The
HISTORY
of
MODERN

MATHEMATICS
Volume III:
IMAGES, IDEAS, AND COMMUNITIES

The
HISTORY
of
MODERN
MATHEMATICS

Volume III:

IMAGES, IDEAS, AND COMMUNITIES

EBERHARD Knobloch
Technische Universität Berlin, Germany

DAVID E. ROWE
Universität Mainz, Germany

ACADEMIC PRESS

Boston San Diego New York
London Sydney Tokyo Toronto

To Dirk J. Struik—the youthful centenarian

This book is printed on acid free paper.

ACADEMIC PRESS

A Division of Harcourt Brace & Company
525 B Street, Suite 1900, San Diego, California 92101-4495

United Kingdom Edition published by
ACADEMIC PRESS LIMITED
24-28 Oval Road, London NW1 7DX

1743193   *7-29-94*

Library of Congress Cataloging-in-Publication Data

The history of modern mathematics / edited by
   Eberhard Knobloch and David E. Rowe.
       p. cm.
   Includes bibliographical references.
   Contents: v. 1. Ideas and their reception—v. 2. Institutions
and applications—v. 3. Images, ideas, and Communities
   ISBN 0-12-599661-6 (v.1: alk. paper).—ISBN 0-12-599662-4 (v.
2: alk .paper) ISBN 0-12-599663-2 (v. 3: alk. paper)
   1. Mathematics—History—Congresses. I. Rowe, David E., Date-
   II. Knobloch, Eberhard, Date-. III. Title.
QA21.S98   ~~1993~~ *1989 vol. 3*
510.9—dc20   89-17766
                                              CIP

PRINTED IN THE UNITED STATES OF AMERICA
94 95 96 97 BC 9 8 7 6 5 4 3 2 1

Contents

Images of Mathematics: Historiographical and Philosophical Issues

Ideas: Differential Geometry and Analysis

Research Communities and International Collaboration

Contents for Volume I

Projective and Algebraic Geometry

Abel's Theorem

Number Theory

Contents for Volume II

Contributors

Numbers in parentheses refer to the pages on which the authors' contributions begin.

Dauben, Joseph W. (263)
PhD. Program in History
City University of New York Graduate Center
33 West 42nd Street
New York, New York, 10036.

Fenster, Della Dumbaugh, (179)
Department of Mathematics, University of Virginia
Charlottesville, Virginia, 22901.

Grattan-Guinness, Ivor (43)
43 Saint Leonard's Road
Bengeo, Hertsfordshire, SG14 3JW, England.

Lützen, Jesper (01)
Mathematics Institute, Copenhagen University
Universitetsparken 5, DK 2000, Copenhagen, Denmark.

Parshall, Karen (179)
Department of Mathematics, University of Virginia
Charlottesville, Virginia, 22901.

Peckhaus, Volker (91)
Institut für Philosophie, Universität Erlangen-Nürnberg
Bismarckstrasse 1, D-91054 Erlangen, Germany.

Purkert, Walter (01)
Binzerstrasse 10, D-04207 Leipzig, Germany.

Tazzioli, Rossana (113)
Dipartimento di Matematica, Università di Bologna
Piazza di Porta S. Donato 5, I-40127 Bologna, Italy.

Ullrich, Peter (139, 155)
Mathematisches Institut, Westfälische Wilhelms-Universität
Einsteinstrasse 62, D-48149 Münster, Germany.

Zhang Dianzhou (263)
Department of Mathematics, East China Normal University
3663 Zhongshan Road Northern
Shanghai 2000062, China.

Preface

When the first two volumes of *The History of Modern Mathematics* appeared five years ago, its editors called attention to the growing interest in mathematical developments of the nineteenth and twentieth centuries. In the intervening period, this interest has engendered a groundswell of activity, much of which has found expression in numerous meetings and symposia that have attracted historians, mathematicians, and philosophers alike. One need but browse through any given issue of *Historia Mathematica*, the official journal of the International Commission on the History of Mathematics, to find numerous reports and announcements pertaining to such gatherings. The last five years have also produced a considerable number of outstanding scholarly studies that have enriched the historical literature devoted to this period. All in all, it seems safe to say that this new trend shows few signs of abating.

The present volume bears further witness to this "boom" of historical activity surrounding modern mathematics. Unlike the first two, which arose in connection with the Symposium on the History of Modern Mathematics, held at Vassar College in June of 1988, this sequel resulted from no real prior plan to publish a similar set of essays. Indeed, most of the nine articles contained herein were originally scheduled for publication in *Historia Mathematica*. As the editors of *Historia Mathematica*, we gradually came to realize that the themes developed in these articles fit in so nicely with those highlighted in the first two volumes of *The History of Modern Mathematics* that preparing such a sequel seemed like a natural and very attractive idea.

The topics covered in the present volume span three different approaches to the history of mathematics that may be considered as, at once, representative and vital for the field. The three contributions by Jesper Lützen and Walter Purkert, Ivor Grattan-Guinness, and Volker Peckhaus deal with diverse "images of mathematics" as seen either through the eyes of the actors themselves (as mathematicians) or through those who sought to interpret their work (as historians or philosophers). On this level, mathematical ideas seem to be accompanied by a sometimes more, sometimes less articulated constellation of associated "meanings." These secondary meanings recede into the background in the contributions of Rossana Tazzioli and Peter Ullrich, which deal with some key episodes in the history of differential geometry and analysis, respectively. In three exemplary case studies, the authors

xiii

trace the, at times, convoluted history of certain fundamental mathematical ideas. In the third and final group of articles, the coauthored contributions of Della Fenster and Karen Parshall and by Diahzou Zhang and Joseph Dauben, the emphasis falls on the larger structure and interaction of mathematical communities. These pioneering studies provide a wealth of new information dealing with the social fabric of two relatively young national communities, those of the United States and China.

In presenting this volume, the editors are pleased to thank the authors, Academic Press, and the International Commission on the History of Mathematics for their mutual support and cooperation. Certainly all nine of these articles would have found an appreciative readership had they appeared in the pages of *Historia Mathematica*. By publishing them separately here, however, we hope to showcase some of the best work currently being done in the field and, at the same time, offer a new proof of the well-known, non-Euclidean theorem that the whole can, indeed, be greater than the sum of its parts.

Eberhard Knobloch
David E. Rowe

Images of Mathematics:
Historiographical and Philosophical
Issues

Moritz Cantor (1829–1920), from the
frontispiece of a *Festschrift* in Cantor's honor,
Abhandlungen zur Geschichte der Mathematik 9 (1899).
A photo of Hieronymus Zeuthen follows page 42.

Conflicting Tendencies in the Historiography of Mathematics: M. Cantor and H. G. Zeuthen

Jesper Lützen* and Walter Purkert†

*Department of Mathematics, University of Copenhagen, Denmark, and †Universität Wuppertal, FB Mathematik, Germany

Moritz Benedikt Cantor (1829–1920) and Hieronymus Georg Zeuthen (1839–1920) were probably the two outstanding historians of mathematics at the end of the 19th century. However, their methods of work differed strikingly. Cantor was an encyclopedist who in his four-volume *Vorlesungen über Geschichte der Mathematik* followed the development of mathematics in a survey of an almost innumerable collection of original and secondary sources from antiquity to the end of the 18th century. He was also an organizer of research in history of mathematics (Vol. 4 of his *Vorlesungen*, Schlömilch's journal, etc.). Zeuthen's papers and books, on the other hand, present deep mathematical analyses of the methods found in classical works mostly from antiquity and from the 16th and 17th centuries in an attempt to capture their fundamental ideas. In this paper we place their works in the context of their lives and mathematical activities (Zeuthen was an outstanding algebraic geometer) and in the context of earlier, contemporary, and later works on the history of mathematics. In particular we compare the differing methods that caused a polemical dispute between them.

1. M. CANTOR (1829–1920)

Moritz Benedikt Cantor belongs to the founders of the historiography of mathematics as an autonomous scientific discipline. His *Vorlesungen über Geschichte der Mathematik* in three volumes [Cantor 1880, 1892, 1894a,b, 1896a, 1898a, 1900, 1901, 1907] (a fourth volume [Cantor 1908] was added by his students and colleagues [1]) long remained the standard work in its field, and even today it has not been supplanted as a detailed account that covers the development of mathematics from its beginnings up to 1758.

Cantor's approach to the history of mathematics was deeply influenced by cultural-historical and cultural-philosophical perspectives; indeed, one of his principal aims was to bridge the gap between the history of mathematics and the general history of human civilization. In the introduction to his first comprehensive work on the history of mathematics, the *Mathematische Beiträge zum Kulturleben der Völker*, Cantor wrote:

> Es gab eine Zeit—und wem wären Werke aus dieser Zeit nicht erinnerlich —in welcher man die Welt fein säuberlich abtheilte und sorgsam Acht gab dass ja nichts aus einem Gefache in das andere hinüberreiche. Da gab es eine Geschichte der Griechen und eine Geschichte der Römer, vielleicht auch eine Geschichte orientalischer Reiche,.... Fast ebenso ging es mit der Geschichte des Mittelalters, der Neuzeit. Noch viel strenger aber war die Sonderung, wenn wir die Geschichte einzelner Kulturzweige betrachten. Es verstand sich fast von selbst, dass eine Geschichte der Religionen sich nicht um die Geschichte der Kunst zu kümmern habe, das diese mit der politischen Geschichte nichts Gemein habe, dass die Geschichte dieser oder jener Wissenschaft wieder ein für sich abgeschlossenes von chinesischer Mauer umgebenes Ganze bilde. In der eigentlich so genannten Weltgeschichte denkt jetzt wohl niemand mehr daran, eine derartige principielle Scheidung für möglich zu halten. Man ist zu der Ueberzeugung gelangt, dass die Menschheit einem Organismus gleicht, dessen einzelne Glieder in immerwährender, wenn auch nicht stets auf den ersten Blick ersichtlicher Verbindung stehen. [Cantor 1863, 1]

For Cantor the goal of describing history within a larger historical context meant above all investigating and comparing the origins of mathematics (elementary geometry, counting, number systems, number symbols, simple arithmetics) among different peoples. These topics were not only the starting point for his research in the history of mathematics; they also formed the main part of his original contributions to the development of our discipline.

From Elementary Mathematics to Its History

Before discussing these matters in detail, we first present a brief synopsis of Cantor's life and academic career. Cantor was born into an orthodox Jewish family in Mannheim in 1829. He thought he was distantly related to the founder of set theory, Georg Cantor, because the latter's ancestors as well as his own came from Denmark. But Georg Cantor did not agree with him on this point [2]. Unlike many Jewish scholars in Germany during the 19th century, Cantor remained true to his Jewish confession all his life. He studied for a year at Heidelberg during 1848, and he continued his studies under C. F. Gauss and M. A. Stern at Göttingen University from 1849 to 1851. He then returned to Heidelberg where he took his Ph.D. with a dissertation on *Ein wenig gebräuchliches Koordinatensystem*. From the beginning Cantor apparently intended to pursue an academic career, since he never bothered to take the "Staatsexamen für das höhere Lehramt" required of all prospective gymnasium teachers. This circumstance also suggests that Cantor's family was wealthy enough to support him in this quest. In 1852 he continued his studies in Berlin under J. Steiner and P. G. Lejeune Dirichlet, and one year later he submitted his *Habilitationsschrift* entitled *Grundzüge einer Elementararithmetik* [3]. Today this book would be regarded as a contribution to mathematical pedagogy rather than as a mathematical work *per se*. After the *Habilitation*, Cantor lectured as a *Privatdozent* at Heidelberg University beginning in May 1853. These lectures were always of an elementary nature, covering topics such as trigonometry, algebraical analysis, analytical geometry, higher curves, and "political arithmetics" [Lorey 1916, 180, 257]. The latter course led to the publication of a small booklet [Cantor 1898b] dealing with mathematical methods for interest and rent calculations, including applications to actuary science and other economic fields. According to Lorey [Lorey 1916, 157], it was well regarded among gymnasium teachers. All of Cantor's 20-odd mathematical publications dealt with elementary topics or recreational mathematics, fields well outside the mainstream of mathematical research during this time. Thus, unlike Zeuthen, Cantor's scholarly reputation was based entirely on his work as a historian of mathematics.

It is not completely clear how Cantor was led to choose the history of mathematics as his main field of research. J. E. Hofmann asserted that the influence of Arthur Arneth was decisive in this regard [Hofmann 1971]. Arneth was a professor at the Lyceum in Heidelberg and also taught as a *Privatdozent* at the university. He wrote a book on the history of mathematics for the *Neue Encyklopädie für Wissenschaften und Künste* [Arneth 1852]. Arneth, for his part, was decisively influenced by the cultural philosopher E. M. Roeth, especially by his *Geschichte der Philosophie* [Roeth 1846, 1858]. Cantor also characterized Roeth as the man "der mich in die historisch-mathematischen Forschungskreise hinüberwies" [Cantor 1863, 84]. Roeth, who had studied science and oriental languages in Paris, taught at Heidelberg from 1840 up until his death in 1858. His main thesis in the history of philosophy was that the roots of modern knowledge can be traced back to Egypt. A modified version of this thesis can also be found in Cantor's early works [4].

The aged Cantor told Cajori [Cajori 1920, 22–23] that he was encouraged to continue his research on the history of mathematics by the very warm response to his lecture [Cantor 1857a] at the 33rd meeting of German Natural Scientists and Physicians. His first paper in the field, entitled "Über die Einführung unserer gegenwärtigen Ziffern in Europa" [Cantor 1856], appeared a year earlier in Volume I of Schlömilch's *Zeitschrift für Mathematik und Physik*. Regarding this topic, M. Curtze wrote that Cantor treated this theme "stets von neuem mit Liebe" and that it resounded "wie ein Leitmotiv durch alle seine späteren Veröffentlichungen" [Curtze 1899, 228]. What is more, his first publication already reveals the principal direction of his later work, which tended to collect numerous scattered accounts and sources related to a certain topic in the history of mathematics and brought them together in a comprehensive form. After mentioning the scholars who had traced the introduction of Indo-Arabic numerals into Europe (A. von Humboldt, Chasles, Libri, Nesselmann), Cantor made the following characteristic remark:

> Wenn wir auch nach diesen Gelehrten versuchen wollen, über denselben Gegenstand Mittheilungen zu machen, so ist unser Zweck weniger der, Neues zu liefern, als vielmehr das vorhandene Material zu sichten und unbeschadet der scheinbaren Widersprüche zu einem klaren Bilde zu vereinigen. [Cantor 1865, 65]

In 1860 Cantor spent some time in Paris, where he was very well received by Chasles, who was then regarded as the doyen among historians of mathematics. Chasles published one of Cantor's articles in the *Comptes Rendus* [Cantor 1860], and he surely encouraged him to continue his work in the field. Cantor cultivated good relations with French historians of mathematics, and in later years he carried on an extensive correspondence with P. Tannery. It was very useful in this regard that he spoke French flawlessly. In 1900, after he delivered an address in French at the Mathematical Congress held in Paris, two Frenchmen came up to him and asked him to settle a wager they had made: one of them had bet that Cantor was a Frenchman, claiming that no foreigner could learn to speak French so well; the other that Cantor was indeed a German who lived in Germany [Cajori 1920, 23].

Around 1860 Cantor began to deliver lectures on the history of mathematics in Heidelberg. Since this field was not an established discipline at the German universities, it was no easy matter finding acceptance. In [Cantor 1863] he alluded to these trying circumstances when, after describing the miserable situation of Pythagoras who at first had to pay his only pupil in order to keep him, he wrote:

> Dieses Dasein, gegen welches die Stellung eines beginnenden Lehres von Nicht-Fach-Gegenständen an deutschen Universitäten eine beneidenswerthe ist, war ihm unerträglich. [Cantor 1863, 67]

His first application for an appointment as *Extraordinarius* in 1860 was turned down by the university authorities [5]. However, this did not discourage him from continuing to lecture on the history of mathematics. From 1875 onward he offered a comprehensive three-semester course covering the whole subject; these lectures formed the basis for his magnificent three-volume work [Bopp 1928]. One of Cantor's auditors, Siegmund Günther, also later went on to become a well-known figure in the field.

A Cultural History of Mathematics

Beginning with his first book, *Mathematische Beiträge zum Kulturleben der Völker* [Cantor 1863], and his investigations on practical arithmetics, numbers, and numeral systems in different cultures, Cantor founded a research tradition that continues to flourish today, especially through the work of Kurt Vogel and his pupils. Cantor himself characterized the history of numeral systems as the thread that tied his work together. He argued that the so-called Indo-Arabic numerals were of Pythagorean origin and came to the Christian Middle Ages through the Romans and Boethius. His argument was based on two manuscripts of the so-called *Geometry II* ascribed to Boethius, one located in Erlangen (*E*), the other in Chartres. His conclusions were as follows:

> Die Geometrie des Boethius, wie sie in der Handschrift *E* enthalten ist, gehört vollständig diesem Verfasser an. Er kannte die im Texte vorkommenden Zahlzeichen und entnahm sie dem Archytas, welchem er auch darin folgt, dass er dieselben *pythagorische Zeichen* nennt. [Cantor 1863, 230]

Thus he traced back Gerbert's knowledge of Indo-Arabic numerals to Graeco-Roman rather than Arabic sources. Today we know that the manuscript *Geometry II*, the earliest Latin work containing Arabic numerals, stems not from Boethius but rather from an unknown scholar who flourished in the first half of the 11th century [Folkerts 1970] [6].

Cantor's first book was not generally well received by his contemporaries. One year after its appearance, Friedlein characterized the question of the origin of the modern numeral system as still being open to debate [Friedlein 1864]. He compared Cantor's results with those published by

Woepcke [Woepcke 1859, 1863] and concluded [7]

> Ich vermochte weder den Resultaten des Herrn Cantor noch denen des
> Herrn Woepcke so beizustimmen, dass ich die Ansichten des einen durch die
> des anderen beseitigt nennen könnte. Ich habe in den Anmerkungen bereits
> einige Punkte angedeutet, in denen mir die aufgestellten Behauptungen
> zweifelhaft erscheinen. Andere lassen sich nur durch eingehende Betrach-
> tungen als unsicher darthun. [Friedlein 1864, 94/95]

Cantor's book is far too speculative to satisfy a modern reader. Moreover, he uncritically presents all the fables surrounding Pythagoras' life as
facts, his account being based on Roeth's history of philosophy. Roeth, for
his part, took for granted notoriously unreliable sources such as Iambli-
chos. Thus in [Cantor 1863] one reads, for example:

> Den unmittelbaren Unterricht des Pherekydes genoss Pythagoras zwei Jahre,
> innerhalb welcher er besonders dessen religiösen Ideenkreis in sich auf-
> nahm, und dann wandte er sich 549 weiter nach Milet zu Anaximander und
> Thales. Schon das dieser damals bereits 90jährige Greis des Jünglings
> Annäherung in vertraulicher Weise zuliess, beweist, wie sehr bereits
> Pythagoras den künftigen grossen Mann verrieth, beweist auf welch günsti-
> gen Boden der Same exacter Wissenschaft fiel, soweit ihn Anaximander und
> selbst Thales ihm anvertrauen konnten.... . Die phönikische Priesterschule
> zu Sidon sollte als Uebergang dienen und dorthin wandte sich Pythagoras
> 548. Ein ganzes Jahr brachte er damit zu, sich die Bekanntschaft mit den
> dortigen Weihediensten zu erwerben, und dann erst betrat er 547 gehörig
> vorbereitet vielleicht zu Naukratis den egyptischen Boden. [Cantor 1863,
> 72–74].

Concerning the critical evaluation of sources, especially ancient ones,
Cantor gradually underwent a process of maturity. In his *Vorlesungen*, for
example, he is much more cautious and critical when dealing with such
sources, and his presentation, on the whole, is much more balanced.

Reviewer and Biographer

In 1863 Cantor was appointed to a position as *Extraordinarius* in Heidelberg. By this time he was also a member of the editorial board of
Schlömilch's *Zeitschrift für Mathematik und Physik*, and in 1875 he was
appointed chairman of the newly established *historisch–literarische
Abteilung* for this journal. He remained on the editorial board until around
1900, when the new editors, Mehmke and Runge, transformed it into an
organ for applied mathematics. Cantor was one of the most diligent
contributors to Schlömilch's journal. He wrote some 650 reviews of historical books and papers, mathematical textbooks, schoolbooks, and mathe-

matical papers and monographs. In each issue one also finds an index of mathematical papers compiled by Cantor and ordered according to the subject matter. Most of his historical papers appeared in Schlömilch's *Zeitschrift*, including [Cantor 1858, 1859, 1865], which are important due to the new sources they contain. From 1877 onward, Cantor edited the *Abhandlungen zur Geschichte der Mathematik*, which appeared as supplementary issues of Schlömilch's journal till 1900 and later as a separate series of monographs.

In 1875 Cantor became *ordentlicher Honorarprofessor* in Heidelberg, but he never attained the position of *Ordinarius*; the assertion to the contrary in [Hofmann 1971] that he became an ordinary professor in 1908 is incorrect (cf. note [5]). Cantor also was engaged by Liliencron's *Allgemeine Deutsche Biographie* in 1875, and over the years he wrote 194 biographies of German scientists, especially mathematicians and historians of mathematics, for this 45-volume work.

Preparing the Vorlesungen

Besides the *Mathematische Beiträge zum Kulturleben der Völker*, Cantor's principal work, the *Vorlesungen über Geschichte der Mathematik*, was preceded by two further preparatory studies which appeared as separate books: *Euclid und sein Jahrhundert* [Cantor 1867] and *Die Römischen Agrimensoren und ihre Stellung in der Geschichte der Feldmesskunst* [Cantor 1875]. In *Euclid und sein Jahrhundert* Cantor dealt with the history of Greek mathematics during the third century B.C., especially with Euclid, Archimedes, Eratosthenes, and Apollonius. His account gives an overview of the contents of their preserved works and discusses the possible contents of lost works such as the *Porismata* of Euclid. Compiling all the ancient sources and information about the lives and works of his heroes, Cantor attempts to give an overall picture of the development of mathematics in Hellenistic Greece during this very fruitful century. A modern historian of science is struck by its lack of reflection on the mathematical ideas themselves and their internal logic. Why, for example, does the theory of proportions in Book 5 of Euclid's *Elements* appear in such a complicated form? Why did Euclid present a second, more elementary theory of proportions, valid only for commensurable magnitudes, in Book 7? Which theory came earlier? Cantor rarely raised such questions. At the end of his booklet he characterized its aim as follows:

> Und somit hätte ich die Darstellung der Fortschritte beendigt, welche, so weit die Spuren zu uns gelangt sind, die Zeit vom Jahre 300 etwa bis 200 vor

Chr. Geb. den mathematischen Wissenschaften hinzufügte. Der Leser wird nicht viel durchaus Neues entdecken, was nicht auch anderwärts schon geboten wäre: das bringt der Gegenstand selbst mit sich. Allein Nichts desto weniger hielt ich die Veröffentlichung einer solchen Skizze eines ein Jahrhundert umfassenden Culturbildes für gerechtfertigt, nachdem seit den letzten zusammenhängenden Darstellungen mathematischer Geschichtsschreiber so manche neue Thatsache entdeckt worden, welche hier und dort zerstreut mitgetheilt ist, ohne zu allgemeiner Kenntnis zu gelangen. Je bedeutender aber die Männer, je interessanter eine Zeit, umso wichtiger ist es für gerechte Würdigung derselben, dass ihre Leistungen uns übersichtlich vor Augen liegen, und gerade dieses habe ich in vorliegender Abhandlung angestrebt. [Cantor 1867, 64]

In his *Die Römischen Agrimensoren*, Cantor not only described the works of Roman engineers and surveyors such as Vitruvius Pollio, Columella, Sextus Julius Frontinus, Hyginus, Balbus, and Sextus Julius Africanus but he also analyzed their Greek and Egyptian sources and the inspiration later scholars (such as Alcuin, Beda, and Gerbert) received from them. He was the first to investigate thoroughly the writings ascribed to Nipsus [8], and he edited and published a Nipsusian manuscript (Codex Erfurt); he also published parts of the Codex Arcerianus (Wolfenbüttel). Cantor himself summarized this work as follows:

Die Römer haben für die Feldmesskunst der Griechen und für unmittelbar oder mittelbar damit Zusammenhängendes, welches ihnen seit dem Beginne der christlichen Aera zufloss, eine aufbewahrende Mittelstelle abgegeben. Sie ähneln darin den Arabern nur dass sie weniger in sich aufnahmen, entsprechend ihrer geringen mathematischen Begabung. Hinzuerfunden haben sie so gut wie Nichts, höchstens einige Operationen wirklicher Feldmesskunst. Weggelassen haben sie von dem, was die sich angeeignet hatten, auch nicht viel; die falschen, meistens altägyptischen Näherungsformeln vor Allen haben sie niemals ausser Uebung treten lassen. Was für die Römer gilt, bleibt wahr für ihre Schüler im Mittelalter. Einzelne hervorragende Geister ausgenommen nimmt das Verständniss des Aufbewahrten immer mehr ab, aber die Menge des Aufbewahrten bleibt. Sie ist nicht gross, aber immerhin erheblicher als man sonst wohl annahm. Dass überhaupt irgend etwas von Geometrie in die wissenschaftliche Barbarei des frühsten Mittelalters hinüber sich retten konnte, ist das unschuldige Verdienst der römischen Agrimensoren. [Cantor 1875, 185].

Cantor's successor in Heidelberg, Karl Bopp, characterized this work as a decisive preliminary study for Cantor's *Vorlesungen*:

Das Jahr 1875 füllte mit dem Buche über die römischen Agrimensoren eine Lücke aus in dem grossen Plane seines Monumentalwerkes; der leitende Faden durch die Geschichte der spätrömischen und mittelalterlichen Mathematik war nach dem sicheren Urteil eines seiner ältesten Schüler, Siegmund Günther, darin gefunden. [Bopp 1928, 510]

The Vorlesungen and their Reception

It is clearly impossible within the scope of this article to give either a synopsis or a critical review of the three volumes of Cantor's *Vorlesungen*. We therefore restrict ourselves here to an account of the successive appearance of the various volumes and the response they received within the academic community. The first volume appeared in 1880 and covers the period from the beginnings up to 1200. A second revised and extended edition appeared in 1894, a third in 1907, and a fourth in 1922. Volume II was published in 1892, with a second edition in 1900, and an unaltered reprint of this second edition appeared in 1913. The second volume covers the period from 1200 to 1668, the year of Newton's appointment at Cambridge. The third volume was brought out in three parts: Part 1 (1668–1699) in 1894, part 2 (1700–1726) in 1896, and part 3 (1727–1758) in 1898. A second edition of volume III (1668–1758) appeared in 1901. At the time of the completion of the third volume, Cantor was 69 years old. He knew that he was no longer equal to the task of writing a fourth volume, but he was also eager to help continue the history. One of the contributors to the fourth volume, F. Cajori, described its origin as follows:

> ... in 1904, at the Congress in Heidelberg, the plan was matured of bringing out a fourth volume on the cooperative plan. Nine men of different countries (V. Bobynin, A.v. Braunmühl, F. Cajori, S. Günther, V. Kommerell, G. Loria, E. Netto, G. Vivanti, C. R. Wallner) were selected to write certain parts under the direction of Cantor as editor-in-chief. Each collaborator was made responsible for his part. As far as possible each man was expected to follow the mode of presentation adopted in the previous volumes. The men were cautioned not to permit themselves to be dominated by preconceived ideas relating to the subjects or men treated in the history. The fourth volume carried the history to 1799, the year of Guass' doctor dissertation. [Cajori 1920, 25]

Cantor's *Vorlesungen* enjoyed an enthusiastic reception. Ohrtmann wrote a 12 page review of the first volume for the *Jahrbuch über die Fortschritte der Mathematik*. It was quite unusual to publish such long reviews in the *Jahrbuch*, but Ohrtmann also emphasized that it was impossible to enter into a detailed criticism in view of the rich contents and the overwhelming variety of material the work contained. Treutlein reviewed the second volume for the *Jahrbuch*, and after summarizing its contents he concluded:

> Der ganze Band und sein Vorgänger stellen eine höchst verdienstliche Leistung ihres Verfassers dar, indem zum ersten Male wieder seit rund 100 Jahren das Ganze der geschichtlichen Entwickelung der (reinen) Mathematik zur Darstellung gelangt. Wir freuen uns und mit uns geniessen die fremden Völker diese Frucht deutschen Fleisses und deutscher Wissenschaft. [Treutlein 1892, 3]

In this review of the second edition of the first volume, Treutlein had this to say about the impact of Cantor's work:

> Das erstmalige Erscheinen des ersten Bandes vorstehenden Werkes (1880) hat die geschichtlich-mathematische Forschung aufs lebhafteste angeregt, eine Fülle von Einzeluntersuchungen sind seitdem erschienen, und was daraus brauchbar war, hat der unermüdliche Verfasser... in der jetzt vorliegenden neuen Auflage des ersten Bandes verwertet. [Treutlein 1894, 1]

Curtze, on the other hand, praised the author's ability to describe the scientific spirit of distant eras:

> Es ist der grosse Vorzug der Darstellung Cantors, dass er sich in den jeweiligen wissenschaftlichen Geist der Zeit, welche er gerade behandelt, hineinzuversetzen vermocht hat, dass er nicht die Arbeiten früherer Epochen mit dem Maßstab des 19. Jahrhunderts misst, und so nicht gegen die grossen Männer der Vorzeit ungerecht wird, ihnen aber auch nicht Sachen und Gedanken unterlegt, welche wir vom Standpunkte unseres Wissens aus in ihren Werken finden können, die ihnen selbst aber niemals gekommen sind. [Curtze 1899, 230]

An English colleague wrote in a review of Cantor's work:

> It hardly requires to be stated that this history is certain to remain for many years the standard work on the subject with which it deals; in completeness, in accuracy, in clearness of arrangement, it stands unrivalled, and for the period which it covers is bound to be a permanent work of reference. [Gibson 1898, 9; quoted according to Cajori]

This opinion was hardly unanimous, however, and with the passage of time Cantor's *Vorlesungen* came to be criticized with considerable severity for its lack of accuracy with regard to historical details. Gustav Eneström devoted much time and energy over the years to recording corrections to Cantor's *Vorlesungen* and publishing them in a special column of his journal *Bibliotheca Mathematica*. Twenty-seven other historians of mathematics also contributed to this regular column, among them such well-known scholars as H. Bosmans, A.v. Braunmühl, M. Curtze, A. Favaro, J. Kürschák, F. Rudio, H. Suter, P. Tannery, G. Vacca, G. Wertheim and H. Wieleitner. H. G. Zeuthen only once made a contribution (a note on Ferrari's solution of the fourth degree equation). [Zeuthen 1900]

Eneström's review of the first part of the second volume was still quite positive:

> Nous attendons avec impatience la fin du 2e tome des *Vorlesungen* et nous espérons que M. Cantor sera en état de continuer encore plus loin son excellente exposition de l'histoire des mathématiques. [Eneström 1891, 118]

The nearer Cantor drew to modern times, however, the more critical Eneström became in his reviews. Analyzing the last part of Volume III, Eneström denies that this is a history of mathematics covering the period from 1727 to 1758. In his opinion the correct title should have been *Beiträge zur Geschichte der Mathematik 1727–1758* [Eneström 1903, 447]. A modern historian of science would fully agree with Cajori, who in 1920 wrote:

> The Herculean efforts of Cantor and the extremely penetrating criticisms of Eneström clearly point out two lessons to scholars of today: (1) The need of a more accurate general history of mathematics, prepared on the scale of that of Cantor's Vorlesungen and embracing the historical researches of the last twenty years; (2) the impossibility of this task for any one man. A history of the desired size and accuracy can be secured only by the cooperative effort of many specialists. [Cajori 1920, 26]

We conclude these remarks on Cantor's main work by quoting the opinion of a renowned authority from the middle of the 20th century, J. E. Hofmann:

> [Cantor's *Vorlesungen*] sind eine für ihre Zeit sehr beachtliche und die bisher eingehendste Gesamtdarstellung von den Anfängen bis 1800, jedoch heute nach Inhalt und Methode überholt. Die internationale mathematikgeschichtliche Forschung hat durch Cantor wesentliche Impulse empfangen, hat seine Hauptergebnisse ergänzt, verschärft und berichtigt und ist in kritischer Auseinandersetzung mit seinem Vorgehen zur ideengeschichtlichen Auffassung übergegangen. [Hoffmann 1957]

Moritz Cantor was a very conscientious academic teacher. By 1913, however, his eyesight had become so poor that he could no longer read his own writing on the blackboard, and so he went into retirement (cf. [5]). He died on April 10, 1920, in his ninety-first year, three days after the death of his only son.

Cantor's Views on Historiography of Mathematics

In 1903 Cantor published a kind of confession of faith as to how one should write the history of mathematics [Cantor 1903]. The following extensive passage from this work reveals that by 1903 he had a more moderate opinion of the kind of historiography which emphasizes mainly the development of mathematical ideas than he had expressed at the time of his controversies with Zeuthen (see later), but he himself would never be willing to write such a history:

> H. Eneström unterscheidet verschiedene Arten, nach welchen Geschichte der Mathematik behandelt werden könne. Er hat darin sicherlich recht. Man kann die verschiedenen Behandlungsweisen selbst in sehr verschiedener Weise schildern, und ich gestatte mir den Gegensatz darin zu finden, dass man von der Wortverbindung Geschichte der Mathematik bald das Wort *Geschichte*, bald das Wort *Mathematik* stärker betont.
>
> Wer das Letztere tut, der könnte vielleicht am zweckmässigsten so verfahren, dass er die einzelnen Sätze der Mathematik in gedrängter Weise mitteilte, bei jedem Satze als Anmerkung beifügend, wann und durch wen er der Wissenschaft einverleibt worden sei. Er könnte durch geschickte Anordnung der Sätze es dahin bringen, dass zwischen den Anmerkungen scheinbar ungewollt, aber die grösste Kunst des Verfassers verratend, ein Zusammenhang sichtbar werde, aus welchem der Leser zu erkennen vermag, wie, wo, durch wen die mathematische Wissenschaft ihre Entwickelung vollzogen hat. Wir haben ein solches Beispiel der Geschichte der *Mathematik* in der grossen im Entstehen begriffenen *Encyklopädie der mathematischen Wissenschaften*, und ohne den übrigen Mitarbeitern an dem monumentalen Werke zu nahe zu treten, möchte ich die von H. Pringsheim bearbeiteten Abteilungen als meiner hier ausgesprochenen Meinung am meisten entsprechend hervorheben.
>
> Etwas ganz anderes ist nach meinem schriftstellerischen Gefühle eine *Geschichte* der Mathematik. In ihr liefert die Mathematik zwar das gesamte Material, aber dessen Benutzung soll nicht ausschliesslich der Mathematik zu gute kommen. Das Bild des gesamten Kulturlebens dient als Hintergrund, von welchem mathematische Charakterzüge sich hell abheben und selbst dazu dienen, jenen Hintergrund zu erhellen. *Mathematische Beiträge zum Kulturleben der Völker* nannte ich 1863, also jetzt vor 40 Jahren, mein erstes geschichtliches Buch und meinte durch diesen Titel mein wissenschaftliches Glaubensbekenntnis und zugleich mein wissenschaftliches Programm zu verkünden. [Cantor 1903, 114/115]

2. H. G. ZEUTHEN (1839–1920)

In 1912 Hieronymus Georg Zeuthen was chosen (probably by Felix Klein) to write the section on *Die Mathematik im Altertum und im Mittelalter* for the collection *Kultur der Gegenwart*. This shows that he was considered one of the leading historians of mathematics at the time. Still Zeuthen always stressed that he made his contributions to this field not as a historian but as a professional mathematician. In order to understand his approach to the history of mathematics it is therefore necessary to start with a brief sketch of his career as a mathematician. (This has been described in more detail in the biographies by M. Noether [1921], J. Hjelmslev [1939], and E. Picard [1922]. Noether's article contains a complete list of Zeuthen's 161 publications.)

Mathematical Career

Hieronymus Georg Zeuthen was born the son of a minister on February 15, 1839 in Grimstrup, West Jutland, Denmark. When he was 10 years old the family moved to Sorø, where he met his schoolmate and later colleague Julius Petersen (1839–1910). As a result of a common interest in mathematics, they both embarked on a study of this subject at the University of Copenhagen (Zeuthen in 1857). Julius Petersen quickly began to systematize and extend the classical methods of geometric construction. Thereby he was led to the problem of constructability by ruler and compass and further to research in Galois theory. Today he is mainly remembered as one of the creators of graph theory (cf. Lützen, Sabidussi, and Toft 1992). Zeuthen, on the other hand, showed such talents in "modern" geometry that when he graduated from the University in 1863, he obtained a scholarship to go to Paris to study enumerative methods with the master geometer Michel Chasles (1793–1880).

As an example of an enumerative problem, consider the following question: What is the number of conics tangent to five given conics? Steiner had argued in 1848 that the number was 7,776, but in 1864 Chasles found that it was "only" 3,264. As this mistake and the size of the numbers indicate, it is no simple matter to solve the corresponding algebraic equations, and therefore Chasles developed a method of "characteristics" to circumvent this problem. In his doctoral thesis, Zeuthen carried these methods and results further. He defended his thesis at the end of 1865, after a less successful attempt to defend his country against the Germans as a soldier in 1864. The thesis dealt with "systems of conics which are subject to four conditions." Before its acceptance, Zeuthen encountered a note in the *Comptes Rendus* in which Chasles had undertaken a study of systems of surfaces of the second degree, and before Chasles published his results, Zeuthen deposited with the Royal Danish Academy a generalization to this situation of his own theory. As it turned out, Zeuthen's treatment contained extensions and corrections of some of the results obtained by his teacher.

Chasles had shown how a family of curves or surfaces could, for enumerative purposes, be characterized by certain numbers. In his 1865 paper and in a subsequent series of remarkable works, Zeuthen developed a method by which one can determine these numbers from the degenerate cases. Moreover, he found an invariant still referred to in modern algebraic geometry as the Zeuthen–Serge invariant. He soon became known (together with his younger contemporary Hermann Schubert) as the master of enumerative methods, and was chosen to write a survey of this subject for the *Encyclopädie der mathematischen Wissenschaften* [Zeuthen 1906]. Eight years later he completed his scientific career with the publication of his monograph *Lehrbuch der abzählenden Methoden der Geometrie* [Zeuthen 1914a], still considered a classic today. In fact, after a period in which algebraic geometers have concentrated mainly on abstract theory and its foundations, a new interest in Zeuthen's concrete ideas reappeared in the mid-1970s. In 1989, a symposium commemorating the 150th anniversary of his birth gathered a great number of experts from all over the world. It was held at the University of Copenhagen, where he taught first as a *docent* from 1871, and then as a professor from 1883 till 1910, ten years before his death in January 1920. [See Kleiman 1991.]

Platonistic Philosophy of Mathematics

When Zeuthen presented his book on enumerative methods [Zeuthen 1914a] to the Royal Danish Academy of Sciences and Letters, whose very active secretary he had been for 25 years, he gave a speech entitled "On the application of calculations and reasoning in mathematics" [Zeuthen 1914b], in which he summarized his philosophy of mathematics and its influence on his historical work. This philosophy was tied to the geometric tradition from which his mathematical works had grown. During the first part of the 19th century, Poncelet and his successors—including Chasles—had developed geometry from a synthetic point of view, claiming that their methods were more natural and intuitive than those of the dominant analysts. The analysts, on the other hand, claimed that their analytic–algebraic methods were more rigorous and better suited to finding solutions to new problems. Zeuthen indirectly referred to these traditions when in his speech he recalled the words of his former teachers at the university, Christian Jürgensen and Adolf Steen. The former had characterized mathematics as a "lazy man's science."

> This appealed to me [Zeuthen said], not because I was particularly inclined to laziness, but, in accordance with the intended meaning of the words, I preferred to get at the results along the shortest path, and, in particular, to avoid complicated calculations with which I was neither as consistent nor as fleet as a mathematician might wish. Therefore, I have often spent days finding a purely geometric solution to a construction problem which could

perhaps be solved by calculation in a quarter of an hour. However, I also sought the solution which, once it had been found, would be considered more simple, direct and enlightening than the one which calculation would give. [Zeuthen 1914a, 1]

Steen, on the other hand, had emphasized that he did not consider a mathematical result to be proved properly until it had been secured by calculation; here "calculation" should be understood in a broad extensive sense to mean any regular operation with symbols. (In his speech Zeuthen implicitly connects this view with Weierstrass' arithmetization of mathematics.) Though Zeuthen understood the need for rigorous proofs, he did not think they could only be attained by calculation. He emphasized that mathematical results are usually discovered by some kind of intuitive reasoning; the proof by calculation only comes later. As an alternative, Zeuthen proposed "that the same certainty can usually be obtained by a thorough examination of the reasoning which has led to the result" [Zeuthen 1914b, 272]. Zeuthen used this approach in algebraic geometry, but he admitted that insufficiencies in the reasoning of his predecessors (Cayley and Salmon) revealed that it is not always easy to make such arguments rigorous. In fact, Zeuthen was aware that some of his own methods were open to criticism, in particular his method of assigning multiplicities. Subsequent investigators often felt compelled to confirm his results by using more rigorous methods.

Zeuthen elaborated on this methodology by referring to the Danish philosopher Harald Höffding's analysis of intuitive knowledge. When a mathematician first guesses a result, he is led by an intuitive feeling, which Zeuthen called the general (or total) impression (*helhedsfornemmelse*). After thorough analysis, the mathematician may finally reach a complete intuitive understanding, which Zeuthen called the general (or total) cognition (*helhedserkendelse*). We may identify this general cognition with Plato's ideal (Zeuthen spoke in Kantian terms of "das Ding an sich"), and, like the ideal objects, the general cognition may attain various symbolic forms (geometric, arithmetic, etc.) A proof expressed in one of these symbolic forms (in particular, in arithmetic terms) may rigorously establish the theorem, but it hides the general cognition or ideal truth. Zeuthen argued that his method was well-suited to reveal this hidden general cognition.

Approach to the History of Mathematics

In order to explain Zeuthen's view of the history of mathematics it is important to note that he considered these pure mathematical ideas as being independent of time, although the symbolism in which they must necessarily be dressed may vary. Thus, through careful analysis, a modern mathematician will be able to uncover the motives and ideas of a writer who lived 2000 years earlier, even though these ideas may be presented in an unfamiliar way. Still, Zeuthen repeatedly emphasized that one cannot evaluate or understand the mathematics of an earlier period on the basis of the form mathematics has taken today. On the contrary, it is indispensable to make oneself acquainted with the techniques and symbolism of former times:

> One can therefore only evaluate the mathematical progress of former times by studying the mathematics of that time as a whole, by acquainting oneself so well with the notions and procedures in use at that time that one is able to judge what *could* be achieved within the area known at the time and under the given conditions, and moreover by examining what these tools have in fact been used for. A thorough understanding will not be obtained until one finds a complete connection between the usability of the tools at their disposal and the range of the results they gained with them. [Zeuthen 1903b, 555]

Yet although technical deductions may appear in different ways, there must be a common logical idea behind them, and in order to uncover this idea mathematical training is necessary. This is the task Zeuthen set for himself when he worked in the history of mathematics. What, then, did he think he could achieve with such an approach? He had both mathematical and historical aims. Mathematicians can in this way learn about forgotten procedures, which though less general or automatic than modern procedures, may lead to a deeper understanding of concrete cases "if only for the reason that the less developed form forced greater brain work and led to observations that easily escape the attention of those who now reach the same result with railroad-like speed" [Zeuthen 1903b, 554]. Moreover, he continued:

> The rules that a mathematician learns today are mostly designed to answer questions formulated in a particular way. He who wants to find something new often has to free himself from existing rules, and he can learn how to do that from those who have not yet converted their procedures into fixed rules. [Zeuthen 1903b, 555]

Finally he asserted that one may be able to streamline the modern approach by removing the unnecessary remains of an earlier presentation.

The aims of Zeuthen's first historical papers were mainly didactic. He chose subjects that shed a new light on a well-known piece of mathematics in order to help, in particular, younger readers "develop their mathematical sense and perception." In fact, he did not bring forward new historical points of view. However, he soon discovered that this approach also allowed him to contribute substantially to the history of mathematics, in particular to the history of ancient Greek mathematics. Indeed the ancient masterpieces were often constructed according to a self-enclosed logical-deductive plan which does not correspond to the way in which the results had been obtained. A critical mathematical analysis combined with the scattered evidence recorded by later commentators may reveal the process of discovery and the more intuitive ideas behind the rigorous façade. This, in turn, may allow us to reconstruct lost works, to understand foggy hints, and to connect seemingly unrelated results.

The clearest explanation of Zeuthen's method in the history of mathematics can be found in his last major work, in which he states at the outset:

> When a professional mathematician wants to cultivate the history of his science, he must, obviously, first of all submit to the rules that apply to everyone who wants to know the historical truth ...
>
> He [the mathematician] will, however, also be able to contribute something of his own. Mathematical truths have been called eternal truths; this, however, is not to say that at any time where they have appeared they should have taken on the same form—and, more, even, that they should have been formulated in the same language. However, in very different expressions one can recognize the same truths. Their concord will often manifest itself in similar applications, and the deductions by which one passes from one truth to the next must, in spite of different appearances, have essentially the same logical foundation, at least if they are mathematically justifiable. In this way, the mathematically educated cultivator of the history of mathematics has the opportunity to interpret texts which otherwise seem to be incomprehensible, or which have been misunderstood, to find connections among historical statements which otherwise may seem to be concerned with different things, to detect the preparation of a discovery which otherwise seems to have been due to the unique gift of prophecy of one single ingenious man, and above all to find and understand the coherence in the research and knowledge of a given period and thereby its connection to the points of view of the earlier and later periods from or to which impulses are given. Not only will historical knowledge thereby increase and be consolidated but one will gain exactly that type of knowledge of old mathematics that will give the greatest benefit for mathematicians and pedagogues. One will learn not only at which times, but also how one has gradually reached the results which are perhaps proved in a completely different way today. One will learn about points of view that have been abandoned in favor of others that as a whole are more profitable but still do not quite render the old superfluous, or conversely one may in modern presentations recognize remnants of methods which have been significant in their time, but which in the modern context are superfluous and therefore ought to be removed. [Zeuthen 1917, 201–202]

The Conic Sections in Antiquity; Geometric Algebra

Where did Zeuthen's approach to the history of mathematics lead him and what convinced him that his method was sound? During the first 10 years of his career, Zeuthen published almost exclusively on algebraic geometry. The first sign of his interest in history came in 1876 in the form of a paper on one of Chasles' favorite subjects: Hindu mathematics (its introduction was quoted above). This was to be the first in a series of papers entitled "From the history of mathematics"; these were written in Danish and published in the *Matematisk Tidsskrift*, which Zeuthen edited. In 1882 and 1883, he published three more notes in this series, dealing with Euclid, Apollonius, and Diophanthus. Zeuthen's interest in ancient mathematics culminated in 1885 with his master work on the theory of conic sections in antiquity (presented to the Royal Danish Academy in October 1884 and published as a monograph in German translation [Zeuthen 1886].

As Zeuthen pointed out in [Zeuthen 1903b], his mathematical works on conic sections naturally led him to study the ancient origins of this theory, and, in particular, Apollonius' surviving seven books. The typical synthetic style of Apollonius' work had caused speculation as to how he and earlier Greek geometers had discovered their results. Several 19th-century mathematicians had advanced the idea that the Greeks had used analytic geometry to find these results, but had kept the methods secret in order to impress their readers. Zeuthen found this hypothesis "tasteless" [Zeuthen 1903b, 557]. Other mathematicians, in particular Descartes, had advanced the opposite idea; namely, the Greeks had no method of discovery, and therefore just collected the theorems they happened to find (cf. [Zeuthen 1885, 3]). Zeuthen, for his part, struck a happy medium, insisting that the Greeks possessed a method which is in many ways equivalent to analytic geometry but which has a different form. His discovery of this method, which he called *geometric algebra*, was one of his most outstanding contributions to the interpretation of the history of mathematics, and, we must add, the most hotly debated among present-day historians of mathematics.

The novelty of Zeuthen's idea must be judged against the background of the general state of affairs in the history of ancient mathematics in the mid-19th century. Many historians believed that while geometry was of Greek origin, arithmetic and algebra were developed by the Hindus, who were supposed to have served as the primary source for Diophantus' work (cf. [Hankel 1874]; mathematical cuneiform texts had not yet been discovered). Cantor, however, wanted to deprive the Hindus of this honor and called attention to Heron as an intermediate link between Euclid's geometric treatment of numbers and Diophantus' arithmetic treatment. Yet already in his 1883 paper on Diophantus, Zeuthen had written:

> "He [Cantor] also points to the fact that much of the real rational basis for the works of Diophantus can be found in Euclid himself; however, in my opinion he does not go at all far enough in this respect." [Zeuthen 1883, 146]

Paul Tannery had already gone further in his paper, "De la solution géométrique des problèmes du second degré avant Euclide" [Tannery 1882]. He pointed out that Theorems 28 and 29 in Euclid's Book VI, and, in a less direct way, Theorems 5 and 6 of Book II, presented geometric solutions of quadratic equations equivalent to Diophantus' and our modern solutions. He further conjectured that already in Pythagorean times the geometric treatment had been derived from an arithmetic procedure, and that the Pythagoreans had made this transformation in order to circumvent the problem of irrational quantities. Moreover, Tannery interpreted Euclid's book X as dealing with the following question: When does a quadratic equation have a rational solution?

Zeuthen characterized Tannery's paper as "both penetrating and brilliant" [Zeuthen 1885, 11] and later described it as the "point de départ d'une grande partie de mes explications" [Zeuthen 1905, 274]. However, in Zeuthen's 1883 paper, where one can see the first traces of his ideas on geometric algebra, he did not refer to Tannery. So it seems unclear if he knew the paper initially.

At any rate, Zeuthen went even further than Tannery. In Euclid's Book II, he found not only quadratic equations, but an entire deductively developed theory, which represents in geometric language many of the same relations that we now express using algebra. For example, the formula

$$(a + b)^2 = a^2 + b^2 + 2ab$$

is equivalent to Euclid's Proposition II, 4:

> If a straight line be cut at random, the square on the whole is equal to the squares on the segments and twice the rectangle contained by the segments.

In a letter to Tannery, dated January 29, 1884 Zeuthen explained the use of this "geometric algebra":

> J'ai reconnu là un instrument remplaçant entièrement une algèbre *du second degré* et satisfaisant, par conséquent, à tous les besoins de la théorie des coniques.... J'ai trouvé, par un peu d'exercice, cet instrument aussi practicable que notre algèbre lorsqu'il s'agit *des propres études et des communications orales où l'on peut montrer les parties* des figures dont il s'agit. Pour

donner de nouvelles inspirations géométriques, il a même eu, selon moi, quelques avantages sur notre algèbre; mais, pour la *communication écrite*, il est très pénible. [Tannery 1943, 628–629]

Once Zeuthen had discovered the Greek geometric algebra, he was then able to understand and analyze Apollonius' conics. Indeed, Zeuthen observed that after Apollonius had derived the *symptoma* of the conic sections (i.e., the equivalents of their equations) the remainder of the work was based entirely on these *symptoma* rather than on the stereometric generation of the curves as sections of cones. Zeuthen even claimed that the curves had their historical origin in their *symptoma* (or similar expressions) as they presented themselves in connection with the Delian problem (or, equivalently, the problem of two mean proportionals) [Zeuthen 1885 Sect. 21]. According to Zeuthen, Menaechmus established the existence of these curves by showing that they were generated by sections in a cone perpendicular to a generator of the cone.

Zeuthen's analysis contradicted Cantors in several ways. First Cantor, following the commentator Geminus, claimed that Apollonius was the first to observe that one did not get more general curves if one made arbitrary sections in arbitrary cones. Zeuthen refuted this by calling attention to the fact that Archimedes explicitly mentions, as something well-known, that all sections of a cone whose plane meet all generators are ellipses (or circles). He then argued that from a mathematical point of view it was unthinkable that Archimedes and his predecessors would have known this property without knowing the related facts about the parabola and the hyperbola. Instead Zeuthen interpreted Geminus' statement to be concerned with the names of the curves only. Secondly, Cantor went so far as to claim that Apollonius was the discoverer of the *symptoma*. Zeuthen's answer is characteristic. First [Zeuthen 1885, 36–37], he tried to reconstruct how Euclid would then have proceeded in his lost book on conics. He tried taking a purely stereometric approach, but since he could not find the preserved traces of such an approach, he rejected it. Secondly, he pointed out that Archimedes had based his study of the conics on proportions that were equivalent to Apollonius' *symptoma* [9].

Already these arguments show the strength of Zeuthen's approach to the history of mathematics. By entering into the mathematical spirit of the time, he was able to find mathematically plausible reconstructions of no longer extant theories, evaluate the surviving sources, and, in particular, show that certain commentators (here, Geminus) were probably mistaken.

In his careful discussion of the work of Apollonius, Zeuthen gave many other examples of this approach. Here we shall only mention that he showed how to connect Apollonius' third book with the three- and four-line loci that came to play such an important role in the invention of analytic geometry [Descartes 1637, 304 ff]. In Zeuthen's modernized language, the four-line locus is the set of points whose distances x, y, z, u from four given straight lines satisfy the equation

$$\frac{xz}{yu} = \text{constant}.$$

Here the distances need not be measured along the perpendicular but along lines given in direction. The three-line locus is found by letting two of the given lines coincide.

In the preface of his work, which is addressed to Eudemus, Apollonius mentioned that Euclid's treatment of these problems had not been entirely satisfactory, but that his own Book 3 provided the basis for a complete solution. Still, this book does not contain any reference to these problems. In his characteristic way, Zeuthen pointed out that "one must not fail to use the means which this information gives concerning the purpose and meaning of these theorems to acquire a better understanding of the work and the considerations which have given the Greek theory of conic sections its development, and the significance of the results it has obtained." [Zeuthen 1885 Chap. 7]

Zeuthen's subsequent reconstruction explains that Euclid's solution might have been incomplete because he only worked with one branch of a hyperbola, and he analyzed in what sense Apollonius' third book contains a solution to the problems. He also conjectured that this solution would be clear to the Greeks because they knew Aristaeus' lost work on solid loci, and further speculated that Euclid's porisms, which had been reconstructed by Chasles, were developed in this connection.

Thus Zeuthen's work on conic sections led him to embark on a comprehensive study of Greek mathematics. In his lectures given at the University of Copenhagen during the following years he continued to study further traces of the use of geometric algebra as well as the use of infinitesimal methods in the Greek texts (cf. [Zeuthen 1903b, 558]). These investigations resulted in his book on the history of mathematics in antiquity and the middle ages [Zeuthen 1893], which was translated into German in 1896 and into French in 1902.

As Zeuthen explained to Tannery in a letter dated June 2, 1893, the mathematical analysis was also at the center of this book:

> J'y borne à un minimum la mention des faits historiques propres, mais je cherche en récompense de représenter clairement *l'état* de la science, les

points de vue, les idées et les manières de les représenter, les *époques les plus importantes*, et je néglige les temps de décadence. [Tannery 1943, 651]

Since it was intended for an audience of mathematicians, mathematics teachers, and their students, all references to the ideas of other historians of mathematics were omitted.

The book contains several new penetrating observations. We mention only one, which Zeuthen also published separately [Zeuthen 1892]; its title reveals the basic idea: *On construction as existence proof in Greek mathematics*. In this vein Zeuthen considered the parallel postulate as a postulate securing the existence of the intersection of two lines. He also gave eye-opening interpretations of other axioms in Euclid and Archimedes, and later in his life he connected these ideas with those of Hilbert and others in a lecture on Euclid's parallel postulate in the light of the modern axiomatic theory given at the Royal Danish Academy [10].

As the title of the book indicates, Zeuthen followed the traces of Greek mathematics into the Middle Ages, and, in particular, gave an account of Indian and Arabic sources. In subsequent years he extended his researches to the European mathematics of the 16th and 17th century. He was particularly interested in studying how geometric algebra was transformed into analytic geometry; therefore he read Viète, Fermat, and Descartes with special care. The development of the infinitesimal calculus, from the ancient method of exhaustion to Newton's and Leibniz's calculi, also attracted his interest, and he reevaluated Barrow's important insight into the inverse nature of tangent constructions and the determination of areas (or differentiation and integration in modern terminology).

Zeuthen first published the results of his research in a series of papers (1893–1900) in French, and then in his third historical book on the history of mathematics in the 16th and 17th centuries [Zeuthen 1903a], which appeared in a German translation the same year in the series *Abhandlungen zur Geschichte der Mathematischen Wissenschaften* founded by Cantor. He continued to publish new contributions to the history of mathematics until one year before his death in 1920. As with his earlier works, these too were mainly aimed at a mathematical audience; however, he also composed expository articles intended for a wider circle. In his capacity of Rector (Wise Chancellor), he wrote a 99-page paper on the development of mathematics as an exact science until the end of the 18th century for the 1896 annual celebration of the University of Copenhagen. Finally, three years before his death, he published a lengthy paper in the Memoirs of the Royal Danish Academy on how mathematics in the period from Plato to Euclid became a rational science, from which we have had occasion to quote the methodological introduction.

Zeuthen and his contemporaries

As Zeuthen often pointed out, his approach to the history of mathematics was influenced by many earlier and contemporary mathematicians. In a certain sense it is a continuation of the long tradition of reconstructing lost Greek works. In this connection Zeuthen only mentioned Chasles. It is probable that Chasles aroused Zeuthen's interest in the history of mathematics during his stay in Paris, but as we have seen, almost 20 years elapsed before the latter began his own historical research. Zeuthen also emphasized the importance of Hankel's unfinished *Zur Geschichte der Mathematik in Alterthum und Mittelalter*, which was published posthumously in 1874. He recognized the similarities in their approaches, although he disagreed with Hankel on a number of particular points.

The two scholars whose work Zeuthen respected most, however, were P. Tannery and J. L. Heiberg. Thus in a letter to Heiberg written October 31, 1893, just after the Danish edition of his *Lectures on the History of Mathematics. Antiquity and Middle Ages* had appeared, he wrote:

> I very much long to hand the book over to you because I consider you to be one of the two most competent judges of its content. The other, P. Tannery, will of course receive it but he cannot read it, as you know. [Zeuthen to Heiberg, correspondence]

In his published works, Zeuthen often praised Paul Tannery (1843–1904) as the true master of the history of mathematics, and the correspondence between the two (published in Tannery's *Mémoires Scientifiques*) bears witness to their mutual respect and, after their first personal encounter in Paris in 1900, also to their friendship.

Tannery, for his part, was very successful in promoting Zeuthen's and his own ideas and his reviews of Zeuthen's works were extremely laudatory. The beginning of his review of *Die Lehre von den Kegelschnitten im Alterthum* is a typical example:

> Le travail du savant professeur de l'Université de Copenhague fera époque, et ce n'est pas assez de dire, pour le louer, qu'il y a longtemps que le sujet qu'il a choisi n'a donné lieu à une étude aussi originale; il faut reconnaître que jusqu'alors l'histoire des coniques dans l'antiquité était incomprise, et que M. Zeuthen, non seulement en donne la clef, mais nous guide de façon à ne plus nous laisser nous égarer.
>
> Il sera impossible, d'ici longtemps, de parler des coniques dans l'antiquité sans avoir eu recours à ce livre magistral. [Tannery 1886]

As we shall see below, Cantor did not agree, but Tannery preferred Zeuthen's work over that of Cantor. Indeed, Tannery was responsible for Zeuthen's first textbook having been translated into French, having first dissuaded the translator from translating Cantor's *Vorlesungen*. He explained the circumstances to Zeuthen in a letter of April 18, 1899:

> Un jeune professeur français, M. Weill, du Lycée de Beauvais, m'ayant récemment manifesté l'intention de traduire en français l'ouvrage de Cantor sur l'histoire des Mathématiques, je l'ai engagé à traduire plutôt de l'allemand vos *Forelaesning over Mathematikens Historie* ...
>
> Je puis m'engager vis-à-vis de vous à surveiller la traduction, de façon qu'elle soit satisfaisante. [Tannery 1943, 656]

As it turned out, a certain Jean Mascart did the translation (which Tannery found too literal), but Tannery kept his promise and even suggested substantial revisions of the content. Moreover, when the Académie des Sciences in Paris in 1903 decided to confer its first Prix Binoux for work done in the history of science to Zeuthen, Tannery wrote to him:

> L'essentiel était pour moi que l'Académie, par son choix de cette année (puisque c'est la première fois qu'elle décerne ce prix) précisât le genre de travaux auxquels elle donnait la préférence, et qu'elle prît une décision éclairée. Malgré toute l'estime que vous savez que j'ai pour Moritz Cantor, je ne puis en particulier m'empêcher de reconnaître votre supériorité sur lui. {Tannery 1943, 667]

In print Tannery's comparisons between his two great contemporaries were more indirect as, for example, in his review of Zeuthen's second textbook on the history of mathematics:

> ... la sûreté de ses informations et la sagacité de sa critique sont telles que cet ouvrage peut encore apprendre bien des choses, même à qui a étudié à fond les *Vorlesungen* de Moritz Cantor, et toute la littérature parue, depuis leur seconde édition, sur l'histoire des mathématiques, à l'époque de leur renaissance. [Tannery 1904]

After Tannery's death, Zeuthen repaid some of this support by agreeing to edit his *Mémoires Scientifiques* together with Heiberg [11].

The Danish philologist Johan Ludvig Heiberg (1854–1928), who gained fame through his still authoritative editions of the Greek mathematicians, was the other person whose judgement Zeuthen valued. Although Heiberg knew enough geometry to understand most of the Greek arguments, he sometimes turned to Zeuthen for help. For example, on November 9, 1890, he wrote to Zeuthen:

> Enclosed please find my questions concerning dubious statements in Apollonios' Book IV, which in clarity and meticulousness fall short of the other books.

After many precise questions Heiberg concluded:

> Please excuse me if I have not expressed myself quite clearly or perhaps even have complained without reason, but on the whole I have found it difficult to master the mathematical content in Apollonios, and moreover I have not seen the fourth book for almost a year.... [Heiberg–Zeuthen, correspondence]

Many subsequent letters from Zeuthen to Heiberg are preserved which discuss various details in Greek mathematics. Considering that the two experts met at least every two weeks in the Royal Danish Academy, there is no doubt that their collaboration was much more intense than the written evidence reveals.

They published two joint papers [Heiberg and Zeuthen 1906 and 1907], the first of which created a great sensation. It contained Archimedes's *The Method*, which had been believed to be lost, but which Heiberg had found in Constantinople. Heiberg provided a translation of the text and Zeuthen wrote the mathematical commentary. The content particularly pleased Zeuthen, for it showed that the ancients had been in possession of an intuitive infinitesimal method by which they found the results that they later proved by the method of exhaustion. This was what Zeuthen and others had guessed on the basis of mathematical analyses of the surviving sources. Thus, as Zeuthen himself pointed out, Heiberg's discovery brilliantly supported Zeuthen's approach (cf. [Zeuthen 1914b, 279]).

Finally, there was a third person who influenced Zeuthen's work greatly, namely Cantor, but the relation between these two scholars was of a different nature.

3. CANTOR AND ZEUTHEN: THE DISPUTE

Cantor and Zeuthen may, in a certain sense, be considered incarnations of two different approaches and ways of thinking in the history of mathematics. On the one hand, we have an encyclopedist who tries to unite a multitude of sources and opinions into a comprehensive picture of the development of mathematics and to integrate these into the evolution of human civilization, underestimating thereby, to a certain degree, an analysis of the inner logical components in the development of mathematical ideas. On the other hand, we have a penetrating mathematician who contributes to the history of mathematics by examining the mathematical content of the works of the most important scholars of the past and by showing the logical connection between them. In the case of Cantor and Zeuthen these different approaches led to a rather polemical scientific dispute which also influenced their personal relations. Zeuthen was the only famous historian of mathematics who did not contribute to the *Festschrift* dedicated to Cantor on his 70th birthday [12].

On a Collision Course

From the very start Zeuthen's interest in the history of mathematics led him on a collision course with Cantor. In his first surviving letter to Heiberg, dated October 20, 1882, he wrote:

> Monday evening I shall give a talk at the Mathematical Society on the theory of conic sections in Greek antiquity. Although I must explain many things that you already know in order to be understood by this audience, my particular aim is to bring forward some remarkable views which mainly contrast with those of Cantor.
>
> I would like to hear your opinion, since you are the one who knows the basic material best, and has made it easier for me to master it. [Zeuthen–Heiberg, correspondence]

In fact, in a letter of November 16, 1886, Zeuthen explained to Tannery that it was the discovery of the inadequacy of Cantor's book that really motivated him to become a historian of mathematics:

> Je finirai par parler d'une suggestion, de nature *négative*, que je ne dois pas à vous, mais à *M. Cantor*. Je ne dis pas que la géométrie grecque m'était inconnue avant ses leçons; j'avais à peu près la même idée du plan du premier Livre d'*Apollonius* que j'ai développée dans mon livre et, en profitant du petit travail de *M. Housel*, que j'apprécie à beaucoup d'égards, j'y avais les mêmes objections à faire qu'à présent; mais c'est le livre de *M. Cantor* qui m'a montré, par ses défauts à cet égard, combien était nécessaire une analyse approfondie de la Théorie ancienne des coniques, et j'espérais d'y réussir en développant ultérieurement les notions que je possédais déjà. [Tannery 1943, 645]

Nevertheless, in his published book *Die Lehre von der Kegelschnitten im Alterthum* [13], Zeuthen treated Cantor with great respect and even called him the leading scholar among those "welche in der neusten Zeit die mathematischen Überlieferungen in einem bis dahin unbekannten Umfange ans Licht gezogen und historisch und chronologisch beleuchtet haben" [Zeuthen 1886, XII]. In many places, he explicitly adopted Cantor's conclusions; indeed he took most of the historical facts from Cantor's *Vorlesungen über Geschichte der Mathematik*. But he also emphasized its

utility, "da ich wegen meines von dem seinen verschiedenen Aus-
gangspunktes an mehreren Stellen genötigt bin, mich seinen Ansichten
polemisch gegenüber zu stellen" [Zeuthen 1886, XII].

As we saw in the previous section, Zeuthen's criticism was directed
against Cantor's dating of the discovery of the *symptoma* and the related
fact that the conics can be generated as oblique planar sections of a cone.
Moreover, Cantor had expressed some doubt as to whether Archimedes
had known the sum of the general arithmetic series, and Zeuthen thought
that this scepticism was unfounded.

When Cantor published the second edition of the first volume of his
Vorlesungen, he mentioned the recent outburst of new results in the history
of mathematics, and partly ascribed that outburst to the publication of the
first edition. (As we saw, this was the case with Zeuthen's contributions.)
He continued:

> dass ich bestrebt gewesen bin, alles mir zugänglich Gemachte aus den letzten
> 13 Jahren zum Theile bis in die Druckzeit selbst, zu verwerthen, bald indem
> ich die neuen Ergebnisse einfach übernahm, bald indem ich sie, ohne mich
> ihnen anzuschliessen, nur erwähnte. [Cantor 1894b, V]

In view of this remark, one would expect to find that Zeuthen's extensive work had been taken into account, but in fact Cantor practically
ignored it. This was not, however, because of a disagreement with Zeuthen
regarding particular points, but rather because of his basic opposition to
the way in which Zeuthen did research in the history of mathematics. In
fact, in his treatment of Apollonius, Cantor failed to mention Zeuthen's
work even once, and one particular passage contains an implicit attack on
Zeuthen's theories:

> Dieses in Kürze der Inhalt des merkwürdigen Werkes, wobei wir uns gegen
> die verlockende Versuchung, noch mehr hineinzulesen als Apollonius gesagt
> hat, zu wappnen gesucht haben. Auch der von uns angegebene nackte Inhalt
> ist sehr wohl geeignet, unsere Neugier anzuregen, inwieweit derselbe Mathe-
> matiker seinen erfinderischen Geist auch noch anderen Gebieten unserer
> Wissenschaft zuwandte. [Cantor 1894b, 327]

In connection with Euclid, Cantor mentioned Zeuthen twice. First he
explained that "Euklid muss mit numerischen quadratischen Gleichungen
zu tun gehabt haben, denn nur daraus lässt sich das Entstehen des X.
Buches der Elemente erklären" and added in a footnote "Dieser feine und
wichtige Gedanke ist zuerst ausgesprochen bei Zeuthen . . . " [Cantor 1894b,
270]. Second Cantor maintained that Euclid did not know the *symptoma* of

the conic sections, but added in a footnote that Arneth and Zeuthen disagreed with him [Cantor 1894b, 276].

The Public Dispute

Without directly attacking it, Cantor's indifference toward Zeuthen's work was tantamount to a pointed rejection of its tendencies. It is therefore no wonder that even the polite Zeuthen responded to this challenge in a rather aggressive note entitled *M. Maurice Cantor et la géométrie supérieure de l'antiquité* [Zeuthen 1894], in which he pointed out that while Cantor had collected and cited many (even conflicting) views in a liberal way he had not profited from his *Kegelschnitte im Alterthum*. Since Cantor had not explained his reasons for ignoring the arguments in this book, Zeuthen felt that the only way he could defend himself was by indicating why Cantor should not be regarded as a leading authority on ancient mathematics. Yet Zeuthen did not deny the virtues of Cantor's book:

> Je me hâte d'ajouter qu'il serait aussi inutile qu'injuste d'essayer de rendre suspects aussi les fruits des recherches plus directement *historiques* de M. Cantor: le soin infatigable dont il a ramassé les faits historiques de tout côté possible, la sage critique dont il les a élaborés et la justesse dont il les a exposés sont trop légitimement respectés pour cela. [Zeuthen 1894, 164]

Even this praise, however, contains an implicit criticism of Cantor as a mere collector. Yet Zeuthen sounds sincere when he continues:

> Il serait même ingrat de ma part d'en élever aucun doute; ce n'est en effet qu'en ayant égard à toutes les remarques historiques de M. Cantor et en consultant les auteurs cités par lui, que je me suis mis à l'abri du danger de donner des auteurs anciens une explication qui ne concorde pas avec le temps de leurs productions. [Zeuthen 1894, 164]

Thus Zeuthen only directed his criticism towards Cantor's "analyse mathématique des grands auteurs ici en particulier de ceux de l'antiquité," and on this point he was rather severe:

> les jugements qu'il porte sur la valeur et sur la connexion de leurs différentes prestations montrent qu'il ne les a pas soumis à l'étude mathématique que méritent les travaux des grands géomètres. [Zeuthen 1894, 164]

In particular, Zeuthen criticized the fact that Cantor "préfère toujours les sources variées de l'historien aux importants documents mathématiques conservés des mains des grands géomètres" [Zeuthen 1894, 166]. As a case in point Zeuthen returned to the argument in his book concerning the discovery of the *symptoma* of the conic sections and the generation of these curves by arbitrary sections of arbitrary cones.

> On le voit, avant tout, à la manière dont il compare Apollonius à ses prédécesseurs. Pour constater les progrès dus à ce grand géomètre il renvoie à des remarques de Geminus et Pappus que sans doute aucun historien ne négligera impunément; mais de son côté M. Cantor néglige (entièrement dans la première édition) d'en contrôler la portée par la comparaison des oeuvres d'Archimède et d'Apollonius. Il laisse échapper de cette façon une occasion de corriger l'erreur généralement répandue qui se rattache à la dénomination de théorème d'Apollonius. [Zeuthen 1894, 166]

In fact, even in the second edition Cantor had maintained that the *symptoma* were discovered by Apollonius; about Archimedes' earlier use of them "M. Cantor n'en dit rien" [Zeuthen 1894, 166]. Zeuthen further noted that Cantor's second edition mentions how Archimedes knew the oblique generation of the ellipse, but he criticized Cantor for having failed to mention his own hypothesis, i.e., that Geminus' statement only concerned the stereometric *definitions* of the conics. Finally, and most serious of all, Zeuthen accused Cantor of having given a bad evaluation of the outstanding nature of Apollonius' Conics and of having completely misinterpreted the *goal* of the work. In fact Cantor had written

> So musste das IV. Buch...gleichmässige Verbreitung mit den 3 ersten Büchern gewinnen, deren Abschluss es gewissermassen für solche Mathematikstudirende bildete, welche von der damaligen höheren Mathematik gerade das in sich aufnehmen wollten, was bis zur Lösung der delischen Aufgabe, diese mit inbegriffen, nothwendig war. [Cantor 1894b, 325]

Zeuthen read this passage as a claim on Cantor's part to the effect that the aim of Apollonius' first four books was a treatment of the Delian problem, and he strongly rejected this point of view. After pointing out some inconsistencies in Cantor's evaluation of the various parts of Apollonius' work, he concluded with a response to Cantor's implicit attack on his methodological approach:

> Il faut convenir à M. Cantor qu'il a très bien évité de trouver [*hineinzulesen*] dans les coniques d'Apollonius quelque chose qu'Apollonius n'a pas dit, sagesse qu'il faut certainement apprécier beaucoup dans un manuel historique; mais nous ne comprenons guère ce qu'il dit de ses tentations à cet égard. Avant d'y venir, il aurait fallu s'occuper de beaucoup de choses que dit Apollonius et que M. Cantor a omises. Il aurait pu, par exemple, donner une idée plus complète des principes des démonstrations, et citer les trois derniers théorèmes du troisième livre qui contiennent de fait la démonstration qu'une conique quelconque est un *lieu à trois droites*. Et M. Cantor a tort en distant qu'il donne le contenu nu (*nackte Inhalt*) de l'Ouvrage

d'Apollonius; en effet, ses remarques citées sur le but des quatre premiers livres en *voilent* les plus grandes beautés...

J'espère avoir dit assez pour montrer que M. Cantor n'ajoute pas à ses autres grands et incontestables mérites celui d'être assez bon interprète d'Archimède et d'Apollonius pour en rendre superflues d'autres interprétations, et qu'il n'a nullement fait voir les qualités nécessaires pour juger de la valeur historique et géométrique des autres interprétations. [Zeuthen 1894, 169]

With this note, the atmosphere became charged and the polite exchange of differing opinions was no longer possible. Cantor replied in this vein in a rejoinder published in the next volume of the *Bulletin des sciences mathématiques*. Already the title contains a subtle attack on Zeuthen's historical approach: "M. Zeuthen et sa [sic!] géométrie supérieure de l'antiquité."

The main thrust of Cantor's counterattack was to point out that Zeuthen had completely distorted the meaning of Cantor's remark concerning the connection between the Delian problem and the first four books of Apollonius' conics. Cantor's goal had been to explain that only the first four books had been preserved in Greek because later scholars and copyists had been interested mainly in those parts of Apollonius' work that could help them treat the Delian problem. This they had found in Books 1–4, whereas the later books were too complicated to be understood by a sufficient number of people.

Mais qu'Apollonius se soit proposé comme but de faire contenir dans ses quatre premiers Livres ce qu'il fallait pour résoudre le problème Délique, c'est ce que je n'ai jamais voulu dire, et j'espère que mes lecteurs, en comparant ma phrase, l'explication que je viens d'en donner et la soi-disante transcription de M. Zeuthen se rangeront de mon avis, que M. Zeuthen me prête des opinions que je n'ai jamais émises...

Certes, l'irritation de M. Zeuthen contre moi a dû être bien grande pour l'aveugler de façon à lui faire faire ce que je nomme *hineinlesen*, c'est-à dire parvenir à lire dans un auteur ce qu'on voudrait y trouver, tantôt pour l'en blâmer, tantôt pour l'en louer. [Cantor 1895, 66]

Zeuthen, who claimed that he could analyze the writings of the ancient Greek authors without projecting his own mathematical ideas into their work, had thus proven himself unable to interpret a text written yesterday without "hineinlesen." Since the irony in this situation was obvious, Cantor did not emphasize it further.

Instead, he went on to explain why he had not referred more explicitly to Zeuthen's ideas in the second edition of his *Vorlesungen*.

Les instincts personnels sont différents. Il y a des personnes qui aiment les débats scientifiques et autres, en un mot la polémique, il y en a d'autres qui la détestent, et je fais partie des derniers. Jamais, dans la carrière scientifique assez longue sur laquelle je regarde en arrière, je n'ait porté les premiers coups, et la polémique me répugne d'autant plus, si elle doit s'adresser à un savant dont j'estime le mérite incontestable sur un terrain qui lui est propre. S'égare-t-il autre part, je me tais d'abord, et je ne parle qu'y étant forcé. C'est ainsi que je me suit tu vis-à-vis de M. Zeuthen le géomètre éminent, et c'est à regret que je me sens obligé à riposter une fois, mais pas davantage, comme je constate dès aujourd'hui. [Cantor, 1895, 67]

Regarding the generation of the conics by oblique planar sections and the *symptoma*, Cantor maintained that he trusted the testimony of an ancient commentator more than the guesswork of a modern mathematician. He then turned to the three and four point loci. Having summarized the facts known from the ancient sources he continued:

C'est ici que M. Zeuthen est entré en lice. Il s'est saisi du problème à trois ou à quatre droites en maître de la Géométrie synthétique moderne. Il a trouvé la conique en question en ne s'appuyant que sur des vérités contenues dans le III Livre d'Apollonius. C'est tout ce qu'il a de plus ingénieux comme étude géométrique, mais ce n'est pas de l'histoire.

Je ne puis pas prouver que la marche d'Apollonius, s'il a mis par écrit ses pensées sur le problème, ce qui n'est sans vraisemblance, ait été différente de celle de M. Zeuthen; nous ne la connaissons pas! Mais M. Zeuthen peut encore bien moins prouver qu'il se trouve sur les pas d'Apollonius. C'est sa Géométrie supérieure de l'antiquité à lui qu'il nous donne.

Avais-je le droit de la passer sous silence dans un Volume gros déjà de cinquante-cinq feuilles et que je devais, par conséquent, m'abstenir de grossir encore, à moins qu'il ne s'agit de nouvelles découvertes historiquement avérées? Je le crois. M. Zeuthen est de l'avis opposé, et c'est ce que je comprends facilement, puisqu'il s'agit d'hypothèses auxquelles il a voué un travail long, consciencieux, et à son opinion fertile. Nous ne différons que sur ce dernier point. [Cantor 1895, 68–69]

After this precise and polemical explanation of his critical view of Zeuthen's historical method, he left it for future historians of mathematics to judge this dispute: "Nous verrons bien, et le public verra aussi, s'ils consentiront à réunir sous le nom d'Apollonius les recherches de M. Zeuthen" [Cantor 1895, 69]. This was clearly an unfair way to pose the question. Of course, no historian, not even Zeuthen, would attribute Zeuthen's reconstruction to Apollonius. It is worth noting, however, that almost all later writers on Apollonius' conics have in fact taken Zeuthen's work into account.

Later in the same year, Zeuthen issued a brief "Réponse aux remarques de M. Cantor" [Zeuthen 1895] in which he apologized for having misinterpreted the passage concerning the Delian problem. Otherwise, however, he maintained his previous points of view, in particular that the last three theorems of the third book of Apollonius' Conics "contient de fait la démonstration qu'une conique quelconque est un *lieu à trois droites.*" He referred to this as a fact, although he admitted:

> ce sont les *démonstrations* de ces théorèmes qui contiennent cette propriété d'une conique quelconque. On la trouve *de fait*, mais non pas énoncée formellement, dans l'édition Heiberg, t. I, 442, ligne 10; 446, 1. 2–3; 448, 1. 24–25. [Zeuthen 1895, 184]

Thus the two adversaries could not even agree about what is meant by a historical fact. Thereby ended the dispute between the encyclopedist and the mathematical analyst.

Cantor's View of Zeuthen's Mathematical Analyses

Before we conclude with the more indirect continuations of the dispute we shall quote a surprising statement from the young Cantor.

> Wenn bei historischen Untersuchungen von einer Verschiedenheit der Schwierigkeiten gesprochen werden kann, so gehören sicher die Forschungen zu den feinsten, bei denen es sich um die Wiederherstellung verloren gegangener Werke aus mehr oder minder sparsamen Resten handelt, und wo es folglich mehr als bei irgend einer anderen Arbeit nötig ist, sich so in den Geist des Autors zu versetzen und den wissenschaftlichen Standpunkt seines Zeitalters so zu erfassen, dass man sich mit demselben identificirt und die Divination des zu behandelnden Werkes mehr eine Production als eine Reproduction wird. [Cantor 1857b, 17]

As an example of a successful reconstruction Cantor mentioned Viviani's reconstruction (1659) of Book V of the Conics of Apollonius, which had been brilliantly confirmed by the discovery of the Arabic version. Cantor continued:

> Nicht so gut ward es den Wiederherstellern sonstiger Werke des Alterthums, die in ihren vielmehr schwierigen als dankbaren Forschungen um so weiter von einander abwichen, je pflichtgetreuer sie sich von allen neueren Kenntnissen entkleidet in die Natur des Autors versenkt hatten. Leicht erklärlich, hatte doch jeder mit seinen Kenntnissen nicht seinen eigenen Geist abstreifen können, und dieser ist zu verschieden, als dass er nicht von demselben Standpunkte ausgehend je nach seiner Individualität unzählige Wege einschlagen könnte, einschlagen müsste, so dass es hier in gutem Sinne sich

bewährt, was unser grösster deutscher Dichter vielleicht etwas boshaft ausspricht: "Es ist der Herren eigner Geist, in dem die Zeiten sich bespiegeln." [Cantor 1857b, 18]

It is hard to reconcile Cantor's early characterization of the art of reconstruction as one of the finest preoccupations of the historian of mathematics with his later criticism of Zeuthen's approach. The only way would be to assume that Cantor did not think that Zeuthen had been able to penetrate far enough into the spirit of the ancient Greeks so that his reconstructions were unsuccessful. However, his lack of direct refutation of Zeuthen's arguments shows already that Cantor simply changed his mind regarding the value of such mathematical arguments. In fact, this is clearly demonstrated in a review of Zeuthen's *Geschichte der Mathematik im Alterthum und Mittelalter* that Cantor wrote one year after the dispute in the *Bulletin*.

After a strange half-page where he argued that publications in minor languages like "böhmisch, dänisch, holländisch, norwegisch, polnisch, portugiesisch, russisch, schwedisch, spanisch, ungarisch" are actually only semipublic, he turned to his most explicit enunciation of his own confession of faith as a historian of mathematics and his clearest repudiation of Zeuthen's approach:

Ein geistvolles Bändchen nannten wir Herrn Zeuthen's Geschichte der Mathematik im Altherthum und Mittelalter, und dieser Bezeichnung thut es keinen Abbruch, dass wir den Ergebnissen mit dem entschiedensten Misstrauen gegenüber stehen. Das kommt wohl daher, dass wir die alten Schriftsteller, über welche wir Bericht zu erstatten haben, anders als Herr Zeuthen lesen. Wir lesen sie ihrem Wortlaute nach, und wo dieser zweifelhaft oder unklar erscheint, fragen wir den Schriftsteller selbst oder seine nächsten Zeitgenossen, ob diese nicht an anderen Stellen Erklärungen abgegeben haben, welche die Zweifel heben können. Nur wenn auch dieses Mittel sich als unzureichend herausstellt, wagen wir es, eigene Hypothesen zu versuchen, die stets nur als solche von uns verwerthet werden. Anders Herr Zeuthen. Er liest mit grösster Genauigkeit die Werke der hervorragendsten Schriftsteller, im griechischen Alterthume etwa Euklid, Archimed, Apollonius, und, wenn er sie gelesen hat, überlegt er, was er selbst sich wohl dabei gedacht haben würde, wenn er genau die gleichen Sätze in gleicher Reihenfolge zu Papier gebracht hätte. Das Ergebniss dieser Nacherfindung ist ihm dann griechische Mathematik. Uns will scheinen, dass dabei nur ein Mangel ist. Der Geist des XIX. Jahrhunderts ist nicht der gleiche wie der der vorchristlichen Zeit. Unsere Denkweise ist, auch wenn wir die gleichen Wege einschlagen, im Laufe von zwei Jahrtausenden eine andere und immer andere geworden. Wir wissen es nicht einmal, wie himmelweit anders unsere Schlüsse sind, als die der Griechen, und verfallen dadurch in nach unserer persönlichen Meinung irrige Auffassungen. Wir glauben uns zu antikisiren, und wir modernisiren die Alten. Wir verfahren auf geometrischem Gebiete nicht viel anders, als wenn wir beim Anblick jener altägyptischen Bilder von Figuren, deren Körper trotz der bedeckenden Kleidung in deutlichen Umrissen gezeichnet ist, an eine Benutzung von X-Strahlen dächten.

Wer dieses unser geschichtliches Glaubensbekenntniss nicht theilt, wird Herrn Zeuthen's kleines aber inhaltreiches Buch mit wahrem Entzücken lesen, wird in Euklid, in Archimed, in Apollonius Mathematiker unserer Zeit wiedererkennen mit gleichem Bestreben nach unverbrüchlicher Strenge, mit fast gleichen Mitteln diese Strenge erzwingend. [Cantor 1896b, 182–183]

Cantor mentioned only one of Zeuthen's ideas in the book, namely that the decline of Greek mathematics in late antiquity was partly caused by its rigorous logical form, which made the mathematics difficult to understand for students who were not taught orally by a master who knew the game.

Die Richtigkeit dieser letzten Thatsache ist nicht anzuzweifeln. Aus ihr zu folgern, dass man nach mehr als zwei Jahrtausenden besser im Stande sei, alte Gedanken noch einmal zu denken, als nach wenigen Jahrhunderten, so weit zu gehen, fühlen wir uns ausser Stande. [Cantor 1896b, 183].

These were the last public words in the dispute between the two great historians of mathematics.

Zeuthen's Personal Views of Cantor's Vorlesungen

Considering the harshness with which Cantor had criticized Zeuthen's work, it is perhaps surprising to see that Zeuthen continued to praise Cantor's contributions as a historian. In the preface to his *Geschichte der Mathematik im XVI. and XVII. Jahrhundert*, he mentioned that he had profited from several contemporary writers:

Unter allen diesen Schriftstellern muss ich wie ein jeder, der grössere oder kleinere Abschnitte der Geschichte der Mathematik behandeln will, zuvörderst dankend Moritz Cantor nennen. Ohne seine in seltenem Grade vollständigen Vorlesungen würde mir sicherlich manches, das mit heranzuziehen war, entgangen sein, und ich würde jegliche Kontrolle vermissen, um erachten zu können, ob ich im wesentlichen auch alles das berücksichtigt habe, dessen ich zu meinem Zwecke bedarf. [Zeuthen 1903c, V]

One might, of course, suspect that Zeuthen wrote this merely to avoid provoking Cantor's anger, but the fact that he published the same praise in the Danish edition makes this implausible. In fact, Zeuthen seems to have sincerely appreciated the great value of Cantor's *Vorlesungen* as an important reference work and as a valuable basis for further progress in the history of mathematics. Thus, when he won the Prix Binoux (cf. above) he wrote back to Tannery:

Cependant il me reste quelque scrupule devant M. Cantor. En tout cas, il ne serait pas difficile de trouver des points de vue pour une comparaison entre lui et moi, d'où sa supériorité serait certaine. En effet, moi je dois beaucoup à M. Cantor, qui m'a fait connaître par ses livres un grand nombre des faits que j'ai essayé d'ordonner, et qui m'a indiqué là les auteurs que je devais étudier, etc. Je pourrais ajouter les suggestions dues à celles de ses explications qui ne m'ont pas satisfait, et qui m'ont excité à en chercher de meilleures*. Cantor, de son côté, ne doit presque rien à moi. Ses idées s'étaient essentiellement consolidées avant mes travaux, ce qui l'a empêché d'en tirer parti même là où cela, peut-être, aurait été utile. [Tannery 1943, 671–672]

In a footnote Zeuthen added:

Je ne dis nullement cela dans un mauvais sens; c'est la voie naturelle des progrès; et je ne dois pas être mécontent si mes propres travaux donneront des suggestions semblables à mes successeurs. [Tannery 1943, 672]

However, although Zeuthen appreciated Cantor's historical contributions, he felt personally hurt by Cantor's attacks. This can be seen from a letter to Heiberg dated December 12, 1896 (i.e., just after Cantor's review of his book). Here Zeuthen writes that "as long as Cantor—to be sure without reason... —so ardently fight against my views," he finds it wisest to answer in the negative to a question Heiberg had asked him. We may guess that the question concerned a possible contribution by Zeuthen to the *Festschrift* celebrating Cantor's 70th birthday. As mentioned above, Zeuthen in fact did not contribute an article to it.

Later, Cantor and Zeuthen corresponded in a more friendly manner than in 1894–1896, as can be seen from the following passage, which, however, also indicates that Zeuthen did not respect Cantor highly as a person.

Que mes avis sont partagés aussi par d'autres, voilà ce que dit la carte ci-joint de M. Cantor [14]. Je lui avais communiqué la décision du ministre, ainsi l'annonce de M. W. de ses premières leçons et mes propres pensées à cet égard (je devais lui répondre à l'invitation à annoncer quelque communication pour Heidelberg).

Vous saurez sans doute faire abstraction des remarques politiques à la fin de la carte—ou vous en amuser. (Zeuthen to Tannery 1904). [Tannery 1943, 683]

POSTSCRIPT

Although the public dispute between Cantor and Zeuthen ended in 1896 it is well known that the controversies between the protagonists of the two methodologies in the history of (particularly ancient) mathematics have continued till this day. The recent disputes are discussed by Berggren [1984, 397–398]. See also Sarton [1955, 368–369] and Neugebauer [1956, 58]

ACKNOWLEDGMENTS

An abbreviated version of this paper was presented at the XVIIIth International Congress of History of Science in the section "Historiography and History of Mathematics." We express our gratitude to the organizers of this section Sergei Demidov and Menso Folkerts and to Hans Wussing who suggested this topic to us.

NOTES

[1] Concerning the origin of this volume, see the report in *Jahresbericht der Deutschen Mathematiker-Vereinigung* **13** (1904), 475–478.

[2] Georg Cantor wrote in a letter to Hermite of Dec. 26, 1895: "Der Heidelberger Professor Moritz Cantor ist *nicht mit mir verwandt*." [Purkert and Ilgauds 1987, 196].

[3] Printed in Heidelberg in 1855 as a separate book (cf. [Cantor 1855]).

[4] Concerning Arneth, F. Cajori denies any influence on Cantor [Cajori 1920, 22]. Cajori's opinion is based on Cantor's criticism of Arneth's book [Arneth 1852] in his short biography of Arneth in the *Allgemeine Deutsche Biographie*. But this criticism was expressed by a mature scholar more than 20 years after his student days and does not concern the methodological and philosophical basis of Arneth's work, but only the poor contents of the part dealing with the history of mathematics from 500 A.D. onward. Arneth's tendency to link the cultural development of different peoples to their particular ways of doing mathematics may well have had an influence on Cantor.

[5] Personalakte M. Cantor (unpaginated). Archive of the University of Heidelberg. We thank R. Siegmund–Schultze (Berlin) for information on its contents.

[6] As to the history of our numerals see [Menninger 1958] and [Ifrah 1986].

[7] Cantor did not mention [Woepcke 1859], and it is very likely that he was not aware of this work when he wrote his book [Cantor 1863].

[8] Cantor was mistaken to ascribe the codex to a Roman land surveyor named Nipsus; this name was invented by a later copyist.

[9] One may argue that from a geometrical point of view there is a great difference between Apollonius' *symptoma* (equality of certain areas) and Archimedes' *symptoma* (equality of certain ratios) but Cantor never adduced this argument.

[10] This talk is mentioned in the archives of the Royal Danish Academy, but like several other talks Zeuthen held at this institution, it does not seem to have been published.

[11] The Zeuthen correspondence gives evidence that Zeuthen first contemplated having Tannery's works published by the Royal Danish Academy, but finally chose a French publisher in order to get a better international circulation. Zeuthen co-edited the first three volumes of Tannery's *Mémoires Scientifiques*. After Zeuthen's death, Madame Tannery insisted that an obituary by Picard be printed as a preface to the fifth volume together with the last letter she had received from Zeuthen.

[12] [Curtze and Günther 1899]. In addition to German scholars the following contributed: V. V. Bobynin, F. Cajori, S. Dickstein, G. Eneström, A. Favaro, E. Gelcich, J. H. Graf, T. S. Heath, J. L. Heiberg, G. Loria, P. Mansion, F. Rudio, H. Suter, and P. Tannery.

[13] In this section we quote the German translations of Zeuthen's works because these were the editions to which Cantor responded.

[14] This post card as well as all other letters between Zeuthen and Cantor seems to have been lost.

REFERENCES

Arneth, A. 1852. *Die Geschichte der reinen Mathematik in ihrer Beziehung zur Geschichte der Entwicklung des menschlichen Geistes*. Stuttgart: Francksche Buchhandlung. [Separate print from the *Neue Encyclopädie für Wissenschaften und Künste*]

Berggren, J. L. 1984. History of Greek mathematics: A survey of recent research. *Historia Mathematica* 11, 394−409.

Bopp, K. 1928. Moritz Cantor. In *Deutsches Biographisches Jahrbuch 1917−1920*, pp. 509−512. Stuttgart: Deutsche Verlags-Anstalt.

Cajori, F. 1920. Moritz Cantor, the historian of mathematics. *Bulletin of the American Mathematical Society* 27, 21−28.

Cantor, M. B. 1855. *Grundzüge einer Elementararithmetik als Leitfaden zu akademischen Vorträgen*. Heidelberg: Bangel & Schmitt.

—— 1856. Über die Einführung unserer gegenwärtigen Ziffern in Europa. *Zeitschrift für Mathematik und Physik* 1, 65−74.

—— 1857a. Petrus Ramus, Michael Stifel, Hieronymus Cardanus, drei mathematische Charakterbilder aus dem 16. Jahrhundert. Vortag, gehalten zu Bonn in der mathematisch-astronomischen Section der 33. Naturforscherversammlung. *Zeitschrift für Mathematik und Physik* **2**, 353–367.

—— 1857b. Über die Porismen des Euklid und deren Divinatoren. *Zeitschrift für Mathematik und Physik* **2**, 17–27.

—— 1858. Ramus in Heidelberg. *Zeitschrift für Mathematik und Physik* **3**, 133–143.

—— 1859. Die Professur des Ramus. *Zeitschrift für Mathematik und Physik* **4**, 314–315.

—— 1860. Sur l'époque à laquelle a vécu le géomètre Zenodore; Lettre à M. Chasles. *Comptes Rendus Académie des Sciences, Paris* **51**, 630–633.

—— 1863. *Mathematische Beiträge zum Kulturleben der Völker*. Halle: H. W. Schmidt.

—— 1865. Über einen Codex des Klosters Salem. *Zeitschrift für Mathematik und Physik* **10**, 1–16.

—— 1867. *Euclid und sein Jahrhundert*. Leipzig: Teubner.

—— 1875. *Die Römischen Agrimensoren und ihre Stellung in der Geschichte der Feldmesskunst*. Leipzig: Teubner.

—— 1880. *Vorlesungen über Geschichte der Mathematik*. Band I. Leipzig: Teubner.

—— 1892. *Vorlesungen über Geschichte der Mathematik*. Band II. Leipzig: Teubner.

—— 1894a. *Vorlesungen über Geschichte der Mathematik*. Band III. Teil 1. Leipzig: Teubner.

—— 1894b. *Vorlesungen über Geschichte der Mathematik*. Band I, 2nd ed. Leipzig: Teubner.

—— 1895. M. Zeuthen et sa géométrie supérieure de l'antiquité. *Bulletin des Sciences Mathématiques* (2. sér.) **19**, 64–69.

—— 1896a. *Vorlesungen über Geschichte der Mathematik*, Band III, Teil 2. Leipzig:Teubner.

—— 1896b. Review of Zeuthen's *Geschichte der Mathematik im Alterthum und Mittelalter*. Zeitschrift für Mathematik und Physik. Historisch-Literarische Abteilung **41**, 182–183.

—— 1898a. *Vorlesungen über Geschichte der Mathematik*, Band III, Teil 3. Leipzig: Teubner.

—— 1898b. *Politische Arithmetik oder die Arithmetik des täglichen Lebens*. Leipzig: Teubner. [1903^2]

—— 1990. *Vorlesungen über Geschichte der Mathematik*, Band II, 2nd ed. Leipzig: Teubner.

—— 1901. *Vorlesungen über Geschichte der Mathematik*, Band III, 2nd ed. Leipzig: Teubner.

—— 1903. Wie soll man die Geschichte der Mathematik behandeln? *Bibliotheca Mathematica, New Series* **4**, 113–117.

—— 1907. *Vorlesungen über Geschichte der Mathematik*, Band I, 3rd ed. Leipzig: Teubner.

—— 1908. *Vorlesungen über Geschichte der Mathematik*, Band IV. Leipzig: Teubner. [With contributions by V. Bobynin, A.v. Braunmühl, F. Cajori, S. Günther, V. Kommerell, G. Loria, E. Netto, G. Vivanti, & C. R. Wallner]

Curtze, M. 1899. Zum siebenzigsten Geburtstage Moritz Cantors. *Zeitschrift für Mathematik und Physik* **44**, 227–231.

Curtze, M. and Günther, S., Eds. 1899. Festschrift Herrn Hofrat und Professor Dr. Moritz Cantor bei der 70. Wiederkehr des Tages seiner Geburt am 23. August 1899 dargebracht von seinen Freunden und Verehrern. *Abhandlungen zur Geschichte der Mathematik*. Supplement to *Zeitschrift für Mathematik und Physik* **44**.

Descartes, R. 1637. *La Géométrie*, Appendix to the *Discours de la méthode*. In *Oeuvres*, Vol. VI, pp. 297–413. Leiden: Ian Maire.

Eneström, G. 1891. Review of [Cantor 1892, first part]. *Bibliotheca Mathematica* **5**, 117–118.

—— 1903. Review of [Cantor 1898a]. *Bibliotheca Mathematica. New Series* **4**, 446–447.

Folkerts, M. 1970. *"Boethius". Geometrie II. Ein mathematisches Lehrbuch des Mittelalters*. Wiesbaden: Steiner.

Friedlein, G. 1864. Zur Geschichte unserer Zahlzeichen und unseres Ziffernsystems. *Zeitschrift für Mathematik und Physik* **9**, 73–95.

Gibson, G. A. 1898. Review of Cantor's "Vorlesungen." *Proceedings of the Edinburgh Mathematical Society* **17**, 9.

Hankel, H. 1874. *Zur Geschichte der Mathematik in Alterthum und Mittelalter*. Leipzig: Teubner.

Heiberg, J. L., and Zeuthen, H. G. 1906. Eine neue Schrift von Archimedes. *Bibliotheca Mathematica* (*3*) **7**, 321–363.

—— 1907. Einige griechische Aufgaben der unbestimmten Analytik. *Bibliotheca Mathematica* (*3*) **8**, 118–134.

Heiberg–Zeuthen correspondence. *One letter from Heiberg to Zeuthen*. Royal Library Copenhagen NKS 4417, 4°.

Hjelmslev, J. 1939. Hieronymus Georg Zeuthen. Foredrag ved Mindefesten i Matematisk Forening i Anledning af 100-Aarsdagen d. 15. Febr. 1939 for H. G. Zeuthens Fødsel. *Matematisk Tidsskrift* (*ser. A*), 1–10.

Hofmann, J. E. 1957. Moritz Benedikt Cantor. In *Neue Deutsche Biographie*, Band III, p. 129. Berlin: Duncker & Humblot.

—— 1971. M. B. Cantor. In *Dictionary of Scientific Biography*, Vol. III, pp. 58–59. New York: Scribner's.

Houzel, C. P. 1858. Les coniques d'Apollonius. *Journal de Mathématiques Pures et Appliquées* (*2. ser.*) **2**, 153–192.

Ifrah, G. 1986. *Universalgeschichte der Zahlen*. Frankfurt: Campus Verlag.

Kleiman, S. L. 1991. Hieronymus Georg Zeuthen 1849–1920. In: Enumerative Algebraic Geometry. Ed. S. L. Kleiman and A. Thorup. Contemp. Math. Vol. 123. AMS Providence 1991, p. 1–13.

Lorey, W. 1916. Das Studium der Mathematik an den Deutschen Universitäten seit Anfang des 19. Jahrhunderts. In *Abhandlungen über den Mathematischen Unterricht in Deutschland*. Bd. III, Heft 9. Leipzig/Berlin: Teubner.

Lützen, J., Sabidussi G., and Toft, B. 1992. *Julius Petersen 1839–1910. A Biography*, Discrete Mathematics **100**, 9–82.

Menninger, K. 1958. *Zahlwort und Ziffer*, Vol. II. Göttingen: Vandenhoeck & Ruprecht.

Neugebauer, O. 1956. A notice of ingratitude. *ISIS* **47**, 58.

Noether, M. 1921. Hieronymus Georg Zeuthen. *Mathematische Annalen* **83**, 1–23.

Purkert, W., and Ilgauds, H.-J. 1987. *Georg Cantor 1845–1918*. Basel: Birkhäuser.

Picard, E. 1922. H. G. Zeuthen. In *Mémoires Scientifiques* (P. Tannery), Vol. V, pp. xv–xix.

Roeth, E. M. 1846, 1858. *Geschichte unserer abendländischen Philosophie*, Band 1 (1846), Band 2 (1862). Mannheim: Bassermann.

Sarton, G. 1955. Review of van der Waerden's "Science Awakening." *ISIS* **46**, 368–369.

Tannery, P. 1882. De la solution géométrique des problèmes du second degré avant Euclide, *Mémoires de la Société des Sciences Physiques et Naturelles de Bordeaux* **4**, 395–416; *Mémoires Scientifiques* I, pp. 254–280.

—— 1886. H. G. Zeuthen—Die Lehre von den Kegelschnitten im Alterthum; Review, *Bulletin des Sciences Mathématiques* **10**, 263–271; *Mémoires Scientifiques* XI, pp. 275–282.

—— 1896. H. G. Zeuthen—Geschichte der Mathematik im Alterthum und Mittelalter; Review. *Bulletin des Sciences Mathématiques* **20**, 105–109; *Mémoires Scientifiques* XII, pp. 11–14.

—— 1904. H. G. Zeuthen—Geschichte der Mathematik im XVIten und XVIIten Jahrhundert; Review. *Journal des Savants*, 364–365; *Mémoires Scientifiques* XII, pp. 255–256.

—— 1943. *Paul Tannery. Mémoires Scientifiques*, Vol. XVI, J. L. Heiberg and H. G. Zeuthen, Eds. Paris: Gauthier–Villars.

Treutlein, J. P. 1892. Review of [Cantor 1892]. *Jahrbuch über die Fortschritte der Mathematik*, 1–3.

—— 1894. Review of [Cantor 1894b]. *Jahrbuch über die Fortschritte der Mathematik*, 1–2.

Woepcke, A. 1859. *Sur l'introduction de l'arithmétique indienne en occident*. Rome: Imprimerie des sciences mathématiques et physiques.

—— 1863. Sur la propagation des chiffres indiens. *Journal Asiatique* 1863, 27–79, 234–290, 442–529.

Zeuthen, H. G. 1876. Fra Mathematikens Historie. I. Brahma-Guptas Trapez. *Tidsskrift for Mathematik* (*3*) **6**, 168–174, 181–191.

—— 1883. Fra Mathematikens Historie. IV. Prøver paa Diofants løsning af arithmetiske Opgaver. *Tidsskrift for Mathematik* (*5*) **1**, 145–156.

—— 1885. *Keglesnitslæren i Oldtiden*. Kjøbenhavn: Bianco Luno. [Videnskabernes Selskabs Skrifter, 6. Række, naturvidenskabelig og mathematisk Afdeling III. 1, 317 pp.]

—— 1886. *Die Lehre von den Kegelschnitten im Alterthum* (German transl. of [1885] by R. von Fischer-Benzon). Kopenhagen: Höst & Sohn.

—— 1892. Om Konstruktion som Existensbevis i den græske Mathematik. *Nyt Tidsskrift for Mathematik* **3A**, 105–113. [Also in German, *Mathematische Annalen* **47**, 222–228.]

—— 1893. *Forelæsning over Mathematikens Historie: Oldtid og Middelalder*. Kjøbenhavn: Høst & Søn. [German transl. 1896. French ed. 1902]

—— 1894. M. Maurice Cantor et la géométrie supérieure de l'antiquité. *Bulletin des Sciences Mathématiques* (*2*) **18**, 163–169.

—— 1895. Réponse aux remarques de M. Cantor. *Bulletin des Sciences Mathématiques* (*2*) **19**, 183–184.

—— 1896. Om den historiske Udvikling af Mathematiken som exakt Videnskab indtil Udgangen af det 18de Aarhundrede. In *Indbydelsesskrift til Kjøbenhavns Universitets Aarsfest i Anledning af hans Majestæt Kongens Fødselsdag d. 8de April 1896*.

—— 1900. Note on Ferrari's solution of the fourth degree equation. *Bibliotheca Mathematica* (*3*) **1**, 270.

—— 1903a. *Forelæsning over Mathematikens Historie i XVI. og XVII. Aarhundrede*. Kjøbenhavn: Høst & Søn.

—— 1903b. Ved Forelæggelsen af Mathematikens Historie i 16. og 17. Aarhundrede. *Oversigt over det Kongelige Danske Videnskabernes Selskabs Forhandlinger* 1903, no. 5, 553–572.

—— 1903c. *Geschichte der Mathematik im XVI. und XVII. Jahrhundert*, (German transl. of [1903a] by Ralph Mayer). Leipzig: Teubner.

—— 1905. L'oeuvre de Paul Tannery comme historien des mathématiques. *Bibliotheca Mathematica* (*3*) **6**, 257–304. [with bibliography by Enestrøm]

—— 1906. Abzählenden Methoden. *Encyclopädie der Mathematischen Wissenschaften* III C3 **3**, pp. 257–312.

—— 1912. *Die Mathematik im Altertum und im Mittelalter*. Kultur der Gegenwart III. Teil I. Abt. Die mathematischen Wissenschaftcn.

—— 1914a. *Lehrbuch der abzählenden Methoden der Geometrie*. Leipzig/Berlin: Teubner.

—— 1914b. Om Anvendelse af regning og af raisonnement i Mathematiken. Ved Forelæggelsen af Lehrbuch der abzählenden Methoden der Geometrie. *Oversigt over det Kongelige Danske Videnskabernes Selskabs Forhandlinger* 1914, 271–286.

—— 1917. Hvorledes Mathematiken i Tiden fra Platon til Euklid blev rationel Videnskab. Avec un résumé en français, *Oversigt over det Kongelige Danske Videnskabernes Selskabs Forhandlinger* (*8*) **1**, 199–379.

Zeuthen–Heiberg correspondence. *111 letters from Zeuthen to Heiberg*. Royal Library Copenhagen NKS 4417, 4°.

Zeuthen–Vilh. Thomsen correspondence. *92 letters from Zeuthen to Vilh. Thomsen.* Royal Library Copenhagen NKS 4291, 4°.

Hieronymus Georg Zeuthen (1839–1920)
Courtesy of Copenhagen University Mathematics Institute

Joseph Fourier (1768–1830)
Courtesy of MIT Press

"A New Type of Question": On the Prehistory of Linear and Non-linear Programming, 1770–1940

I. Grattan-Guinness

Middlesex University at Enfield, England

To the memory of Cecily Tanner, 1900–1992

Linear programming developed as a substantial branch of mathematics after the Second World War; but it has an extensive, though scrappy, prehistory. It appeared in embryo form at various times between the late 18th and the mid 19th centuries, with a very clear formulation put forward by Fourier in the 1820s, but then it fell largely into desuetude. During the first four decades of this century various studies concerning linear inequalities and/or convexity were carried out, in many different branches of pure and applied mathematics, but they did not often lead to linear programming, where progress remained slow. The whole history is a disjointed affair, with several opportunities missed or only partly taken; possible causes of the long gestation are discussed in the closing section. In an epilogue some tendencies toward nonlinear programming are detected, largely in connection with mechanics.

Modern mathematics is unthinkable without inequality and the inequality signs.

Tanner [1961, 293]

1. TERMS OF REFERENCE

The phrase "linear programming" will refer to a mathematical theory in which the optimum value(s) of some "objective function" (as it is now called), specified as a linear combination of variables of the problem context involved and with all coefficients usually taking the same sign, is to be found subject to further constraints expressed as linear equations and/or inequalities. If interpreted geometrically, these constraints define a convex (sometimes concave) region in an n-dimensional space with linear edges and faces, and the optimal value(s) may correspond to the vertices of the convex region. In some contexts no objective function as such is involved: the task set is just to determine a (convex or concave) region of "feasible solutions" of the equations and inequalities involved. In addition to the *conception* of the theory and attendant *applications*, the possibility of *effective methods of solution* needs to be considered.

The prehistory of this discipline is the concern of this essay. (Throughout a distinction is made between a "topic" and a "discipline," where the latter is understood as the state achieved by establishment as an individuated branch of mathematics). Further refinements or extensions of the umbrella discipline now often called "mathematical programming," such as dynamic or integer programming, will not be discussed, for they do not seem to have arisen in any significant way before the establishment of linear programming as a discipline after the Second World War. However, a line of thinking leading towards nonlinear programming will be described in Section 7.

The account of the prehistory is oriented around the mathematical aspects of the topic, although influences from other fields (such as economics) will be mentioned. Further, it is limited to work in which *substantive* doctrine of some kind approximating the above theory, or involving parts of it, was introduced or used; passing remarks and speculations are not recorded. Similarly, standard ways of using linear systems of equations in (proto-)linear algebra or inequalities in mathematical analysis, geometry, mechanics, or mathematical statistics are not considered.

The story is quite lengthy and rather disconnected, with different selections of the required conceptions and methods in place in a variety of mathematical contexts, often at different times (but, on interesting occasions, present contemporaneously): thus hindsight shows a generous collection of missed opportunities. These points are pondered over in Section 6. Before that, Sections 2 and 3 deal with work carried out between 1770 and 1860; they are followed by a lengthy Section 4 on various parallel and partly distinct developments executed between 1900 and 1940. Section 5 briefly records the establishment of linear programming as a discipline after the Second World War; however, the *content* of the founding works of this time will not be discussed in detail.

The bibliography includes all the main historical items, but only the most important or illustrative works from the primary sources. The dates of principal figures are given at their first main entry in the text. Braces are used to indicate sets: the use of set theory as such will be noted in Subsection 4.2.

2. BOSCOVICH'S CRITERION, AND FRENCH APPROPRIATIONS

2.1. *The Slow Dissemination of Boscovich's Criterion*

During the second half of the 18th century developments in celestial and planetary mechanics involved an improvement in the quality of instruments, and thus of observations, from which parameters, distances and angles could be calculated. However, although accuracy may have increased, observation error and equipment malfunction meant that individual data could not be relied on, to that sets of data had to be amassed from runs of readings. Methods of calculating parameters from the pertinent data available were then needed. Mathematical statistics was born in substantial part out of such tasks.

This was the context for the first significant foray in the direction of linear programming. It was carried out by R. J. Boscovich (1711–1787), as a consequence of an account of a geodetic survey of the Papal States which he wrote with the mathematician and theologian C. Maire. The task involved the preparation of a map of the States, which led them both to consider the shape of the earth also to note the "irregularities" in the data which they wished to deploy [Maire and Boscovich 1755, especially 498–502].

In the late 1750s Boscovich developed the following criterion [Eisenhart 1961, Sheynin 1973]. He calculated the length l of a degree of meridian at a given latitude u by a method which bears some similarity with our topic. He determined the coefficients A and B in the well-known formula

$$l = A \sin^2 u + B \tag{1}$$

by defining the correction c_i:

$$c_i := l_i - A \sin^2 u_i - B \tag{2}$$

for each set of data $\{(u_i, l_i)\}$ taken, and imposing the conditions

$$\sum_i c_i = 0, \quad \text{and} \quad \sum_i |c_i| \text{ to be minimal.} \tag{3}$$

The relationship between this criterion of absolute deviation and that of least squares regression will be considered in Subsection 7.3.

As if to presage the long gestation of linear programming, the history of Boscovich's criterion is complicated and lengthy. He stated it first in a paper placed with the Bologna Academy summarising the 1755 book; but the formulation was buried fathoms deep there, at least for any other reader [Boscovich 1757, 391–393], and not surprisingly it did not gain attention. He gave it a more substantial presentation in 1760, but in an unusual location: a series of prosodic supplements to a versified commentary on Newtonian philosophy written by the poet and philosopher B. Stay [Boscovich 1760, esp. 420–425]. All these texts were in Latin.

At the time of writing his supplements Boscovich was visiting London. He described his criterion to the English mathematician and statistician Thomas Simpson, who promptly found a partial algebraic solution which pleased Boscovich greatly [Stigler 1984, Farebrother 1990]. However, Simpson died during the following year and did not publish his result; and in the supplement Boscovich presented only his own geometrical solution. This did not employ a convex region; it was based upon plotting the data $\{(u_i, l_i)\}$ in the (u, l)-plane, dropping perpendiculars $\{p_i\}$ from each member datum onto a straight line passing through their centroid, and considering the effect of varying the slope of the line upon the value of $\sum_i |p_i|$ [Boscovich 1760, Fig. 44].

Stay's book did not catch the general interest of scientists upon its publication; thus Boscovich's criterion lay dormant for another decade until there appeared a translation of [Maire and Boscovich 1755] into French. Boscovich added an appendicial "Note," that included a translation of part of this supplement for Stay and some related material [Maire and Boscovich 1770, esp. 477–483, 501–508].

2.2. *Laplace and de Prony*

Boscovich's criterion was now more firmly placed in the scientific domain (so that I prefer to take 1770 as the effective starting point for this essay) [1]. In particular, it came to the attention of P. S. Laplace (1749–1827). Drawing upon Boscovich's work in a paper [1793, art. 9–14] and then in the second volume of his *Mécanique céleste* [Laplace 1799, Book 3, arts. 39–41], he replaced (3) by the minimax condition

$$\sum_i c_i = 0, \quad \text{and} \quad |\max\{c_i\}| \text{ to be minimal.} \tag{4}$$

He also developed a weighted version of this condition. With typical disdain, he acknowledged Boscovich only once, and then to say that Boscovich had "uselessly complicated" his solution "with the consideration of figures" [1793, art. 11], and he gave an analytical presentation of the solution to his own version (4), which assumed linear "equations of condition" between the $\{c_i\}$ and devising a method of elimination to give the minimax error.

Laplace's method was taken up by the engineer G. C. F. M. de Prony (1755–1839). This time the context was water-flow: in a monograph on this subject he proposed, as an empirical law, that the gradient P of a body of water was a quadratic function of its mean velocity u, and he wrote it in the linear form

$$P/u = au + b, \tag{5}$$

where a and b were constants [de Prony 1804, xvi–xxxii]. (I have omitted various other parameters from (5).) In order to determine a and b he used Laplace's approach but turned the tables on Laplace's insult to Boscovich by following "geometrical constructions [, which] are much more familiar than abstract analysis to a great number of engineers." He set up a Cartesian axis system in (P/u) and u, and used Laplace's method to form a convex region that enclosed the experimental data and to construct a

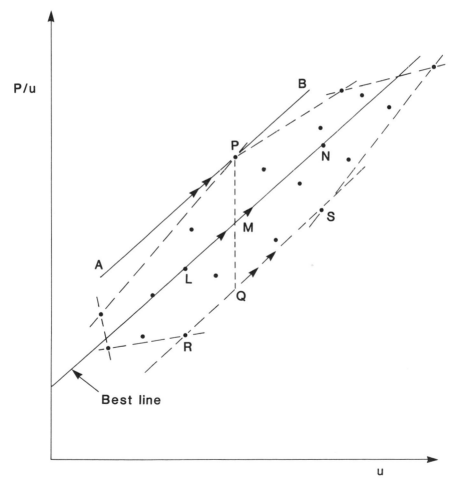

Fig. 1.   Geometrical representation of de Prony's method.

line of "best fit" through the region according to (4). Figure 1 represents his line of thought which, surprisingly, he did not provide: he presented his theory in the preface of his monograph, and maybe the plates had already been prepared [2]. Presumably he was unaware of Boscovich's work, for his own construction was quite different in deploying a convex region.

3. FOURIER'S INSIGHT, AND BOOLE'S FORAY

[T]his new branch of indeterminate analysis [. . .] comprises a very extended class of questions, susceptible to the most varied applications, and which are resolved by a uniform calculation, analogous to the algebraic method.

Fourier [1827b, 326]

3.1. From "Virtual Velocities" to "The Analysis of Inequalities"

The most explicit anticipation of linear programming also has its origins around 1800; the author involved was J. B. J. Fourier (1768–1830). He worked under the influence of Lagrange's recent treatise [1788] on variational mechanics, in which Lagrange used the principle of virtual work (to use the modern name) as an axiom. The question had arisen as to whether or not the "axiom" could actually be proved [Lindt 1904], and to this end Fourier published an important paper in 1798 containing some proofs. In the course of this study he stated the principle itself in terms of an inequality for the work expression instead of the usual equality to zero, on the grounds that the equilibrium of the mechanical system required that the work expression (which he called "the sum of moments") had to be nonnegative, not just zero. He also emphasised the point that the sign of a term in the work expression depended upon whether the force increased or decreased with increasing displacement [1798, arts. 20–21].

These remarks were not developed into a pukka theory, but Fourier worked further on inequalities from time to time, in between his other concerns [3]. At last he published his ideas during the 1820s: in parts of two notices of the work for 1823 and 1824 of the *Académie Royale des Sciences* (of which by then he was a *secrétaire perpétuel*) [1827a, 1827b], and also in a paper [1826]. Much of this material reappeared in the lengthy preface to his incomplete and posthumous book on the theory of equations [1831, 75–84], and all of it was reprinted in the second volume of his collected works [Fourier 1890, 315–328] [4].

Fourier had a complete understanding of the basic ideas of linear programming, which he regarded as "a new type of question" referred to as "the analysis of inequalities." In the paper [1826] he took a problem with six inequalities, which he illustrated by drawing the corresponding three pairs of lines in Fig. 2. These lines produced the convex polygon (for him, "the extent of the question") of the feasible solutions (as we now call them), and he indicated in verbal terms how to find the desired minimax case [5].

In the second self-notice [Fourier 1827b] gave a very general presentation, outlining both a method of vertical descent to the minimax point of a convex region and the essentials of a method of solution by elimination. (A more highly developed form is attributed to [Motzkin 1936], on whom see Subsection 4.5.) He knew of various applications: not only mechanics but also statistical contexts such as the errors of observations [6], and elections.

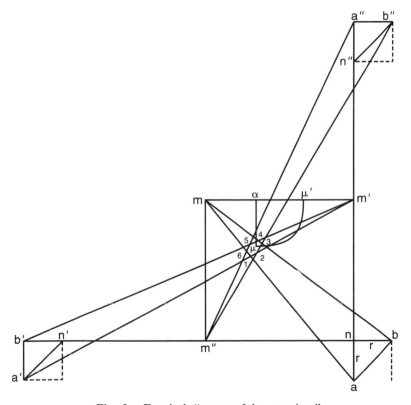

Fig. 2.   Fourier's "extent of the question."

3.2. Fourier's Followers

By the 1820s Fourier was an influential figure in French science, but his ideas stimulated interest among (only) two followers [7]. C. L. M. H. Navier (1785–1836) was one of Fourier's great admirers: he was to edit for publication Fourier's posthumous book [1831] on equations mentioned above. Working principally on elasticity theory and engineering mechanics in the mid 1820s, he also discussed Fourier's new idea in two papers. In [1825a] he outlined the theory (including the method of elimination) as presented in [Fourier 1827b].

At the end Navier referred to a well-known problem in indeterminate statics "which Fourier has communicated to us" (and was to publish in [Fourier 1827a]), giving the details in a succeeding paper [Navier 1825b]. Called the "paradox of statics," the problem concerned the calculation of the loads on a table which had more than three supporting legs, or a bar with more than two: the "paradox" arose from the fact that the balance of forces and balance of moments yielded only three (or two) equations, so that extra conditions were required to calculate the loads on the supports. Various proposals had been made during the 18th century, always expressed as equations; but (Fourier and) Navier introduced inequalities to express the breaking loads on the supports in the case of a loaded bar lying on three supports. Figure 3 is his illustration of the convex region $2cdef$ within which feasible solutions would be found: the line $\alpha\beta$ was shown as a typical example of a minimal value of the variable representing the inverse of the load, set at the place on the bar corresponding to the point α on the line $0b$ [8].

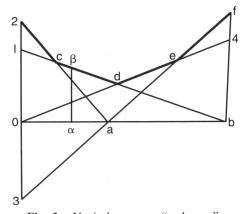

Fig. 3. Navier's concave "polygon."

Fourier's other follower at the time was A. A. Cournot (1801–1877), remembered today as a major pioneer of mathematical economics and as an important career educationalist in the *Université Royale de France*; in the mid 1820s he was a young man exploring various ideas of mathematics and making a living partly by reviewing. He wrote three short papers on our topic in the comprehensive abstracting serial *Bulletin universel des sciences et de l'industrie:* as with Navier, a survey [1826] of "Fourier's conditions of inequality" as presented in Fourier's first self-notice [1827a], but also including examples concerning the roots of polynomial equations; [1827] on the principle of virtual work including mechanical systems acting under constraints, in which he considered cases where the terms of the generalised work expression did not all take the same sign; and [1828] on indeterminate statics, largely following Fourier and Navier.

Then, in one of the theses written for the *Université* doctorate taken at the Paris *Faculté des Sciences*, Cournot passed from statics to dynamics and used inequalities to study "the motion of a rigid body, supported by a fixed plane" [1829a] but receiving both forces and impacts. For (relative) simplicity he worked mostly on the case of a cuboid. Starting out from Newton's second law and Euler's equations for the motion of a rigid body, he invoked inequalities to state that an impact was positive and thus the velocity of the body normal to the fixed plane was zero, or vice-versa. He placed a summary [1829b] of his thesis in the *Bulletin*, and then republished it in full the following year in Crelle's (new) journal. He quickly put out a two-part sequel paper [1830–1832] there, in which he considered effects of friction: this time inequalities stated that the frictional forces could or could not overcome forces of reaction. He noted that consideration of inequalities alone could show that "certain hypotheses on the values of friction are inadmissible" (p. 6: we might now say "infeasible").

The failure of Cournot to continue these studies is quite poignant to modern eyes, for later he was to be contributor to the theory of probability and a major founder of mathematical economics; moreover, he saw mechanics both as a model and a standard to be emulated in economics (compare Subsections 5.1 and 7.2). However, he seems not to have discussed his work again—not even in his major innovative study [1838] of economics, despite using there inequalities and convex or concave functions in his (new) theory of demand and supply functions; nor in his study [1847] of "the limits of the correspondence between algebra and geometry"; not even in his posthumously published memoirs [1913].

Similarly, Navier did not present his own contributions in his textbook on mechanics, which was published in editions during the 1820s and 1830s

[Navier Mechanics]. Saint-Venant did note the contents of the articles in the introductory material of his own edition of the volume of Navier's textbook devoted to elasticity theory, but he obviously did not realise their significance [Saint-Venant 1864, lxix–lxx, cix]. Thus, when editing Fourier's works in 1890, Darboux could say in this preface with apparent justice that "it is permitted to find now a little exaggerated" Fourier's enthusiasm for the topic [Fourier 1890, vi].

3.3. Boole's Logic and Probability

Another use of linear inequalities was made in Ireland by G. Boole (1815–1864). In a paper concerning "propositions numerically definite," probably written around 1850 but published posthumously by the Cambridge Philosophical Society as [Boole 1868], he deployed inequalities to express upper and lower bounds on the number $\{n(s_i)\}$ of members of "numerically definite" classes (that is, classes containing known numbers of members). Later, in his second book on logic, he applied these ideas in harness with his new Boolean algebra to probability theory in an exercise which was interestingly entitled "Of statistical conditions" [1854a, ch. 19]. He determined bounds upon the probability of a compound event (W, say) as a logical combination of the probabilities of the simple events $\{S_i\}$. Deploying Boolean algebra to express W as a linear combination of the $\{S_i\}$ with coefficients $\{a_i\}$, he worked with the numbers $\{n(s_i)\}$ of members of the classes $\{s_i\}$ corresponding to the $\{S_i\}$, defining the probability of S_i as the ratio $n(s_i) \div n(1)$ relative to the "universe of discourse" 1. The character of his procedures is of interest here, for he used not only these linear combinations, but also linear inequalities of the type $f \le 1$ or $g \le 0$ on the $\{a_i\}$ expressing bounds upon either the numbers of members of classes or on probability values (see [1854a, 310]). In the next chapter he applied his results to some "problems relating to the connexion of causes and effects."

Soon afterwards, in a paper [1854b] in the *Philosophical Magazine*, Boole proposed "an easy and general method" for finding the conditions, relative to which "I am not sure that [the previous one] is equally general." To our eyes his new treatment was oriented around the dual programming problem inasmuch as the $\{a_r\}$ become the subjects of equations and inequalities expressing relationships between probability measures, and his method involved elimination of variables, a technique which Fourier had already tried.

Boole seems not to have known of any of his predecessors. Similarly, although he published his ideas in a book which became well-known and in journals already then of wide reputation, many future workers in, and historians of, linear programming have not been aware of his contribution. For example, A. De Morgan corresponded with him about his theory as it was in genesis (see [G. Smith 1982, 58–62]) and arranged for the publication of his posthumous paper [1868]; but he clearly did not grasp its import. It gained some attention in the late 1920s, in connection with the strong law of large numbers [Seneta 1992], and the problem of finding bounds on the probability of the occurrence of at least one of a set of events given the probabilities for each of the members was also studied occasionally [Hailperin 1991, arts. 5–6], especially in [Fréchet 1935] on the best possible bounds. But Boole's contribution was brought to the general attention of mathematicians only in [Hailperin 1965]: see also his [1986, 36–43, 338–350].

4. CONVEXITY, LINEAR INEQUALITIES, AND MINIMAX PROPERTIES, 1900s–1930s

4.1. A Growth in Interest

The uses of inequalities described in the previous section seemed to die away almost entirely for the rest of the century, and no others emerged. For example, a crystallographer might have perceived the bearing of Fourier's theory of inequalities upon his concerns; but this does not seem to have happened, although the field was in active development and mathematical crystallography was even emerging as a discipline [Scholz 1989].

However, around 1900 interest grew quite quickly in a number of topics in which major roles were played by convexity and/or linear inequalities (often of combinations of terms of some kind with coefficients taking the same sign). These researches are the subject of this section. Linear programming was only one of these topics (a point to be mulled over in Section 6); it appears in the penultimate of the five Subsections into which I have divided the rest of this section.

The story reveals a rich growth in interest in inequalities and in convexity (sometimes both together) in several areas of mathematics, and is worth the telling in its own right. A full account awaits the writing, especially concerning the interactions between the different branches of mathematics involved, and relationships with other branches: I cite only the main papers and a selection of the more minor contributions as typical of the kinds of continuing investigation being undertaken. Further references and information can be gleaned from the excellent survey in [Dines and McCoy 1933] and the fine bibliographies in [Bonnesen and Fenchel 1934, Motzkin 1936, Valentine 1964, and Prékopa 1980].

4.2. Convex Sets

During the last fifteen years of his life H. Minkowski (1864–1909) devoted much time to properties of convex sets (see especially the material grouped in the second volume of his collected works at [1911, 103–229]; and compare Subsection 4.4 below). He concentrated upon convex regions bounded by (usually) planar faces, and was naturally drawn to express membership of the region in terms of inequalities. (I use set-theoretic language here deliberately, as he drew on point set topology.) His concerns included conditions for the types of solution of a set of inequalities, the unique specification of a convex region, and inequalities and other relationships concerning its surface area and volume (such as the smallest volume containing a given collection of point locations). He came closest to our concerns when he considered extreme points of a region [1911, 157–161]. One of his main inspirations came from number theory, as we shall note in Subsection 4.4.

Minkowski's work provided a basis for the treatment of convexity in general. It was also drawn on by several of the mathematicians to be mentioned in the next two subsections. Convexity received a comprehensive treatment in the monograph [Bonnesen and Fenchel 1934]; it was quickly followed by a survey of the "Elementary theory of convex polyhedra" [1935] by H. Weyl (1885–1955), which was later to become a classic reference source for linear programming and the theory of games (the link here is described in Subsection 4.6).

4.3. Inequalities in Mathematical Analysis

Several analysts, mostly with German backgrounds, worked in this area shortly before the First World War. Their studies often involved linear combinations of functions, sometimes under the influence of the newly emerged discipline of functional analysis (on which see especially [Siegmund-Schultze 1982]).

Quick off the mark was A. Hurwitz (1859–1919), who in his paper [1902] applied Fourier series to Minkowski's theory of curves and surfaces. Later, and conversely, C. Carathéodory (1873–1950) used inequalities in [1911] to state conditions satisfied by the normalised coefficients of the nth partial sum of a Fourier series for all n; he interpreted them as specifying a point A in a $2n$-dimensional space, and showed that good behaviour (such as regularity) of the series required that A lie within the smallest convex region containing the curve parametrically defined by the $2n$ trigonometric terms of the series. In a sequel paper [1911] written with L. Féjer (1880–1959), the same approach was applied to harmonic functions.

Féjer's compatriot F. Riesz (1880–1956) rapidly used the idea in a study of integral equations: the context involved conditions for the safe conversion of a convergent power series $\sum_s c_s z^s$ in a complex variable z into a monotonic power series $\sum_s d_s |z|^s$, of bounded variation, where d_s took the form $k_s \frac{\sin}{\cos} (\arg sz)$. The integral equations arose from the Fourier forms for the $\{k_s\}$: this time the point $\{(k_s r^s)\}$ was confined to convex barracks in a $2n$-dimensional space [Riesz 1911, art. 9].

E. Steinitz (1871–1928) deployed inequalities in a long study [1913–1914] of another department of potential disorder in mathematical analysis, namely, conditionally convergent series. Here the inequalities were used for expressing majorant conditions on partial sums and remainder terms as well as for other purposes. As divergence was a possibility, he allowed for unbounded convex regions. This point was recognised at the end of a paper [1913] by E. Schmidt (1876–1959) concerned with illuminating Hilbert's recent difficult analytical proof of Waring's problem in number theory (that every positive integer can be represented as the sum of the nth powers of a certain number of nonnegative integers). Schmidt adapted Carathéodory's idea of coefficients as coordinates by interpreting the q coefficients of the $2n$th power of a general linear form in five variables $\{x_i\}$ as a point in a q-dimensional space which should lie within the smallest convex region containing the coefficients of the $2n$th power of the sum of the squares of the $\{x_i\}$.

Far away, the Japanese mathematician S. Kakeya (1886–1947) found various uses for inequalities in a long series of short papers, of which I note two examples. In [1914a], working from Riesz's method, he sought conditions on the constants $\{a_i\}$ in the system of integral equations

$$\int_\alpha^\beta \phi_i(x)f(x)\,dx = a_i, \qquad 1 \le i \le n,$$

$$\text{with} \quad \int_\alpha^\beta f(x)\,dx = 1,$$

(6)

under which solutions for f, continuous and nonnegative over (α, β) in x, could be found. In [1914b], and now aware of Steinitz, Kakeya used convex regions to study linear combinations of functions satisfying given inequalities.

A different use of "the Minkowski geometry," as he called it, was made by the Hungarian mathematician A. Haar (1885–1933) in a paper [1918]. This time the context was the Chebysheff approximation to any continuous function $f(P)$, defined on a set $\{P\}$ of values, by a linear combination of functions $\{f_i(P)\}$. He found necessary and sufficient conditions for there to be just one set of coefficients $\{a_i\}$ to give the best such approximation to $f(P)$. Inequalities came in precisely in connection with the required minimal condition of best fit; and he also cited Carathéodory for the idea of letting $\{a_i\}$ define a point in a multi-dimensional space.

4.4. Linear Systems of Inequalities

From the late 1870s onwards linear algebra had advanced sufficiently far for questions of rank and signature of a matrix to be considered, often in connection with the (non)uniqueness of the solution of systems of linear equations. Minkowski was again a pioneer, for in his book on geometrical number theory he studied "nowhere concave surfaces" and related volumes, and when he permitted them to become convex he found conditions for solutions of sets of linear inequalities as linear combinations of particular solutions, should these exist [Minkowski 1896, especially 39–45].

In the 1920s some North American mathematicians realised that certain of the conditions could be applied to the study of the (non)existence of solutions of systems of linear inequalities in which all the coefficients of some or all of the inequalities took the same sign; they are considered cases of strict inequality. The main contributors were W. B. Carver (1892–1961) with his paper [1922] and L. L. Dines (1885–1964) in [1925], and Dines extended his results to "functions of a general variable" [1927], in the tradition established in the USA by E. H. Moore. Then in a paper [1933] based on her doctoral dissertation, R. W. Stokes combined elements of her compatriot predecessors' work by jointly treating solutions of systems of linear inequalities, strict inequalities and equations.

At the same time I. J. Schoenberg (1903–1990) extended these considerations in two different ways. In [1932–1933] he treated "linear inequalities in infinitely many variables," sketching out the notion of a convex region in an infinite-dimensional space and also making use of systems of functions. His work thereby enriched knowledge of infinite matrices, which was gaining ground at that time in connection with functional analysis in the hands of von Neumann (who will be an important figure in our essay from Subsection 4.6 onwards). It was a fairly new topic [Bernkopf 1968], although once again Fourier had been a pioneer long before and Riesz a recent contributor.

In [1933] Schoenberg continued the links with systems of functions when he sought conditions on the change of sign of "linear combinations of continuous functions" over "convex domains." For methods he drew upon [Carathéodory 1911] (though not Haar, whom he had cited in his previous paper); his results involved integral equations, and in an addendum he reported his recent sighting of [Kakeya 1914a].

A few years later Minkowski's geometrical number theory itself received publicity in the USA, with the survey [Hancock 1939], Chapter 2 on "surfaces that are nowhere concave" being the most pertinent to our discussion. It was also given chapters in some of the treatises and textbooks on topology that were then beginning to appear.

4.5. Linear Proto-Programming

Inequalities and convexity were the staple components of the work described earlier in this section, but linear programming was not necessarily involved; that is, attempts were not usually being made to emphasise the feasible region as such or to determine optimal values of some objective function in a convex region. For such procedures the principal figure for this period was the Hungarian mathematician J. Farkas (1847–1930) [see Brentjes 1976a, Prékopa 1980]. He produced several papers from the mid 1890s onwards, mostly in Hungarian: the best known was [1902], written in German. Motivated by questions in variational mechanics, he was interested in interpreting as linear inequalities both the principle of virtual work and also the constraint conditions upon a mechanical system. He knew of Fourier's paper [1798], perhaps from its reprint in [Fourier 1890].

Farkas' main result stated that if the n variables $\{x_j\}$ satisfied the m inequalities

$$\sum_{i=1}^{m} a_{ij}x_j \geq 0, \quad \text{with} \quad 1 \leq j \leq n \tag{7}$$

and the $\{a_{ij}\}$ constants, it also held for, and only for, any vector composed of a linear combination of the m vectors $\{(a_{ij}), 1 \leq j \leq m\}$ with nonnegative coefficients. His theorem is somewhat similar to Minkowski's of Subsection 4.2, which he did not know; but he advanced further in that he included mixed systems of equations and inequalities. (Although on pp. 3–4 he considered the maximum possible number of independent inequalities, he did not make clear how many inequalities were involved, or whether the total system was square or rectangular; I have assumed the latter in (7).) The obvious interpretation is that certain variables (which we now call "slack") could be eliminated from the system; but a reading in terms of optimising inequalities given the equations as defining boundaries bestows upon his work the flavour of linear programming. Curiously, despite his awareness of Fourier's contribution, he did not stress or develop this aspect.

Farkas wrote a succeeding paper [1906] on mechanics, which we shall note in Subsection 7.5. In view of the quality of his work it is a pity he did not write further on inequalities until the early 1920s, and then his papers from that time were generally written in Hungarian, and so did not create much attention. But language may not have been the only obstacle; interest in the topic still seems to have been modest.

Evidence for this state of affairs is furnished by Farkas' compatriot Haar, who returned to the topic with a valuable paper [1924] "On linear inequalities," published in German, as a translation from a paper in Hungarian. Here he extended the rectangular version of Farkas' theorem to "inhomogeneous inequalities," where the linear combination $(7)_1$ was joined by a constant term $a_{i(j+1)}$ (art. 2); he also related his result to a question in integral equations (art. 4), though without citing [Kakeya 1914a]. However, little attention was paid to his paper, even by those who were to use Farkas. He published his paper in an early volume of the new journal of Szeged University, which presumably did not gain much circulation at first.

The importance of Farkas' work for linear programming (and the question of the rigour of its proofs) did not gain proper recognition until the German-Jewish mathematician T. Motzkin (1908–1970) gained his doctorate at the University of Basel in 1933 with his important "Contributions to the theory of linear inequalities," which was published as a short book [1936] in Jerusalem [9]. Here he presented his "transposition theorem" on types of solution of a matrix inequality and of the inequality associated with the transposed inequality [art. 73], and the "simplex theorem" on the rank of a smallest system of the "definitely dependent" columns of an $(r + 1) \times r$ matrix and its r-th order "simplex" submatrix whose rows were linearly dependent with positive coefficients [arts. 80, 55, 70]. He presented his elimination method in terms of nonsingular submatrices of the matrix defining the given problem, and their pertaining solution spaces [arts. 28–34, 44–48].

Motzkin also knew his prehistory very well, for he prefaced his results with valuable historical remarks and an excellent bibliography (arts. 5–13): he cited most of the figures mentioned earlier in this essay as well as several others. Substantial reviews appeared in the German abstracting journals, both the old *Jahrbuch über die Fortschritte der Mathematik* [Behrend 1940] and the new *Zentralblatt für Mathematik* [Ridder 1936], despite the unwelcome nationality of its author. (Ridder gave the place of publication as Basel!) However, the manner and place of publication reduced the chances of publicity, and its place in the history of linear programming has not been sufficiently recognised. Nevertheless, his work appeared just before the time when world events were focusing the attention of others on the kinds of methods he had studied.

4.6. *Minimax Theorems and the Theory of Games*

Our last subsection takes us back to the concerns of Subsection 2.2, firstly with a paper [1911] on "minimum approximation" by the Belgian mathematician C. de la Vallée Poussin (1866–1962). Define a "residual" r_j as

$$r_j := \sum_{i=1}^{n} (a_{ij} x_j - b_j), \quad \text{with} \quad 1 \le j \le m \quad \text{and} \quad n > m. \tag{8}$$

In general the system of equations obtained by setting each residual to zero does not admit a solution for the $\{x_j\}$; so the value $\min\{\max |r_i|\}$ was sought for which a solution could be found. Aware of Laplace's condition (4), he built upon some work of his colleague P. J. E. Gredseels on errors of observation to find more efficacious procedures than Laplace's; thus, the orientation was around numerical linear algebra rather than linear programming, and his work made no special impact on our later story.

A much more significant result involving minimaxisation, this time in mathematical economics, was produced by another Hungarian, and a key figure in our next section: J. von Neumann (1903–1957). In his paper [1928a] "On market games," he proved his minimax theorem on the existence of strategies and studied certain special cases, mostly in game situations with a small number of players. Naturally he deployed inequalities; but at this stage his thinking about his result was oriented around fixed-point theorems. This connection was emphasised in a succeeding paper [1937], "On an economic equation system," where the use of inequalities was still clearer but where fixed-point theorems were stressed even in the rest of the title.

The link with convexity was made only by J. Ville, in a "Note" [1938] added to his edition of some lectures on probability theory given by his master E. Borel, and published in a series of volumes on this subject that were produced under Borel's direction. In order to present a new proof of von Neumann's minimax theorem, Ville considered a convex region and drew upon a lemma which stated (in the notation of the previous subsection) that if at least one of the p linear forms

$$\sum_{j=1}^{n} a_{ij} x_j, \qquad 1 \leq i \leq p, \tag{9}$$

was nonnegative when $\{x_j \geq 0\}$, there was at least one collection of multipliers $\{X_j \geq 0\}$ for which

$$\sum_{i=1}^{p} X_i = 1 \quad \text{and} \quad \sum_{i=1}^{p} \left(X_i \sum_{j=1}^{n} a_{ij} x_j \right) \geq 0 \tag{10}$$

for all $\{x_j \geq 0\}$.

There are shades of Farkas, Motzkin and especially some North Americans here; but Ville mentioned only von Neumann (and even then without giving a reference). The omission is rather bizarre, for Borel was already aware of von Neumann's theorem; he had communicated to the Paris *Académie des Sciences* a French summary [1928b] of von Neumann's first paper [1928a]. To Ville's edition he added a brief note [Borel 1938] of appreciation of the result, and a link between linear programming and the theory of games was more in the air.

Yet even then connections with mathematical economics were not fully grasped. According to his own testimony there, von Neumann wrote the essentials of his [1937] five years before its publication, and in the meantime various economists had studied Walrasian models of economic pricing. Now in these studies demand quantities of goods were expressed as linear functions of their prices, and the inverse systems of equations were also taken; it was found that if side conditions on some of the variables were taken in the form of inequalities, then a unique solution could be found. As mathematics these analyses are once again redolent of Farkas, of North America, and even of [von Neumann 1928a] to some extent; yet none of these sources was cited, not even [Wald 1935, 1936], who published his important analyses *in the same journal* that was to take [von Neumann 1937].

Further to the East, the penchant for planning evident in the Soviet Union led to certain other aspects of mathematical economics being studied with profit. At the onset of the Second World War the economists A. N. Tolstoy and V. V. Novozhilov, and especially the mathematician L. V. Kantorovich (1912–1986), published works involving models of industrial economic planning [Brentjes 1976b]. Kantorovich's study [1939] of "methods of organizing and planning production" became the most influential source from our point of view, for he really grasped the linear programming problem and found the solution in terms of determining as "multipliers" the values of the variables for optimising the objective function. He gave a wide range of examples (with a nice use of diagrams, when suitable), all based on finding the best balances of measure between the various goods or means involved: mixed-means freight shipment, distribution of arable land, manufacture of parts of a compound piece of equipment, and so on. Citing none of our earlier sources, he may have started out afresh mathematically. His and his colleagues' work were to help convert our topic at last into a discipline.

5. THE ESTABLISHMENT OF LINEAR PROGRAMMING

The mathematical foundations of game theory are found in [von Neumann 1928a, 1937] and less conclusive contributions by Borel [...] Actually, until 1944 there were almost no papers. [...]

In the summer of 1949 at the University of Chicago, a conference was held [on linear programming...] What was most surprising was that the research had taken place during the preceding two years.

Dantzig [1963, pp. 277, 1]

5.1. The Emergence of a Discipline

The 'arrival' of linear programming has two independent strands. The prior but slower and less influential one occurred in the Soviet Union. Novozhilov and Kantorovich continued their work, with others, during and after the Second World War [Brentjes 1976b]. However, progress seems not to have been very rapid, partly because the notion of production centres producing more than one product could infringe Communist planning (for discussion of such issues, see [Novozhilov 1967, chs. 9–10]) [10]. This laggardness was deplored by Kantorovich himself in 1959 [Dantzig 1963, 23]. Further, his work became generally known in the West only at that time, when his [1939] was translated into English by the Management Society of America in 1960 and admired by [Koopmans 1960, Charnes and Cooper 1962] in their journal *Management science*; and Tolstoy and Novozhilov are still largely unrecognised outside the former socialist countries. Thus the Soviet pioneers did not contribute to the second though major strand, to which we confine the rest of this section.

The second strand developed largely in the USA, during the Second World War; credit belongs to T. C. Koopmans and G. B. Dantzig, for research largely related to questions of military organisation. The story has been outlined by Dantzig in his treatise on the discipline [1963, ch. 2] and in greater detail later in his overview [1970] and reminiscences [1982], and more recently it has been sketched in [Dorfman 1984] with comments from [Schwartz 1989, Gass 1989] [11].

Dantzig attributes the name "linear programming" to Koopmans, although he gives the date 1951 in [1963, 6] and 1948 in [1982, 45]; in the latter reference he also credits R. Dorfman with devising the name "mathematical programming" in 1949. This latter name is now nicely confused with "computer programming." (In the course of my library research I handed over several volumes for recataloguing!) It is worth mentioning that the new techniques in computing soon made feasible many of the solutions proposed to problems in linear programming (and, conversely, that computing was thereby encouraged in certain aspects of its development). However, while both disciplines had military and economic applications in common [Rees 1980, Dantzig 1982, Cohen 1988], neither of them seems to have been a major factor in the *inauguration* of the other one.

A leading figure in the establishment of computing was von Neumann. He was also crucial for the development of aspects of linear programming within the theory of games, especially in his book on that subject "and economic behaviour" written with O. Morgenstern (1902–1977) [1944: see especially art. 16 on "Linearity and convexity"]. The interaction between linear programming and the theory of games was of great and mutual immediate benefit to both fledgling disciplines. However, for historical purposes we should still keep the theory of games rather apart from linear programming. For example, in their book von Neumann and Morgenstern did not mention Fourier, Farkas, or Motzkin, not even in the new editions of 1947 and 1953 [12].

Among early theoretical innovations connected with the establishment of the discipline, von Neumann's conception of the dual programme and Dantzig's simplex algorithm were particularly significant; yet even these notions were not quite new, for, apart from such related touches in [Motzkin 1936] and occasionally elsewhere (such as Boole; cf. Subsection 3.3), they had been partly foreshadowed by the American mathematician F. L. Hitchcock in his war-time study [1941] of "the distribution of a product from several sources to numerous localities." Although mathematically his linear model resembled the linear algebra of his compatriots in Subsection 4.4 and some of the economists of Subsection 4.6, he cited nobody; and he received the same treatment himself until well after the discipline had been established.

The two new disciplines were also adjoined to other emerging ones like location theory, and also older ones such as mathematical economics. In addition, operational research began to develop during the War, as an umbrella subject partly involving all these disciplines [13]. A tangle of subjects and techniques is involved here, especially in the USA, and their mutual relationships need much clarification (and not only historically!). For example, optimisation was one important linking thread, and analogies from mechanics constituted another; locational equilibrium has a prehistory of its own going back to Fourier's time, again partly inspired by the principle of virtual work [14]; meanwhile, economics drew also on other mechanical inspirations, such as conservation principles [Franksen 1969]. Fortunately it is not our task here to unravel all these threads, for our concern is centered around linear programming.

The fact that different mathematical contexts were involved in its establishment as a discipline is also a feature of its prehistory. The contacts with linear algebra and some aspects of convexity have been maintained and even enriched; however, those with mathematical analysis and mechanics seem rather to have fallen away. The prominence of economics in the establishment of the discipline, and its continuing importance (especially with the link to the theory of games), is another change from previous sloth (on which see also Subsection 6.4 below). This impulse was no doubt the main cause for the extraordinary speed with which both disciplines were then to develop worldwide [15]. Another fascinating historiographical question, the converse of the one addressed in the next section, awaits some future student: how does a discipline develop when it goes very *quickly*?

5.2.   A Very Partial Heritage

After linear programming had become established, some posterior recognition of the past workers was made. For example, D. R. Fulkerson translated Motzkin's thesis into English in 1952, and Dantzig mentioned Motzkin and Fourier (and several other figures briefly) in [1963, 12–13 and 21–25] and in [1982]. But how much did the new work *owe* to the past?

Much of the history outlined earlier in this essay seem to have played little role in the establishment of linear programming. The apparent lack of awareness of the linear algebraists (Subsection 4.4) is particularly striking, for their work was not only recent and relevant but it had even been executed in the USA, which was to be the principal country for the discipline. Again, the arrival of integer programming later in the 1950s apparently owed little or nothing to Minkowski's geometry of numbers (Subsection 4.4); indeed, his work on convexity in general exercised only a modest and indirect influence on the discipline even after it had been established, although it was well established in mathematics in general.

In particular, among the precursors Fourier deserves special credit or perceiving that the topic of linear programming constituted "a new kind of question," for conceiving of two methods of solution, and for having in mind a variety of applications. Yet neither his nor anyone else's efforts seem to have played important roles in the establishment of the discipline. This history of missed opportunities is *especially* interesting because of Fourier: not only the clarity of his vision but also his status as a major figure in the mathematical world at the time, who nevertheless failed to galvanise activity among others in a way that one might have expected. It is meet, therefore, to consider possible causes of the 120-year delay before his vision was realised.

6. QUESTIONS OF CONTEXT: THE GESTATION OF LINEAR PROGRAMMING

One of the most intriguing aspects of the history of mathematics is the 'record' of developments which could have happened but did not, or which occurred only in bits and pieces. Causes of the non-events will be examined here. The historical context could be broadened to take in (for example) much more detail on the general context of optimisation in probability and statistics, and in economics. However, I have no doubt that a more detailed tale would only reinforce the considerations (and perplexities!) which will be presented.

6.1. A Scattered Background

A common feature of scrappy developments like this one is the difficulty of being *aware* of the pertinent past. At the time the founders of linear programming would have had great difficulty in ascertaining many of the relevant texts, even if they had wanted to find them. Few workers in any branch of mathematics could or can claim the bibliographical tenacity of Motzkin noted in Subsection 4.5! This point has bearing upon this essay itself, for I have faced the same problem in tracking down the relevant primary literature in my historical research. As was mentioned in Section 1, not all items consulted are listed here, and doubtless others have been missed in the research.

But *why* is this background scattered? In linear programming we treat of a subject which draws on four general *aspects* of mathematics—namely, optimisation, linearity, convexity and inequalities—which made themselves in several of its *branches*, either singly or in some combination. Here we have the source of the importance of the subject, and the frequency of the partial appearances; but we find also a major cause of the missing of opportunities, when work executed in one branch is not noted by those active in another one. In this case, a wide range of branches was involved: linear algebra, mathematical economics and mathematical analysis were the major ones, and others, such as mechanics, mathematical statistics, probability theory, number theory, functional analysis and infinite matrices occasionally came on stage. As a result workers in one area were not necessarily aware of colleagues elsewhere. I have mentioned some cases of mathematicians not citing each other, and the list could be lengthened considerably.

Further, in areas such as the mathematical analysis described in Subsection 4.3, convexity and/or inequalities were being used as tools, and so (quite understandably) were not pursued for their intrinsic interest. (For example, Carathéodory's condition in Subsection 4.3 for the regularity of Fourier series stated in effect that the convex region specified for the parametrically defined curve identified a feasible region for the location of the point defined by the Fourier coefficients; but it would have been very surprising if he, or one of the mathematical analysts who cited his paper, had formed that interpretation.) Further, linear programming does not necessarily contain deep examples or applications of these tools (for instance, the convexity as such is normally fairly straightforward); thus such an interpretation of its results would not have seemed to offer attractive avenues of study.

6.2. Persons and Contexts

Concerning individual mathematicians, only Fourier and Minkowski, out of all the principal figures, could be described as eminent (remember from Subsection 4.6 that von Neumann was explicitly *not* a contributor during the period of gestation). No major schools took a coordinated interest in the work.

As often happens in the history of mathematics (whether scrappy or connected), national differences are evident; the contributors came from a certain number of countries, sometimes with different interests in different contexts. Germans (and some Austro-Hungarians) focussed upon mathematical analysis (Subsection 4.3), while the North Americans liked systems of linear inequalities (Subsection 4.4): the French largely dropped the topic after Fourier's death, and Ville was unlucky to have his link between convexity and the theory of games (Subsection 4.6) rather buried in a *fascicule* of a series of volumes, and to appear just before the Second World War. Among other nationalities, no contributions of note were made by the British after Boole, the Scandinavians, the Russians (apart from the Soviet contributions noted in Subsections 4.6 and 5.1), the Poles (who became so active in mathematical analysis and set theory from the 1920s), or the Italians after their resident Boscovich (although [Ascoli 1932] used Minkowskian "convex fields" in a paper on functional analysis).

6.3. Conceptions and Techniques

Let us turn now to the strictly mathematical impediments to more rapid and connected progress. One of these must have been a lack of powerful methods of numerical calculation. But a major further difficulty was surely the absence of linear algebra, to serve as a practical mathematical technique to effect the theories envisaged at Fourier's time (and for long after). For example, in his posthumous book [1831] he did not repeat the method of elimination which he had presented in [1827b], and of course he could not convert his method of vertical descent into (a version of) the simplex algorithm.

Matrix theory was not to emerge until quite late in the 19th century, and became prominent only from the 1920s. Indeed, its history forms nearly as scrappy and tardy a tale as that of linear programming itself—and moreover, linear inequalities was a (minor) part of the discipline [16]. Among the literature discussed in this essay, [Farkas 1902], [de la Vallée Poussin 1911] and even as late as [Haar 1924] are typical examples of the continuing slowness, for they explicitly worked with determinants but wrote out the matrix equations and inequalities longhand.

Hence the lack of awareness (noted in Subsection 5.2) of the work of the linear algebraists (described in Subsection 4.4) can be partly understood. It is striking to note that, well after the establishment of linear programming itself, the authors of a major monograph felt the need to devote an appendix to matrix theory, upon which "there are *now* available several excellent textbooks" [Dorfman, Samuelson and Solow 1958, 470: italics inserted].

However, these comments overlook the possibility that an earlier development of linear programming might have accelerated the establishment of matrix theory *itself*. Indeed, an irony of history has to be noted; for exactly at the time of Fourier's enthusiasm for "the analysis of inequalities," has compatriot A. L. Cauchy published a paper [1829] on the simple latent root problem (to use modern terms) which could have inaugurated the spectral theory of matrices. However, Cauchy did not realise the importance of his work. Neither did the young Swiss mathematician J. C. F. Sturm, who worked on the generalised latent vector problem at exactly the same time in a paper [1829d] and also reviewed Cauchy's paper in his [1829b] (in the Paris abstracting journal *Bulletin universel des sciences et de l'industrie*, of which he was then the editor for the series of mathematical sciences); he also dropped the matter [17].

But even the eventual arrival of matrix theory, while necessary for the establishment of linear programming, was not sufficient: another necessary component was geometrical thinking, which Fourier had so clearly exhibited. Yet of all his successors, only Navier seems to have recognised the benefit of interpreting the problems and techniques geometrically. Even though Cournot held that "the calculus of inequalities" was "intimately linked to geometrical considerations" and held geometry to be "indispensable for interpreting the results handily" [1830, 135], he did not provide any diagrams in his papers. Similarly, Minkowski, for all his concern with the "geometry of numbers," did not give geometry a major place in his presentation—a fact noted at the head of [Haar 1924], who reworked Minkowski's and Farkas' theorems in more geometrical vein in his art. 1.

Much more typical of the mathematics surveyed in earlier sections is the North American work on linear inequalities noted in Subsection 4.4, which normally concentrated on algebraic issues and methods. Indeed, in their survey [1933] Dines and McCoy stated quite explicitly that the geometrical side, "despite its simplicity and suggestiveness, has not proven so directly useful" as the more algebraic approaches [1933, 39]. In another irony, in that same year [Stokes 1933] appeared, containing a "geometrical theory of linear inequalities"; but while she lived up to her title by laying emphasis upon "flats" (a much crisper word than "hyperplanes"!), geometry still played servant to her algebra. Finally, Schoenberg began his [1932–1933] with a promise to publish a geometrical treatment, but it seems not to have appeared.

Geometrical thought is related to, but does *not* necessarily require, the use of diagrams, especially as in this topic a diagram would often be planar sections of multidimensional situations. However, Fourier and Navier showed their utility; but the practice was not adopted by successors. Even Motzkin deployed diagrams only on occasion, although effectively [1936, arts. 60, 64, 84]. By contrast, the treatment of "Linearity and convexity" in [von Neumann and Morgenstern 1944, art. 16] was positively Fourieran in its use of pictures.

6.4. *"Only Connect"*

Even allowing for the (absence of) effective algebraic tools and geometrical intuition, the length of the delay is still perplexing. Three possibilities deserve some emphasis, in chronological order of concern.

Firstly, a primary use of linear programming is in economic optimisation; but this was a growing interest in the increasingly industrialised societies of Fourier's time (the new-fangled railway versus other forms of transport, the best number of locks to construct in a sequence in a canal, and so on). Even though the mathematics did not usually entail more than the usual first-order optimisation condition $f'(x) = 0$ on some appropriately constructed function $f(x)$, more realisation might have been made of Fourier's proposals, especially among the followers he had gathered about him by the late 1820s [18].

Secondly, tasks set by the Second World War played major roles in establishing linear programming; so why had the First World War, and/or its immediate aftermath, failed to do the same? At that time the same kind of logistical problems were evident, convexity was being actively studied in several contexts, and linear inequalities were known; even just about enough matrix theory was in general circulation, thanks to pioneer textbooks such as [Bôcher 1907] [19]. True, the methods of calculation were far less powerful than those available by the mid 1940s (Subsection 5.1), but at least some forays could have been effectively attempted and solutions found.

Perhaps the crucial difference between the situations of the two wars was that during the First World War optimal modelling was dominated by deterministic strategies such as those developed by F. W. Taylor and put into widely publicised practice by Henry Ford: rather impersonal in character, they imposed or accepted hierarchical structures in the workplace. By contrast, in the Second World War *in*deterministic theories were proposed, in and for more democratic working conditions: they led to great (though naive) expectations of the *value* of planning and organisation within a capitalist system, as a means of avoiding the social and economic disasters that followed the First—and linear programming was one of the newborn disciplines mentioned in Subsection 5.1 to be so encouraged. The similarities but also contrasts with the Soviet predicament mentioned in Subsection 4.6 are striking.

Thirdly, von Neumann might have helped to establish the topic as a discipline after his paper [1928a] on market games; but he did not make the essential connections, and neither did anyone else for nearly two decades—not even in the 1930s when von Neumann, Ville, Borel and Wald were tackling very similar problems and using such kindred techniques to solve them (Subsection 4.6). Indeed, it is puzzling that earlier motivations in Western work were not drawn from mathematical economics. Clearly, that topic/discipline could have played a major role in the establishment of linear programming during the early decades of this century, and various of its forays in optimisation can now be so interpreted (compare [Dorfman, Samuelson and Solow 1958, ch. 14] and [Baumol and Goldfield 1968] for a valuable assembly of original texts). However, despite the poignant case of Cournot from Subsection 3.2 and the use made by him and various successors of convex or concave functions in mathematical economics, no significant insights of this kind seem to have been made; even Hitchcock's transportation model, proposed in wartime of all times (Subsection 5.1), did not make an impact. Finally, the use of convexity initiated by Minkowski was uninfluential, although it had been taken up in several branches or aspects of mathematics, including in other kinds of optimisation.

7. EPILOGUE. INEQUALITIES IN THE PRINCIPLES OF MECHANICS: BACKGROUNDS FOR NONLINEAR PROGRAMMING

7.1. The Kuhn-Tucker Theorem

In the paragraph containing the quotation at the head of Section 3 Fourier even saw that "the question does not change its nature" if the inequalities were not linear [1827b, 326]. Expressed in the modern terms of Section 1, Fourier envisioned the possibility that the objective and/or constraint functions were not linear in form. Among applications, presumably he had in mind the restatement of the principle of virtual work (which involved a linear combination of nonlinear terms) in terms of inequalities. In the same spirit, some of the prehistory of linear programming described in earlier sections related both to linear and nonlinear modelling; in particular, the conception of convexity in Minkowski and its applications to mathematical analysis (Subsections 4.2–4.3).

In addition, however, there is a specific matter to be treated here. When nonlinear programming was established in the early 1950s (Subsection 7.6), the basic theorem was found by H. W. Kuhn and A. W. Tucker in their [1950] [20]. They studied the optimisation of a convex function $f(\mathbf{x})$ of a point $\mathbf{x}$ $(= \{x_i\})$ in an n-dimensional region specified by m concave differentiable functions $\{g_s(\mathbf{x})\}$. By using variational arguments, conditions were found under which optimisation of f would occur at a point $\mathbf{z}$ in the region and m multipliers $\{k_j\}$ could be found such that

$$\sum_{i=1}^{n} \left\{ (x_i - z_i) \left[\partial \left\{ f(\mathbf{x}) + \sum_{s=1}^{m} k_s g_s(\mathbf{x}) \right\} \middle/ \partial x_i \bigg|_{\mathbf{x}=\mathbf{z}} \right] \right\} \geq 0. \quad (11)$$

The long, though broken, ancestry of cousins of this theorem is our chief concern here.

7.2. Fourier's Followers Again

Fourier did not pursue the consequences of his insight about nonlinearity, though it excited the interest of his followers. Cournot's study [1827] of the principle of virtual work (Subsection 3.2) was concerned with it, and in his paper on frictional effects on a rigid body he extended his analysis from the cuboid to a "convex and continuous figure" [1830–1832, 152–158]. But such extensions were taken much further a few years later by M. A. Ostrogradsky (1801–1862), a young Russian mathematician who had spent the middle 1820s in Paris. He discussed the use of inequalities in variational mechanics in two papers published by the Saint Petersburg *Académie des Sciences*.

In [1838a] Ostrogradsky went beyond Fourier and Cournot in discussing the general use of multipliers in the analysis of a mechanical system of mass-points acting under time-free constraints. Strangely, he cited only John Bernoulli and Lagrange (whose approach in [1788] he followed); there was only an oblique allusion to Fourier on p. 130, and no mention of [Cournot 1827], despite several similarities of thought. He also distinguished cases where multiplier and constraint function took the same or different signs (so affecting the sign of their contribution to the generalised work expression involving constraints); he also brought d'Alembert's principle into account. He showed (none too clearly) that if the constraint conditions were stated as equations rather than inequalities, then they might not be satisfied by some virtual or even actual displacements of the system from equilibrium.

In a sequel paper [1838b] Ostrogradsky extended the analysis to cases where the constraint functions involved time. Here his use of "linear inequalities" (p. 573) was much more extensive. He took inequality forms of the generalised work expression (E, say) to both first and second degree of the time interval dt, and obtained systems of linear inequalities as components of E in given axis directions. (Today we would write them in terms of matrices and vectors.) From our point of view a striking passage dealt with conditions when these component expressions could be nonpositive given that the corresponding displacements were nonnegative (pp. 589–590): he took for granted that his conditions were necessary and sufficient, a matter upon which Farkas was to ponder more deeply (Subsection 4.5). One of Ostrogradsky's main general conclusions was that "effective" displacements were distinguishable from "virtual" ones on the grounds that only the former would not affect the equilibrium of the system (p. 586). And on p. 570 he referred to another use of inequalities in mechanics, to which we now turn.

7.3. Gauss' New Principle of Mechanics

The criterion of absolute deviation which we saw Boscovich propose in Subsection 2.1 was soon rather eclipsed in mathematical statistics by least-squares regression. This was introduced in 1805 and A. M. Legendre, and it soon became the subject of an unlovely priority controversy between him, Laplace, and C. F. Gauss (1777–1855) [Plackett 1972]. In the course on one of his own discussions, Gauss added a (somewhat careless) statement of Boscovich's criterion and Laplace's version of it [Gauss 1809, art. 186]. His own preference for least-squares regression was to bear fruit for our theme two decades later, with a paper [1829] in which, acting in apparent independence of Fourier's circle, he formulated a principle of "least constraint" for mechanics.

In this approach to dynamics, the motion of a system of mass-points moving under the action of impulses and under restrictions of some kind was asserted to occur under the smallest value of "constraint" ("Zwang"). The word "constraint" as not used in the same way as in variational mechanics; Gauss defined it for the system as the sum of the product of each mass and the square of the distance D between the point to which it would have moved during an infinitesimal time interval dt if able to do so free of the restrictions, and the point where it actually arrived. The form of the constraint expression was the same as that used in least-squares regression.

Gauss's proposal used inequalities to express the least value held to pertain to the motion. In addition, it related to two other principles of mechanics: d'Alembert's, which was also concerned with D in equilibrate situations of the system; and virtual work in its usual equational form, which could be read as stating the first-order condition of complete differentials for minimising the constraint function. The relationships between these various principles, and also with that of least squares, were matters of study.

Sturm briefly reviewed Gauss's paper in his [1829d], in the abstracting *Bulletin* noted in Subsection 6.3; however, the principle did not gain interest in France. But a few of Gauss' later compatriots, such as C. G. J. Jacobi, A. F. Möbius, C. Neumann, R. O. S. Lipschitz, and H. Herz took note of it, and a quiet tradition of interest ran through the century (see [Jourdain 1908, Stäckel 1919, Appell 1925], sources discussed in the next two subsections; and, for a modern account, [Cyganova 1983–1984]). Yet it was a modest strain. For example, G. A. G. Ritter (1826–1908) wrote his thesis [1853] on Gauss's principle under the master's direction, but his later textbook on analytical mechanics contained only seven pages out of 271 on it [Ritter 1873, 216–222], with no attention paid to Fourier's French idea. Again, neither principle seems to have gained currency in engineering mechanics, although generalised principles of work/energy exchange gained considerable prominence there [Grattan-Guinness 1990, ch. 16]: influential German texts such as [Weisbach 1855] and French ones like [Poncelet 1870] were similar in their silence.

7.4. Some Anglophonic Interest

The tradition of using inequalities in mechanics continued intermittently in the late 19th century. An interesting case is J. W. Gibbs (1839–1903) in the USA, who pondered, in [1879], upon "the fundamental formulae of dynamics," especially in its variational tradition. He allowed for, and even preferred, variations to be imposed on the velocities or accelerations of the mass-points rather than on their positions. He also formulated Fourier's inequality idea, applying it especially to Newton's second law of motion, and when allowing impulses to act on the mass-points, he came to Gauss' principle; but of his predecessors he cited only Gauss, and then in a footnote implying that his paper had already been thought out.

Again, Gibbs' work gained little response; but some other slight Anglo-Saxon interest in these principles is evident in the 1900s. For example, in his influential English textbook on dynamics, E. J. Routh devoted a few pages to (only) Gauss' principle [Routh 1905, 311–313, 350–351]. Three years later the historian of mathematics P. E. B. Jourdain included [Gauss 1829] in an edition of texts on the foundations of mechanics edited for the series *Ostwalds Klassiker der exakten Wissenschaften* [Jourdain 1908]; it contained many valuable historical notes of his own.

7.5. Germanic Concerns, 1900s–1930s

Jourdain's edition was, of course, in German. As he showed in his notes, concern with the foundations of mechanics had increased considerably in German literature around the turn of the century (see also [Voss 1902]). The main interest was the relationship between Hamiltonian and least-action formulations in generalised coordinate systems, and so were usually equational in form; but some note was taken of inequality principles. A most distinguished commentator in this context was L. Boltzmann (1844–1906), in a substantial discussion of inequality principles in his lectures on mechanics [1897, 209–241]. Starting out from Gauss' principle, he related it to others, such as that of virtual work and d'Alembert's principle; he noted Gibbs' preference for it (and, like Gibbs, was apparently unaware of Fourier).

Another contributor to this tradition was Farkas, our hero in Subsection 4.5. In his paper [1902] he cited [Ostrogradsky 1838a] as a predecessor, and in an interesting passage on "infinitesimal systems," he considered linear combinations of functions and their first-order derivatives and found inequality relationships for such functions over regions and on their surfaces [1902, 17–20]. In a succeeding paper [Farkas 1906] he discussed "the foundations of analytical mechanics"; here he deployed a mixture of equations and inequalities, which were used to express the "Relationships of the constraint," where Gauss's word, "Zwang," was used. He examined conditions for the continuity of the constraint, and also its discontinuity in the case of impulses (where he cited both Gibbs and Boltzmann on p. 179), followed on pp. 181–184 by a consideration of "multiplicatorial equations" (where the usual multipliers were included); he treated both friction-free and frictional systems. On p. 166 he recorded his awareness of the work of Minkowski for the first time.

The history of Gauss' principle was also recorded at this time. It was noted in an important article on the history of the principle of virtual work which has already been cited in its own right in Subsection 3.1 [Lindt 1904, 172–173], and it appeared in the *Encyklopädie der mathematischen Wissenschaften*, the vast German enterprise which reported on the development of all areas of mathematics up to that time. Two articles included a short statement: [Voss 1901, 84–87], in a discourse on "the principles of rational mechanics"; and [Stäckel 1905, 460] in a survey of "elementary dynamics."

A learned historian of mathematics, P. Stäckel (1862–1919) briefly also noted the work of Fourier and Cournot in this reference, and later he gave the topic more historical attention. In 1917, as an editor of a volume of the edition of Gauss' works, he reproduced and discussed some texts prepared by Gauss' student Ritter (whose work was discussed in Subsection 3.3), part of Ritter's own doctoral thesis of [1853], and his edition of some pertinent lectures by Gauss on the principle of least squares [Gauss 1917, 468–482] [21]. Two years later, Stäckel sent his own "Remarks on the principle of least constraint" [1919] to the Heidelberg *Akademie der Wissenschaften*, in which he discussed in more detail both the historical literature and also the logical relationship between Gauss' and d'Alembert's principles.

Stäckel's paper and its considerations received some more general publicity in 1927, in two articles in a volume on mechanics published in the large series *Handbuch der Physik*. Inequality formulations of both the principle of virtual work and d'Alembert's principle were given in a study of "the foundations of mechanics" ([Hamel 1927, 17, 24], building upon earlier work), and in a survey of "the principles of dynamics" ([Nordheim 1927, 49–52] on Fourier and later work, and [1927, 62–70] on Gauss and later work). However, neither author discussed these formulations at any length. Indeed, neither Fourier's nor Gauss' principles seemed to gain a prominent place in mechanics. For example, they gained a modest place in a "general formulation of the equations of dynamics" [Appell 1925, 30–32, 37–43] (a noteworthy source, as this pamphlet inaugurated the French series *Mémorial des sciences mathématiques*); and they did not appear at all in the survey of "Dynamical systems" which G. D. Birkhoff published in the USA as his [1927] (exactly the same time as the *Handbuch* volume), although variational mechanics was featured prominently there. Publicity for them was only occasional and usually historical, such as two items from 1933: a (fine) study of Gauss' contributions to mechanics and potential theory published as a supplement to the edition of his works [Geppert 1933, 16–32]; and, more briefly, in a long survey of analytical mechanics, one of the last articles for the great German *Encyklopädie* [Prange 1933, 541–547]. Thus the principles did not play a significant role in the inauguration of nonlinear programming.

7.6. On the Establishment of Non-Linear Programming

The origins of this sub-discipline have been admirably recorded in [Kuhn 1976], and I need only note a few points from him here. The pioneering paper [Kuhn and Tucker 1950] (Subsection 7.1) was the major initial source, although a few authors had been in the vicinity since the mid 1930s (such as F. John, and Kuhn and Tucker themselves). The contexts had included the calculus of variations, network theory and geometrical inequalities, and of course the later establishment of linear programming itself and its link with the theory of games.

But all this was *recent* work: apart from the bearing of studies in the calculus of variations, and the *general* influence of mechanics upon linear programming noted in Subsection 5.1, the principles of mechanics discussed in this section do not appear to have guided its birth. [Kuhn and

Tucker 1950] exemplifies the point, for the authors cited only [Farkas 1902] among the older sources, and just for its main theorem (which bears upon nonlinear as well as linear programming), not for its treatment of "infinitesimal systems." Yet their theorem (which was stated in Subsection 7.1) is reminiscent of other predecessors, especially the deployment of multipliers in [Ostrogradsky 1838b] and [Farkas 1906]. Indeed, Kuhn discovered later that even the Kuhn-Tucker theorem (11) of 1950 had been proved 12 years earlier in a master's dissertation of 1939 on the calculus of variations written by W. Karush, who did not publish it.

Thus, as with linear programming, we find in this section an excellent case study of nonheritage. A significant difference needs to be recorded, however; in this case almost all the potential sources lay in one branch of mathematics [22], in contrast to the wide range of branches discussed in the earlier sections of this paper. But, as before, the opportunities were largely missed. One of the main causes may have been the strongly algebraic cast of much of the previous work, which took most of its leads from Lagrange's variational tradition in mechanics: Ostrogradsky's papers (Subsection 7.2) are particularly good examples of this genre. A second cause is the fact that the nonlinear functions deployed were not necessarily required to specify convex or concave regions. Finally, one must note national differences again: the establishment and its immediate background were very much North American affairs, but the prehistory was dominated by Germanic literature.

The history of mathematics reveals many "sleepers" such as linear and nonlinear programming (and matrix theory also), due in large part to mathematicians not knowing enough of the pertinent past. Which major topics in the mathematics of the 21st century are sleeping today?

ACKNOWLEDGMENTS

Thanks are due to colleagues at the Middlesex Polytechnic Mathematics Seminar: their reactions to a presentation of some of this material there directed me towards finding much of the rest of it. For comments on draft material, I thank O. I. Franksen and T. Hailperin, and two anonymous referees.

NOTES

[1] Some of the historical literature gives credit for linear programming to [Quesnay 1758], for his attempt to relate the roles of the various socio-eco-

nomic classes relative to their use of "productive" and "sterile" expenditures. Without taking away from the originality of his work (especially concerning the distinction just mentioned), it cannot really be interpreted as linear programming.

[2] On these and some other developments, see [Tilling 1973, esp. 113–139]; Figure 1 is based on hers at p. 138. For a briefer study of Laplace and de Prony, see [Grattan-Guinness 1990, 344–347, 557–559].

[3] From 1802 to 1815 Fourier was employed as a Prefect of two French *départements*, and was also much involved in Egyptology. During these years he also produced his mathematical theory of heat diffusion, his version of Fourier analysis, and much work on the theory of equations [see Grattan-Guinness and Ravetz 1972]. L. Charbonneau (Montreal) is completing an exercise on dating the manuscripts of Fourier in the *Bibliothèque Nationale* in Paris: mss. fr. 22501–22529: he has kindly told me that he identifies material pertinent to inequalities from the periods 1806–1811 and 1818–1830. There are quite a few manuscripts, and they deserve a decent study.

[4] On Fourier's work, with references, see [Grattan-Guinness 1970]; on the subject matter of this subsection see [Prékopa 1980, 534–536] and [Grattan-Guinness 1990, 1140–1146]. [Fourier 1827b] is translated, with a short commentary, in [Kohler 1973].

[5] Fourier expressed the problem verbally, but he seems to have (mistakenly) thought of the inequalities in terms of " $<$ " instead of " $\le$ ". Navier so represented his theory in [1825b, 66–67]; and Darboux also followed him when rendering the problem of Fig. 2 in algebraic form [see Fourier 1890, 321, 320].

[6] Since 1818 Fourier had become concerned with statistics, as the chief of the Bureau of Statistics for the *département* of the Seine. His main papers are reproduced in [1890, 523–588]. Pertinent manuscripts (see [3]) also survive; once again, the ensemble is worthy of detailed examination.

[7] In 1827 Fourier tried, but without success, to stimulate the interest of S. Germain (see his letter in *Bibliothèque Nationale* (Paris), *manuscrits français, nouvelles acquisitions*, ms. 4073, fol. 8).

[8] Gergonne claimed priority in his [1826] in the use of inequalities in mechanics. Unfortunately his work seems to have been uninfluential even at the time. On the general history, and bearing, of inequalities upon mechanics and also thermodynamics, see [Franksen 1981, 1985]; and Section 7 below.

[9] Motzkin also wrote some short papers on convex sets at this time. In the "curriculum vitae" at the end of his thesis [1936], he reported that he had moved to Jerusalem around 1930. Among his professors at Basel, he paid especial thanks to A. Ostrowski, who might have indicated the topic to him. Dantzig mistakenly described the thesis as proffered at the University of Zürich [1963, 609].

[10] Developments seem to have been still slower in other communist countries; for example, [Hua "and others" 1962] is an early case of linear programming in China (involving the harvesting of wheat).

[11] A curious historical detail emerged during research on this paper. The *American Mathematical Monthly*, 33 (1926), 487–490, records the abstracts of lectures given at the annual meeting for 1926 of the Mathematical Association of America. C. F. Gummer spoke on linear systems of equations and inequalities, within the American tradition recorded in Subsection 4.4; and T. Dantzig, best remembered today as an historian of mathematics but also the father of a future pioneer of linear programming, spoke on "Mathematics in modern engineering practice."

[12] Compare [Dantzig 1963, 277, 286–290]. I am somewhat surprised by his later reminiscence that after conversations with von Neumann in 1947 he "heard of Farkas' Lemma [. . .] for the first time" [Dantzig 1982, 45]; perhaps von Neumann had found the theorem for himself and then communicated it to Dantzig without using or knowing of Farkas's name. Anecdotal history on the origins of mathematical programming is furnished in [Lenstra, Rinnooy Kan and Schrijver 1991].

[13] R. Rider (Berkeley) is examining the origins of operational research, especially in Britain and the USA.

[14] A very remarkable anticipation of locational equilibrium was achieved by G. Lamé and B. Clapeyron, French scientists then working in Saint Petersburg, in a wonderful paper [1829] published there. Sadly, it had no impact beyond a review [Sturm 1829a] in the abstracting *Bulletin* in Paris (Subsection 3.2), not even on its own authors. [Franksen and Grattan-Guinness 1989] contains an English translation of the paper and an extensive commentary on the later history of this topic; see also [D. M. Smith 1981, ch. 4]. Curiously, Minkowski was not far away from it in his study of geometrical number theory (Subsection 4.4) with his examination of the "economy of the smallest radial distance" of a set of points [1910, 176–179, 189–191]. For another important missed opportunity of 1829, in which Sturm was again involved, see Subsection 6.3 below.

[15] For evidence of the rapid rate of early growth of linear programming in mathematical economics, see, for example, the bibliographies in [Dorfman, Samuelson and Solow 1958] and [Dantzig 1963]. In 1975 Koopmans and Kantorovich shared the Nobel Prize for economics.

[16] This judgement on the slow and patchy development of matrix theory is based largely on my own examination of the literature. Some of the characteristics of scrappy histories discussed in this section apply; perhaps major mathematicians were more significant, even if not frequently so. The story has not received the integrated study that it deserved, although various papers by T. W. Hawkins have filled several parts of the gap; his [1975b] contains an overview of progress in the 19th century.

[17] We saw in Subsection 3.2 that Cournot used this *Bulletin* for his articles; he was another of its regular reviewers. On these papers by Cauchy and Sturm, see [Hawkins 1975a]; and [Grattan-Guinness 1990, 1150–1154].

[18] Fourier's followers included J. M. C. Duhamel, J. Liouville, Navier, Sturm and G. Lamé: they also drew on other senior contemporaries as well as on Fourier. Lamé's involvement in locational equilibrium was noted in [14]; so was Sturm's review, and he appears in this role again in Subsection 7.3. On the work of these followers in the 1820s and 1830s, and its general context, see [Grattan-Guinness 1990, chs. 17–18, esp. pp. 1206–1212 on economic optimisation].

[19] From an Anglo-Saxon viewpoint, the failure was a blessing in disguise; for the high proportion of German contributors to the work described in Sections 4 and 7 suggests that the Kaiser's High Command might have been the first to take such an initiative! After all, even Bôcher's textbook was available in German three years after its first appearance [Bôcher 1910].

[20] On this theorem see also [Abadie 1967]. An excellent account of the general significance and development of nonlinear programming is given in [Avriel 1981].

[21] In view of the concern with n-dimensional spaces evident in some of the later work that was discussed in Section 4, it is very striking to see Ritter refer in 1853 to "manifolds of n dimensions" [Gauss 1917, 470]. Presumably some reverberation from Grassmann is evident [compare 436–437].

[22] The prehistory of nonlinear programming marginally includes F. Y. Edgeworth, who used the calculus of variations and multipliers in his work on mathematical economics of the late 19th century [see his texts in Baumol and Goldfield 1968, 188–228]. Similarly, the studies made by G. Monge from the 1770s onwards of "cuts and fills," on the most efficient means of transporting earth from one location to another, were handled nonlinearly, as an application of descriptive geometry [see Taton 1951, 193–204]; only [Kantorovich 1942] perceived this application among the major figures discussed here.

REFERENCES

Abadie, J. 1967. On the Kuhn-Tucker theorem. In *Nonlinear programming*, J. Abadie, ed., pp. 19–36. Amsterdam: North Holland.

Appell, P. 1925. *Sur une formulation générale des équations de la dynamique*. Paris: Gauthier-Villars (*Mémorial des sciences mathématiques*, no. 1).

Ascoli, G. 1932. Sugli spazi lineari metrici e le loro varietà lineari. *Annali di matematica pura ed applicata* (4)10, 33–81, 203–232.

Avriel, A. 1981. Non-linear programming. In *Mathematical programming for operational researchers and computer scientists*, A. G. Holzman, ed., 272–367. New York: Dekker.

Baumol, W. J. and Goldfield, S. M., eds. 1968. *Precursors in mathematical economics: an anthology*. London: London School of Economics.

Behrend, F. 1940. [Review of Motzkin 1936]. *Jahrbuch über die Fortschritte der Mathematik* **62** [for 1936], 54–55.

Bernkopf, M. 1968. A history of infinite matrices. A study of denumerably infinite linear systems as the first step in the history of operators defined on function spaces. *Archive for History of Exact Sciences* **4**, 308–358.

Birkhoff, G. D. 1927. *Dynamical systems*. New York: American Mathematical Society.

Bôcher, M. 1907. *Introduction to higher algebra*. New York: Macmillan. [Reprinted New York: Dover, 1964; 2nd ed. 1922; German translation 1910.]

Bôcher, M. 1910. *Einführung in die höhere Algebra*. Leipzig and Berlin: Teubner.

Boltzmann, L. 1897. *Vorlesungen über die Principe der Mechanik*, pt. 1. Leipzig: Barth.

Bonnesen, T. and Fenchel, W. 1934. *Theorie der konvexen Körper*. Berlin: J. Springer. [Reprinted New York: Chelsea, 1964(?).]

Boole, G. 1854a. *The laws of thought* .... London: Walton and Maberley. [Reprinted New York: Dover, 1958; new edition, P. E. B. Jourdain, ed., Chicago and London: Open Court, 1916.]

Boole, G. 1854b. On the conditions by which the solutions of questions in the theory of probabilities are limited. *Philosophical Magazine* (4)8, 91–98. Also in [1952], 280–288.

Boole, G. 1868. Of propositions numerically definite. *Transactions of the Cambridge Philosophical Society* **11**, 396–411. Also in [1952], 167–186.

Boole, G. 1952. *Studies in logic and probability*. R. Rhees, ed. London: Watts.

Borel, E. F. E. J. 1938. Observations sur la note précédente [Ville 1938]. In *Traité du calcul des probabilités*, E. F. E. J. Borel, gen. ed., Vol. 4, pt. 2 (J. Ville, ed.), 115–117. Paris: Gauthier-Villars.

Boscovich, R. J. 1757. De litteraria expeditione per Pontificiam ditionem. *Bononiensi scientiarum et artium Instituto atque Academia commentarii* **4**, 353–396.

Boscovich, R. J. 1760. Supplementum ad librum quintum. In B. Stay, *Philosophiae recentioris versibus traditae libri X*, Vol. 2, pp. 385–472. Rome: Palearini. (See also [Maire and Boscovich]).

Brentjes, S. 1976a. Bemerkungen zum Beitrag von Julius Farkas zur Theorie der linearen Optimierung. *NTM-Schriftenreihe zur Geschichte der Naturwissenschaften, Technik und Medizin* **13**, 21–23.

Brentjes, S. 1976b. Der Beitrag der sowjetischen Wissenschaftler zur Entwicklung der Theorie der linearen Optimierung. *NTM-Schriftenreihe zur Geschichte der Naturwissenschaften, Technik und Medizin* **13**, 105–110.

Carathéodory, C. 1911. Über den Variabilitätsbereich der Fourier'schen Konstanten von positiven harmonischen Funktionen. *Rendiconti del Circolo Matematico di Palermo* **32**, 189–217. Also in [1955], 78–110.

Carathéodory, C. 1955. *Gesammelte mathematische Schriften*, vol. 3. *Bayerische Gesellschaft der Wissenschaften*, Ed. Munich: Beck.

Carathéodory, C. and Féjer, L. 1911. Über den Zusammenhang der Extremen von harmonischen Functionen mit ihrem Koeffizienten und über den Picard-

Landauschen Satz. *Rendiconti de Circolo Matematico di Palermo* **32**, 218−239. Also in Carathéodory [1955], 111−138. Also in Féjer, *Gesammelte Arbeiten*, Vol. 1, P. Turán, ed., 693−715. Basel and Stuttgart: Birkhäuser, 1970.

Carver, W. B. 1922. Systems of linear inequalities. *Annals of Mathematics* (2)**23**, 212−220.

Cauchy, A. L. 1829. Sur l'équation à l'aide de laquelle on détermine les inégalités séculaires des mouvements des planètes. *Exercices de mathématiques* **4**, 140−160. Also in *Oeuvres complètes*, ser. 2, vol. 9, *Académie des Sciences*, Ed., 174−195. Paris: Gauthier-Villars, 1891. [Review in Sturm 1829b].

Charnes, A. and Cooper, W. W. 1962. On some works of Kantorovich, Koopmans, and others. *Management science* **8** (1961−1962), 246−263.

Cohen, I. B. 1988. The computer: a case study of support by government, especially the military, of a new science and technology. In *Science, technology and the military*, E. Mendelsohn, M. R. Smith and P. Weingart, eds., Vol. 12, 119−154. Dordrecht: Kluwer.

Cournot, A. A. 1826. Sur le calcul des conditions d'inégalité. *Bulletin universel des sciences et de l'industrie, sciences mathématiques* **6**, 1−8.

Cournot, A. A. 1827. Extension du principe des vitesses virtuelles au cas où les conditions de liaison du système sont exprimées par des inégalités. *Bulletin universel des sciences et de l'industrie, sciences mathématiques* **8**, 165−170.

Cournot, A. A. 1828. Sur la théorie des pressions. *Bulletin universel des sciences et de l'industrie, sciences mathématiques* **9**, 10−22.

Cournot, A. A. 1829a. *Mémoire sur le mouvement d'un corps solide, soutenu par un plan fixe*. Paris: Hachette. Also in *Journal für die reine und angewandte Mathematik* **5** (1830), 133−162.

Cournot, A. A. 1829b. [Summary of 1829a.] *Bulletin universel des sciences et de l'industrie, sciences mathématiques* **11**, 264−266. [Authorship attributed.]

Cournot, A. A. 1830−1832. Du mouvement d'un corps sur un plan fixe, quand on a égard à la résistance du frottement. *Journal für die reine und angewandte Mathematik* **5** (1830), 223−249; **8** (1832), 1−12.

Cournot, A. A. 1838. *Recherches sur les principes mathématiques de la théorie des richesses*. Paris: Hachette. Also *Oeuvres complètes*, Vol. 8, G. Jorland, ed. Paris: Vrin, 1980.

Cournot, A. A. 1847. *De l'origine et des limites de la correspondance entre l'algèbre et la géométrie*. Paris and Algiers: Hachette.

Cournot, A. A. 1913. *Souvenirs (1760−1860)*, E. P. Bottinelli, ed. Paris: Hachette.

Cyganova, N. J. 1983−1984. Die Entwicklung des Prinzips des kleinsten Zwanges in den Arbeiten deutscher Forscher des 19. Jahrhunderts. *NTM-Schriftenreihe zur Geschichte der Naturwissenschaften, Technik und Medizin* **20** (1983), no. 2, 25−38; **21** (1984), no. 1, 79−90.

Dantzig, G. B. 1963. *Linear programming and extensions*. Princeton: University Press.

Dantzig, G. B. 1970. Linear programming and its progeny. In *Applications of mathematical programming techniques*, E. M. L. Beale, ed., 3–16. London: English Universities Press.

Dantzig, G. B. 1982. Reminiscences about the origins of linear programming. *Operations research letters* **1**, 43–48. [Slightly revised version in *Mathematical programming. The state of the art*, A. Bachem, M. Grotschel and B. Corte, eds., 78–86. Berlin: Springer, 1983.]

de la Vallée Poussin, C. 1911. Sur la méthode de l'approximation minimum. *Annales de la Société Scientifique de Bruxelles*, pt. 2, 1–16.

de Prony, G. C. F. M. 1804. *Recherches physico-mathématiques sur la théorie des eaux courantes*. Paris: *Imprimerie Impériale*.

Dines, L. L. 1925. Definite linear inequalities. *Annals of mathematics* (2)**27**, 57–64.

Dines, L. L. 1927. Sets of functions of a general variable. *Transactions of the American Mathematical Society* **29**, 463–470.

Dines, L. L. and McCoy, N. H. 1933. On linear inequalities. *Transactions of the Royal Society of Canada* (3) **27**, sec. 3, 37–70.

Dorfman, R. 1984. The invention of linear programming. *Annals of the history of computing* **6**, 283–295.

Dorfman, R. Samuelson, P. A. and Solow, R. M. 1958. *Linear programming and economic analysis*. New York: McGraw Hill.

Eisenhart, C. 1961. Roger Boscovich and the combination of observations. In *Roger Joseph Boscovich*, L. L. Whyte, ed., 200–212. London: Allen and Unwin. Also in [Kendall and Plackett 1977], 88–100.

Farebrother, R. W. 1990. Further details of contacts between Boscovich and Simpson in June 1760. *Biometrika* **77**, 397–400.

Farkas, J. 1902. Theorie der einfachen Ungleichungen. *Journal für die reine und angewandte Mathematik* **124**, 1–27.

Farkas, J. 1906. Beiträge zu den Grundlagen der analytischen Mechanik. *Journal für die reine und angewandte Mathematik* **131**, 165–201.

Fourier, J. B. J. 1798. Mémoire sur la statique, contenant la démonstration du principe de vitesses virtuelles, et la théorie des momens. *Journal de l'École Polytechnique* (1)**2**, cah. 5, 20–60. Also in [1890], 475–521.

Fourier, J. B. J. 1826. Solution d'une question particulière du calcul des inégalités. *Bulletin des sciences, par la Société Philomatique de Paris*, 99–100. Also in [1890], 315–321 [including note by G. Darboux. Anonymous review in *Bulletin universel des sciences et de l'industrie, sciences mathématiques* 7 (1829), 1–2.]

Fourier, J. B. J. 1827a. [Part of] Analyse des travaux de l'Académie Royale des Sciences, pendant l'année 1823. Partie mathématique. *Histoire de Académie Royale des Sciences 6*, xxix–xli. Also in [1890], 321–324 [cited here]. [Part in [1831], 75–82.]

Fourier, J. B. J. 1827b. [Part of] Analyse des travaux de l'Académie Royale des Sciences, pendant l'année 1824. Partie mathématique. *Histoire de Académie*

Royale des Sciences **7**, xlvii–lv. Also in [1890], 325–328 [cited here]. {Part in [1831], 82–84. English translation: [Kohler 1973].}

Fourier, J. B. J. 1831. *Analyse des équations déterminées*, C. L. M. H. Navier, ed. Paris: Firmin Didot.

Fourier, J. B. J. 1890. *Oeuvres*, Vol. 2, *Académie des Sciences* [G. Darboux], ed. Paris: Gauthier-Villars.

Franksen, O. I. 1969. Mathematical programming in economics by physical analogies. *Simulation*, June, 297–314; July, 25–42; August, 63–87.

Franksen, O. I. 1981. The virtual work principle—a unifying system concept. In *Structures and operations in engineering and management systems*, Ø. Bjørke and Franksen, eds., 17–152. Trondheim: Tapir.

Franksen, O. I. 1985. Irreversibility by inequality constraints. *Systems analysis in modelling and simulation* **2**, 137–149, 251–273, 337–359.

Franksen, O. I. and Grattan-Guinness, I. 1989. The earliest contribution to location theory? Spatio-economic equilibrium with Lamé and Clapeyron, 1829. *Mathematics and computers in simulation* **31**, 195–220.

Fréchet, R. M. 1935. Générlisations du théorème des probabilités totales. *Fundamenta mathematicae* **25**, 379–387.

Gass, S. 1989. Comments on the history of linear programming. *Annals of the history of computing* **11**, 147–151.

Gauss, C. F. 1809. *Theoria motus corporum coelestium in sectionibus conicis solem ambientium*. Hamburg: Perthus and Besser. Also in *Werke*, Vol. 9, 1–282. Leipzig: Teubner, 1903 [reprinted Hildesheim: Olms, 1973]. [English transl. C. H. Davis: Boston: Little, Brown, 1857; reprinted New York: Dover, 1963.]

Gauss, C. F. 1829. Über ein neues allgemeines Grundgesetz in der Mechanik: "Die Bewegung eines Systems, materieller, auf was immer für eine Art unter sich verknüpfter Punkte, deren Bewegung zugleich an was immer für äussere Beschränkungen gebunden sind, geschieht in jedem Augenblick in möglich grösster Übereinstimmung mit der freien Bewegung, oder unter möglich kleinsten Zwange, in dem man als Mass des Zwanges, den das ganze System in jedem Zeittheilchen erleidet, die Summe der Produkte aus dem Quadrate der Ablenkung jedes Punktes von seiner freien Bewegung in seiner Masse betrachtet." *Journal für die reine und angewandte Mathematik* **4**, 232–235. Also in *Werke*, Vol. 5, 23–28. Leipzig: Teubner, 1877 [reprinted Hildesheim: Olms, 1973]. Also in [Jourdain 1908], 27–30. [Review in Sturm 1829c.]

Gauss, C. F. 1917. *Werke*, Vol. 10, pt. 1, *Göttingen Gesellschaft der Wissenschaften*, Ed. [including P. Stäckel]. Leipzig: Teubner. [Reprinted Hildesheim: Olms, 1973.]

Geppert, H. 1933. *Über Gauss' Arbeiten zur Mechanik und Potentialtheorie*. Berlin: J. Springer [as Gauss *Werke*, Vol. 10, pt. 2, no. 7].

Gergonne, J. D. 1826. Note sur le calcul des conditions d'inégalité. *Annales des mathématiques pures et appliquées* **17** (1826–27), 134–137.

Gibbs, J. W. 1879. On the fundamental formulae of dynamics. *American journal of mathematics* **2**, 49–64. Also in *Scientific papers*, Vol. 2, H. A. Bumstead and R. G. van Name, eds., 1–15. London: Longmans, 1906.

Grammel, R. 1927. (Ed.) *Grundlagen der Mechanik. Mechanik der Punkte und starren Körper*. Berlin: J. Springer (*Handbuch der Physik*, Vol. 5).

Grattan-Guinness, I. 1970. Joseph Fourier's anticipation of linear programming. *Operational research quarterly* **21**, 361–364.

Grattan-Guinness, I. 1990. *Convolutions in French mathematics, 1800–1840. From the calculus and mechanics to mathematical analysis and mathematical physics*, 3 vols. Basel: Birkhäuser, and Berlin: Deutscher Verlag der Wissenschaften.

Grattan-Guinness, I. and Ravetz, J. R. 1972. *Joseph Fourier 1768–1830. A survey of his life and work, based on a critical edition of his monograph on the propagation of heat, presented to the Institut de France in 1807*. Cambridge, Mass.: MIT Press. (See also [Franksen and Grattan-Guinness].)

Haar, A. 1918. Die Minkowskische Geometrie und die Annäherung an stetige Funktionen. *Mathematische Annalen* **78**, 294–311. Also in [1959], 403–420.

Haar, A. 1924. Über lineare Ungleichungen. *Acta litterarum ac scientarum regaie Universitatis Hungaricae Francisco-Josephinae, sectio scientarum mathematicam* **2** (1924–1926), 1–14. Also in [1959], 439–452.

Haar, A. 1959. *Gesammelte Arbeiten*, B. Szökefalvi-Nagy, ed., Budapest: Akademie der Wissenschaften.

Hailperin, T. 1965. Best possible inequalities for the probability of a logical function of events. *American Mathematical Monthly* **72**, 343–359.

Hailperin, T. 1986. *Boole's logic and probability*, 2nd ed. Amsterdam: North Holland.

Hailperin, T. 1991. Probability logic in the twentieth century. *History and philosophy of logic* **12**, 71–110.

Hamel, G. 1927. Die Axiome der Mechanik. In [Grammel 1927], 1–42.

Hancock, H. 1939. *Development of the Minkowski geometry of numbers*. New York: Macmillan. [Reprinted New York: Dover, 1964.]

Hawkins, T. W. 1975a. Cauchy and the spectral theory of matrices. *Historia mathematica* **2**, 1–29.

Hawkins, T. W. 1975b. The theory of matrices in the 19th century. In *Proceedings of the International Congress of Mathematicians, Vancouver, 1974*, Vol. 2, 561–570. [n.p.]: Canadian Mathematical Congress.

Hitchcock, F. L. 1941. The distribution of a product from several sources to numerous localities. *Journal of Mathematics and Physics* **20**, 224–230.

Hua, L. K. "and others" 1962. Application of mathematical methods to wheat harvesting. *Chinese Mathematics* **2**, 77–91.

Hurwitz, A. 1902. Sur quelques applications géométriques des séries de Fourier. *Annales de l'École Normale Supérieure* (3)**19**, 357–408. Also in *Mathematische Werke*, Vol. 1, G. Polya, ed., 509–554. Basel: Birkhäuser, 1932.

Jourdain, P. E. B. 1908. (Ed.) *Lagrange, Rodrigues, Jacobi und Gauss. Abhandlungen über die Prinzipien der Mechanik.* Leipzig: Engelmann (*Ostwalds Klassiker der exakten Wissenschaften,* no. 187).

Kakeya, S. 1914a. On some integral equations. *Tôhoku mathematical journal* **4**, 186–190.

Kakeya, S. 1914b. On some positive forms. *Tôhoku mathematical journal* **6**, 27–31.

Kantorovich, L. V. 1939. *Matematicheskie metody organizatsiy i planirovaniya proizvodstva.* Leningrad: University Publishing House. [English translation: *Management science* **6** (1959–1960), 366–422.]

Kantorovich, L. V. 1942. O peremeshcheniyu mass. *Doklady Akademii Nauk SSSR* **37**, 199–201. [English translation: *Management science* **6** (1958–1959), 1–4.]

Kendall, M. G. and Plackett, R. L., eds. 1977. *Studies in the history of statistics and probability.* London: Griffin.

Kohler, D. A. 1973. Translation of a report by Fourier on his work on linear inequalities. *Opsearch* (India) **10**, 38–42. [Translation of Fourier 1827b.]

Koopmans, T. J. 1960. A note about [Kantorovich 1939]. *Management science* **6** (1959–1960), 363–365.

Kuhn, H. W. 1976. Nonlinear programming: a historical view. In *Nonlinear Programming,* R. W. Cottle and C. E. Lemke, eds., 1–26. Providence: American Mathematical Society.

Kuhn, H. W. and Tucker, A. W. 1950. Nonlinear programming. In *Proceedings of the Second Berkeley Symposium on Mathematical Statistics and Probability,* J. Neyman, ed., 481–492. Berkeley: University of California Press.

Lagrange, J. L. 1788. *Méchanique analitique.* Paris: Desaint. [Not in *Oeuvres.*]

Lamé, G. and Clapeyron, B. P. E. 1829. Mémoire sur l'application de la statique à la solution des problèmes relatifs à la théorie des moindres distances. *Journal des voies et communications* no. 10, 26–49. [Review in [Sturm 1828a]. English translation and commentary: [Franksen and Grattan-Guinness 1989].]

Laplace, P. S. 1793. Sur quelques points du système du monde. *Mémoires de l'Académie Royale des Sciences,* (1789), 1–87. Also in *Oeuvres complètes* Vol. 11, *Académie des Sciences,* ed., 477–558. Paris: Gauthier-Villars, 1910.

Laplace, P. S. 1799. *Traité de mécanique céleste,* Vol. 2. Paris: Duprat. Also *Oeuvres complètes,* Vol. 2, *Académie des Sciences,* ed. Paris: Gauthier-Villars, 1880.

Lenstra, J. K., Rinnooy Kan, A. H. G. and Schrijver, A. 1991 [Eds.] *History of mathematical programming, a collection of personal reminiscences.* Amsterdam: Elsevier.

Lindt, R. 1904. Das Prinzip der virtuellen Geschwindigkeiten. Seine Beweise und die Unmöglichkeit seiner Umkehrung bei Verwendung des Begriffes "Gleichgewicht eines Massensystems." *Abhandlungen zur Geschichte der Mathematik* **18**, 145–195.

Maire, C. and Boscovich, R. J. 1755. *De litteraria expeditione per Pontificiam ad dimetiendos duos meridiani gradus et corrigendam mappam geographicam,* Rome: Typographio Palladis.

Maire, C. and Boscovich, R. J. 1770. *Voyage astronomique et géographique, dans l'état de l'église* [...] *pour mesurer deux degrés du méridien, et corriger la carte dans l'état ecclésiastique.* [?] Hugon, trans. Paris: Tilliard. [French translation/edition of 1755.] (See also [Boscovich].)

Minkowski, H. 1896. *Geometrie der Zahlen*, 1st ed. Leipzig and Berlin: Teubner. [Reprinted New York: Chelsea, 1964.]

Minkowski, H. 1910. *Geometrie der Zahlen*, 2nd ed. Leipzig and Berlin: Teubner.

Minkowski, H. 1911. *Gesammelte Abhandlungen*, Vol. 2. D. Hilbert, ed. Leipzig and Berlin: Teubner.

Motzkin, T. 1936. *Beiträge zur Theorie der linearen Ungleichungen.* Jerusalem: Azriel. [English translation: Rand Corporation report T-22 (1952); also in *Selected papers*, Vol. 2, D. Carter, B. Gordon and B. Rothschild, eds., 1–80. Basel: Birkhäuser, 1983. Reviews in [Ridder 1936] and [Behrend 1940].]

Navier, C. L. M. H. Mechanics. *Résumé des leçons données à l'École Nationale des Ponts et Chaussées, sur l'application de la mécanique à l'établissement des constructions et des machines*, pt. 1. 1st ed. Paris: Didot, 1826. 2nd ed. Paris: Carilain-Goeury, 1833. [3rd ed.: see Saint-Venant.]

Navier, C. L. M. H. 1825a. Note sur les questions de statique dans lesquelles on considère un corps pesant supporté par un nombre de points d'appui surpassant 3. *Nouveau bulletin des sciences, par la Société Philomatique de Paris*, 35–37. {Anonymous review in *Bulletin universel des sciences et de l'industrie, sciences mathématiques* **5** (1826), 166–167.}

Navier, C. L. M. H. 1825b. Sur le calcul des conditions d'inégalité. *Nouveau bulletin des sciences, par la Société Philomatique de Paris*, 66–68, 81–84. (See also [Fourier 1831].)

Nordheim, L. 1927. Die Prinzipe der Dynamik. In [Grammel 1927], 43–90.

Novozhilov, V. V. 1967. *Problemy izmereniya zatrat i resul'tatov pri optimal'nom planirovanii.* Moscow: Ekonomika. [English translation: *Problems of cost-benefit analysis in optimal planning.* White Plains, New York: International Arts and Sciences Press, 1970.]

Ostrogradsky, M. A. 1838a. Considérations générales sur les moments des forces. *Mémoires de l'Académie des Sciences de St. Pétersbourg* (6)**3**, pt. 1, 129–150.

Ostrogradsky, M. A. 1838b. Mémoire sur les déplacemens intantanés des systèmes assujettis à des conditions variables. *Mémoires de l'Académie des Sciences de St. Pétersbourg* (6)**3**, pt. 1, 565–600.

Plackett, R. L. 1972. The discovery of the method of least squares. *Biometrika* **59**, 239–251. Also in [Kendall and Plackett 1977], 279–291.

Poncelet, J. V. 1870. *Introduction à la mécanique industrielle, physique ou expérimentale*, 3rd ed. X. Kretz, ed. Paris: Gauthier-Villars.

Prange, G. 1933. Die allgemeinen Integrationsmethoden der analytischen Mechanik. In *Encyklopädie der mathematischen Wissenschaften*, Vol. 4, sec. 2, 505–804 (article IV 12–13).

Prékopa, A. 1980. On the development of optimisation theory. *American mathematical monthly* **87**, 527–542.

Quesnay, F. 1758. *Tableau économique avec son application, ou extrait des économies royales de Sully*, Versailles: privately printed. [Various reprints and translations: for example, part in English in *Early Economic Thought*, A. E. Monroe, ed., 339–348. Cambridge: Harvard University Press, 1924 (reprinted 1965).]

Rees, M. 1980. The mathematical sciences and World War II. *The American Mathematical Monthly* **87**, 607–621.

Ridder, T. 1936. [Review of Motzkin 1936]. *Zentralblatt für Mathematik* **14**, 246–247.

Riesz, F. 1911. Sur certains systèmes singuliers d'équations intégrales. *Annales scientifiques de l'École Normale Supérieure (3)***28**, 33–62. Also in *Oeuvres complètes*, Vol. 2, A. Császár, ed., 798–827. Budapest: Akademie der Wissenschaften, 1960.

Ritter, G. A. G. 1853. *Über das Princip des kleinsten Zwanges*. Göttingen: dissertation. (Part in [Gauss 1917], 468–472.)

Ritter, G. A. G. 1873. *Lehrbuch der analytischen Mechanik*. Hannover: Rümpler.

Routh, E. J. 1905. *The elementary part of a treatise on the dynamics of a system of rigid bodies*, pt. 1, 2nd ed. London: Macmillan.

Saint-Venant, A. J. C. Barré de. 1864. Notice sur les ouvrages de Navier [and] Historique abrégé des recherches sur la résistance et sur l'élasticité des corps solide. In C. L. M. H. Navier, *De la résistance des corps solides*. Saint-Venant, ed., lv–lxxxiii and xc–cccxi. Paris: Dunod. (The book is the 3rd ed. of Navier [Mechanics], pt. 1.)

Schmidt, E. 1913. Zum Hilbertschen Beweise des Waringschen Theorems. *Mathematische Annalen* **74**, 271–274.

Schoenberg, I. J. 1932–1933. On finite-rowed systems of linear inequalities in infinitely many variables. *Transactions of the American Mathematical Society* **34** (1932), 594–619; **35** (1933), 452–478.

Schoenberg, I. J. 1933. Convex domains and linear combinations of continuous functions. *Bulletin of the American Mathematical Society* **39**, 273–280.

Scholz, E. 1989. *Symmetrie, Gruppe, Dualität. Zur Beziehung zwischen theoretischer Mathematik und Anwendung in Kristallographie und Baustatik des 19. Jahrhunderts*, Basel: Birkhäuser, and Berlin: Deutscher Verlag der Wissenschaften.

Schwartz, B. L. 1989. The invention of linear programming. *Annals of the History of Computing* **11**, 145–147.

Seneta, E. 1992. On the history of the strong law of large numbers and Boole's inequality. *Historia Mathematica* **19**, 24–39.

Sheynin, O. B. 1973. R. J. Boscovich's work on probability. *Archive for History of Exact Sciences* **9**, 306–324.

Siegmund-Schultze, R. 1982. Die Anfänge der Funktionalanalyse und ihr Platz im Umwälzungsprozess der Mathematik um 1900. *Archive for History of Exact Sciences* **26**, 13–71.

Smith, D. M. 1981. *Industrial location. An economic geographic analysis*, 2nd ed. New York: Wiley.

Smith, G., ed. 1982. *The Boole-De Morgan correspondence*, Oxford: Clarendon Press.

Stäckel, P. 1905. Elementare Dynamik der Punktsysteme und starren Körper. In *Encyklopädie der mathematischen Wissenschaften*, Vol. 4, sec. 1, 435-684 (article IV 6).

Stäckel, P. 1919. Bemerkungen zum Prinzip des kleinsten Zwanges. *Sitzungsberichte der Heidelbergschen Akademie der Wissenschaften, mathematisch-physikalische Klasse*, paper no. 11, 25 pp. (See also [Gauss 1917]).

Steinitz, H. E. 1913-1914. Bedingt konvergente Reihen und konvexe Systeme. *Journal für die reine und angewandte Mathematik* **143** (1913), 128-175; **144** (1914), 1-40.

Stigler, S. 1984. Boscovich, Simpson and a 1760 manuscript note on fitting a linear relation. *Biometrika* **71**, 615-620.

Stokes, R. W. 1933. A geometric theory of linear inequalities. *Transactions of the American Mathematical Society* **31**, 782-807.

Sturm, J. C. F. 1829a. [Review of Lamé and Clapeyron 1829]. *Bulletin universel des sciences et de l'industrie, sciences mathématiques* **11**, 327-328.

Sturm, J. C. F. 1829b. [Review of several papers by Cauchy, including 1829]. *Bulletin universel des sciences et de l'industrie, sciences mathématiques* **12**, 301-303.

Sturm, J. C. F. 1829c. [Review of Gauss 1829]. *Bulletin universel des sciences et de l'industrie, sciences mathématiques* **12**, 304.

Sturm, J. C. F. 1829d. Sur l'intégration d'un système d'équations différentielles linéaires. *Bulletin universel des sciences et de l'industrie, sciences mathématiques* **12**, 313-322.

Tanner, R. C. H. 1961. Mathematics begins with inequality. *The Mathematical Gazette* **44**, 292-294.

Taton, R. 1951. *L'oeuvre scientifique de Monge*. Paris: Presses Universitaires de France.

Tilling, L. 1973. The interpretation of observational errors in the eighteenth and early nineteenth centuries. University of London Ph.D.

von Neumann, J. 1928a. Zur Theorie des Gesellschaftsspiele. *Mathematische Annalen* **100**, 295-320. Also in [1963], 1-26. (Summary in [1928b].)

von Neumann, J. 1928b. Sur la théorie des jeux. *Comptes rendus de 'Académie des Sciences* **186**, 1689-1891. (Summary of [1928a].)

von Neumann, J. 1937. Über ein ökonomisches Gleichungssystem und eine Verallgemeinerung des Brouwerschen Fixpunktsatzes. *Ergebnisse eines mathematischen Kolloquiums* **8**, 73-83. (English translation: *Review of economic studies* **13** (1945), 1-9; also in [von Neumann 1963], 29-37; and in [Baumol and Goldfield 1968], 296-306.)

von Neumann, J. 1963. *Collected works*, Vol. 6, A. Taub, ed. Oxford: Pergamon.

von Neumann, J. and Morgenstern, O. 1944. *The theory of games and economic behaviour*, 1st ed. Princeton: University Press. [2nd ed. 1947, 3rd ed. 1953.]

Valentine, F. A. 1964. *Convex sets*. New York: McGraw Hill.

Ville, J. 1938. Sur la théorie générale des jeux où intervient l'habilité des joueurs. In [Borel 1938], 105–113.

Voss, A. 1901. Die Prinzipien der rationellen Mechanik. In *Encyklopädie der mathematischen Wissenschaften*, Vol. 4, sec. 1, 3–121 (article IV 1).

Wald, A. 1935. Über die eindeutige positive Lösbarkeit der neuen Produktionsglei-chungen. *Ergebnisse eines mathematischen Kolloquiums* 6, 12–18. (English trans-lation: [Baumol and Goldfield 1968], 281–288.)

Wald, A. 1936. Über die Produktionsgleichungen der ökonomischen Wertlehre. *Ergebnisse eines mathematischen Kolloquiums* 7, 1–6. (English translation: [Baumol and Goldfield 1968], 289–293.)

Weisbach, J. 1855. *Lehrbuch der Ingenieur- und Maschinen-Mechanik*, pt. 1, 3rd ed. Braunschweig: Vieweg.

Weyl, H. 1935. Elementare Theorie der konvexen Polyeder. *Commentarii mathe-matici Helvetici* 7, 290–306. Also in *Gesammelte Abhandlungen*, Vol. 2, K. Chandrasekharan, ed., 517–533. Berlin: Springer, 1968. [English translation: *Annals of mathematical studies* 24 (1950), 3–18.]

David Hilbert (1862–1943)
Courtesy of Handschriftenabteilung der Niedersächsische
Staats-und Universitätsbibliothek Göttingen

Ernst Zermelo (1871–1953)
Courtesy of Handschriftenabteilung der Niedersächsische
Staats-und Universitätsbibliothek Göttingen

Hilbert's Axiomatic Programme and Philosophy

Volker Peckhaus

Institut für Philosophie der Universität Erlangen-Nürnberg, Germany

It was only after the publication of the set-theoretic paradoxes by Russell and Frege (1903) that the problem of ensuring a consistent axiomatization of set theory and logic fell within the scope of David Hilbert's early axiomatic programme (1899–1914). The paradoxes affected the consistency proof for the axioms of arithmetic, a central concern of the programme. The "direct" proof Hilbert had hoped for seemed now clearly to be impossible by means of a logic proved to be inconsistent. This necessitated a fundamental revision of the concept of axiomatization, which led Hilbert to pursue a "partly simultaneous" development of the laws of logic and arithmetic, an idea he expressed for the first time in 1904. Hilbert's somewhat vague ideas can be clarified by referring to his explanations in the hitherto unpublished lecture course *Logische Principien des mathematischen Denkens* (1905). This lecture course also reveals that Hilbert regarded the philosophical foundations of his axiomatics as a desideratum to be answered by systematically oriented philosophers. Hilbert himself did not work in detail on the axiomatization of set theory and logic, but he expected this task to be carried out by the mathematician Ernst Zermelo and the philosopher Leonard Nelson. He tried to secure for both men jobs at Göttingen, thereby hoping to create stable long-term working conditions for them there.

1. INTRODUCTION

It is not the aim of this paper to give a philosophical analysis of David Hilbert's ideas concerning the foundations of mathematics, or even his axiomatic programme. Studies of this kind have been presented, for example, by Philip Kitcher [1976] and Wolfgang Schüler [1984]. These studies concur on at least one central point: Hilbert's preference for Kantianism was largely incompatible with Kant's philosophy. Rather than analyzing Hilbert's philosophical views directly, I intend, however, to inquire into the motivation behind them. This essay will, therefore, be concerned above all with Hilbert's *attitude* towards philosophy as science, and the rôle of philosophy in the kind of mathematics which he propounded. This orientation was decisively shaped by the philosophical implications and complications surrounding his early programme for establishing an axiomatic foundation of mathematics. Hilbert's *early* axiomatic programme was implicit already in his "Grundlagen der Geometrie" [Hilbert 1899] but first explicitly formulated in his lecture, "Mathematical Problems," which he delivered at the Paris International Congress of Mathematicians in August 1900 [1900b]. This programme became the basis for virtually all axiomatic research over the next 20 years.

In elucidating the relations between Hilbert's axiomatic programme and philosophy, I set forth the following two theses:

First, the discussion of the paradoxes of set theory evoked by the publications of Bertrand Russell (*Principles of Mathematics* [1903]) and Gottlob Frege (the postscript to his *Grundgesetze der Arithmetik* [1903]) gave rise to a "philosophical shift" in Hilbert's axiomatic programme. This shift can be seen from the fact that set theory became a central focus of his attempts to axiomatize mathematics, whereas the foundations of logic were included as a new area of investigation.

Secondly, in his revised programme Hilbert regarded the actual foundations of axiomatic systems, which were, in his opinion, based on metaphysical conceptions, as a true philosophical concern. Moreover, he expected that its solution would be found, if not by mathematicians, then by mathematically-trained philosophers. The axiomatization of mathematics thus became a common task for both mathematicians and philosophers, who would establish a natural division of labour based on the actual concerns of the two disciplines.

After discussing these theses, I shall show how Hilbert tried to set up the personnel and administrative requirements for this collaboration between philosophers and mathematicians through his activities in academic politics. Before turning to these matters directly, however, I begin with a sketch of the early axiomatic programme, as it appeared at the end of the year 1900.

2. HILBERT'S AXIOMATIC PROGRAMME

In his *Festschrift* contribution, "Grundlagen der Geometrie" [1899], published for the unveiling of the Gauß-Weber monument on 17 June, 1899, in Göttingen, Hilbert proceeded from three imagined "systems of things" (points, straight lines, planes), and described their interrelations in a set of 20 axioms. In addition he investigated this set of axioms *as an object in itself*, proving its completeness, the independence of the axioms, and its consistency. It is known that he reduced the consistency of the geometrical axioms to the presupposed consistency of arithmetic. Therefore, a complete consistency proof was, in fact, postponed. At the same time, he set forth a new task: to find a consistent set of axioms for arithmetic.

Hilbert presented his ideas concerning the foundations of arithmetic three months later, in September 1899, at the annual meeting of the *Deutsche Mathematiker-Vereinigung* held in Munich. In his lecture, "Über den Zahlbegriff" [1900a], he elaborated the foundations of arithmetic—in his opinion the fundamental discipline of mathematics—independent of set-theoretic considerations. Again, Hilbert proceeded from an imagined system of things, which he now called numbers. He constructed the field of the real numbers in the same way he founded geometry. The system of axioms he proposed for arithmetic was largely equivalent to the 17 propositions on the nature of real numbers in §13 of his "Grundlagen der Geometrie," although in the latter work these were employed for calculations with straight lines. Entirely new was only the axiom of completeness which he added as an 18th axiom. In accordance with the sketchy character of the lecture, Hilbert did not attempt to carry out the meta-axiomatic investigations of the independence of the axioms or the completeness of the system and its consistency. Still, he did assert that the "necessary task" of proving consistency required "only a suitable modification of known methods of inference" [1900a, 184].

These optimistic words from September 1899 clearly indicate that Hilbert initially underestimated the enormity of the task at hand. However, he soon altered these views, and in August 1900, less than one year later, he included the consistency proof for arithmetic as the second among his 23 famous unsolved mathematical problems presented before the Second International Congress of Mathematicians in Paris. In this lecture, Hilbert explained that he did not see his foundational conception as restricted to pure mathematics. He wrote:

> Wherever, from the side of the theory of knowledge or in geometry, or from the theories of natural or physical science, mathematical ideas come up, the problem arises for mathematical science to investigate the principles underlying these ideas and to establish them upon a simple and complete system of axioms in such a manner [1] that the exactness of the new ideas and their applicability to deduction shall be in no respect inferior to those of the old arithmetical concepts [2].

Hilbert thus expanded his axiomatic programme to cover all of mathematics, including its applications, in his words introducing problem two even to all sciences:

> When we are engaged in investigating the foundations of a science, we must set up a system of axioms which contains an exact and complete description of the relations subsisting between the elementary ideas of that science [3].

Hilbert's remarks concerning the problem of the consistency proof were, in fact, taken from his earlier lectures. He noted that the consistency of the axioms of geometry could be deduced from the consistency of the arithmetical axioms. Of course, for the axioms of arithmetic a similar deduction was not possible. For these, a direct proof would be required. He expressed, however, his conviction

> that it must be possible to find a direct proof for the consistency [4] of the arithmetical axioms, by means of a careful study and suitable modification of the known methods of reasoning in the theory of irrational numbers [5].

I have presented this short sketch to indicate the pre-eminent position Hilbert assigned to the consistency proof already in the year 1900. Three years later, on 27 October, 1903, he spoke to the Göttingen Mathematical Society on the foundations of arithmetic. He then emphasized that he aimed "to work out the 'axiomatic' standpoint clearly." On the rôle of consistency, he coined the brief formula: "the principle of contradiction[:] the pièce de résistance" [6].

3. PARADOXES AND THE AXIOMATIC METHOD

The principle of (excluded) contradiction as the "main course" of all attempts to axiomatize mathematics was also the topic of another lecture Hilbert delivered before the Göttingen Mathematical Society on 8 November, 1904 [7]. On this occasion, Hilbert called the consistency proof the "regulating principle" ("ordnendes Prinzip") of axiomatic foundations, and he informed his Göttingen colleagues of a lecture he had given at the Third International Congress of Mathematicians held in Heidelberg in August of that year. This Heidelberg lecture, "Über die Grundlagen der Logik und der Arithmetik," revealed a fundamental change in Hilbert's attitude towards the foundations of mathematics. In fact, it marked the beginning of a new, second phase in Hilbert's programme, since here, for the first time, he took the philosophical and logical implications of the axiomatic programme into consideration. In his early contributions Hilbert used the words "logic" or "logical" without any clear signification. Up until his lecture, "Mathematical Problems," he regarded the axiomatization of arithmetic as a purely mathematical problem, independent of logical and set-theoretic investigations. It was therefore even more surprising that in his Heidelberg lecture Hilbert demanded a "partly simultaneous development of the laws of logic and arithmetic" ("teilweise gleichzeitige Entwicklung der Gesetze der Logik und der Arithmetik" [Hilbert 1905a, 176]). Thus, between 1900 and 1904, a significant change in Hilbert's views took place, as he moved toward the conviction that the foundations of arithmetic—and, therefore, of all mathematics—required investigations from both the mathematical and philosophical sides. In the 11-page published version of his Heidelberg lecture, Hilbert only hinted at the nature of the joint construction of logic and arithmetic he had in mind, aiming merely to indicate the approximate direction of the research that would be required. He was apparently unable to attain even this limited aim. For, according to Hilbert's biographer, Otto Blumenthal, his lecture remained "completely misunderstood, and its indications proved insufficient, when studied more thoroughly" [1935, 422].

Fortunately, we are in possession of another source which helps to clarify Hilbert's intentions during this crucial period: his lecture course, "Logische Principien des mathematischen Denkens," held in Göttingen during the summer term of 1905. Two sets of notes from these lectures have survived: the "official" notes taken by Ernst Hellinger, located in the library of the Mathematics Institute of the University of Göttingen [Hilbert 1905b]; and a second manuscript, written by the Nobel Laureate, Max Born, stored among Hilbert's papers in the State and University Library of Göttingen [Hilbert 1905c]. These documents make it clear that Hilbert's ideas concerning the philosophical foundations of mathematics were still quite provisional. Nevertheless, this lecture course provides remarkable testimony to Hilbert's views on the philosophical foundations of axiomatics in general.

It is generally known that the paradoxes of set theory, which gained enormous publicity through Russell's and Frege's publications in 1903, had already been discussed in Göttingen somewhat earlier. Walter Purkert has documented the extant correspondence between Hilbert and Georg Cantor (only Cantor's letters survived) [1986] from the years 1897 to 1900. These letters dealt, above all, with the systems (or sets) of all cardinals and ordinals, i.e., with precisely those problems that were independently formulated in Burali–Forti's paradox. Purkert's findings support the results of

Gregory H. Moore and Alejandro Garciadiego [e.g., 1981], who concluded that Cantor was aware of these paradoxes, but that he did not regard them as logically unsolvable. Returning to the discussion in Göttingen, it should be pointed out that already in April 1902 Zermelo had produced a proof showing that the concept of a set M which contains all of its subsets as elements leads to contradictions [Rang and Thomas 1981]. In his lecture course on the logical principles of mathematical thinking Hilbert dealt in detail with a paradox involving all sets constructed by self-occupancy ("*Selbstbelegung*") and set union. This paradox, known in Göttingen as "Hilbert's paradox," was probably one of the contradictions Hilbert claimed to have found before the turn of the century, as he wrote to Frege on 7 November, 1903 [Frege 1976, 79 *f.*].

The question has been discussed as to why Hilbert and Zermelo did not publish the paradoxes which they had found (e.g., [Moore 1978]). In my opinion, Hilbert, like Cantor before him, failed at first to appreciate the logical relevance of the paradoxes. In his lecture, "Über den Zahlbegriff" [1900], and later in his discussion of Cantor's set theory in "Mathematische Probleme," Hilbert stated that a proof of the consistency of arithmetic would simultaneously prove that the system of real numbers forms a consistent and complete set ("fertige Menge") in Cantor's sense. He emphasized, however, that any attempt to prove the existence of the set of all of Cantor's Alephs would necessarily fail [Hilbert 1900a, 184; 1900b, 265].

It should be stated clearly that in his early writings Hilbert did not advance an axiomatic approach to set theory. For him, set theory was a programme that concurred with his axiomatic foundations for arithmetic, and he claimed that within the framework of his axiomatics all of the problematic concepts of set theory are either not expressible or their existence can be proved. Around 1900 Hilbert did not regard set theory as an independent mathematical subdiscipline but rather as an alternative methodological approach to arithmetic. In accordance with his "pragmatic" viewpoint, the appearance of contradictions was nothing to be alarmed about so long as the stock of accepted mathematical knowledge could be preserved by other means.

To illustrate Hilbert's pragmatic orientation I would like to quote a metaphorical passage on the general development of science that he invoked in opening his discussion of the paradoxes of set theory in the lecture course of 1905.

> It has, indeed, always been so in the historical development of knowledge that one begins to cultivate a discipline without many scruples, and, pressing onwards as far as possible, one thereby runs into difficulties (often, however,

only after some time) which, consequently, forces one to turn back and reflect on the foundations of the discipline. The house of knowledge is not erected like a dwelling where the foundation is first well laid-out before the erection of the living quarters is begun. Knowledge prefers to obtain comfortable rooms as quickly as possible in which it can rule, and then only afterwards, when it appears that, here and there, the loosely joined foundations are unable to support the completion of the rooms, does it proceed to prop up and secure them. This is no shortcoming but rather a correct and healthy development [8].

By 1905, however, Hilbert had revised his attitude towards set theory completely. Thus, he followed up on the above remarks by noting that the strictest methods developed in mathematics had led to "unsolvable problems" and evoked "very grave contradictions" in new mathematical fields (i.e., set theory). These domains had been established naively on the no longer sufficient fundamentals of such mathematical disciplines as the infinitesimal calculus or geometry. Their foundations, therefore, urgently needed to be strengthened and their methods thoroughly investigated. The situation in set theory differed from that of the other mathematical disciplines essentially because its problems were such

> [...] which tend substantially further towards the theoretical philosophical side than those which occurred earlier. As in new disciplines, one always uses the old customary laws which are, at first, considered obvious, so likewise for set theory, one took the traditional Aristotelian logic and used its definitions without scruples [9].

These paradoxes, Hilbert noted, arose from purely logical operations (above all, from comprehending conceptions under general terms), and they could not be solved by hitherto existing devices. This circumstance reveals the essential cause of the "philosophical shift" in Hilbert's axiomatic programme. By 1904, Hilbert realized that it was impossible to axiomatize mathematics with purely mathematical methods. Moreover, the paradoxes affected the germ of his programme, the consistency proof for the axioms of arithmetic, because it was impossible to obtain the desired "direct" proof by means of an inconsistent logic. Hilbert was convinced that he had found the source of these contradictions: the unrestricted comprehension of generalities as new sets. Still, nothing was gained from this diagnosis, as he knew, "since all thinking is based on just such comprehensions, and it remains a problem to separate that which is allowed from that which is not allowed" [1905b, 215]. Attempts were made, from both the mathematical and the philosophical sides, to grapple with these questions, and Hilbert hoped that these things "would henceforth receive more and more attention—especially from philosophers as well" [1905b, 216].

In his 1905 lecture course, Hilbert constructed an axiomatized logical calculus based upon contemporary algebraico-logical examples. His calculus was clearly—although not really justifiably—set apart from earlier studies of contemporary logicians:

> I would characterize all that which has been achieved up till now as too formal, because these logical calculi presented themselves with no far-reaching aims or tasks but were rather concerned only with the formal construction of a calculus for the presentation of the old logical inferences [10].

Hilbert's logical calculus constitutes a propositional logic without quantification based upon the principle of identity [11] and intended as an instrument for mathematical operations. As a second step in this direction, Hilbert introduced number signs and the notions of the "things of thought," "everything," and "infinite," thus making it possible to reform the definition of a set, a definition which should serve to build a bridge between logic and arithmetic.

An examination of the differences between his Heidelberg lecture of 1904 and the lecture course he began a few months later reveals the underlying dynamic in the development of Hilbert's ideas. In 1904 Hilbert still construed the problematic term "everything," which was, in his view, responsible for the known contradictions, as a shorthand expression representing infinite conjunctive and adjunctive connexions of propositions. By employing this process he produced quantifiers in the Peirce–Schröder tradition [12]. In 1905, however, he relinquished this presentation. His vague hints at this time seem to indicate that he intended to abandon the admission of infinite operations. These operations he later tried to avoid in his finitistic proof theory by introducing the epsilon axiom. In the 1905 lectures, Hilbert reinterpreted his programme of 1904, which vaguely called for the "simultaneous development of the laws of logic and arithmetic." He now sought to successively introduce logical and mathematical operations according to pragmatic criteria, in the sense of a construction of mathematics in stages: axiomatized calculus of propositional logic, axiomatic systems for mathematical operations, axiomatic systems for set theory and arithmetic.

4. PHILOSOPHICAL FOUNDATIONS OF AXIOMATIZATION

I now turn to discuss the second aspect of the connection between Hilbert's axiomatic programme and philosophy: the philosophical foundations of axiomatization itself. I deal here only with one part of this aspect, the "problem of the initial first step." Hilbert began his axiomatization of geometry in his "Grundlagen der Geometrie" [Hilbert 1899] with the words "we imagine three different systems of things." One year later, in "Über den Zahlbegriff," he introduced the arithmetical axioms in a similar way: "We imagine a system of things." In his 1904 Heidelberg lecture he wrote: "An object of our thought processes is to be called *a thing of thought (Gedankending)* or, in brief, a *thing*, and will be designated by a sign."

For the foundations of logic it appeared necessary to justify the first step itself, the "we imagine." This was accomplished by means of an "axiom of thinking" ("Axiom des Denkens"), or by an "axiom of the existence of an intelligence," not really belonging to the logical calculus:

> I have the ability to think *things*, and to designate them by simple signs $(a, b, \ldots, X, Y, \ldots)$ in such a completely characteristic way that I can always recognize them again without doubt. My thinking operates with these designated things in certain ways, according to certain laws, and I am able to recognize these laws through self-observation, and to describe them perfectly [13].

In a marginal note to Hellinger's lecture notes, Hilbert called this axiom "the philosophers' *a priori.*" This remark confirms the standard view that Hilbert was a Kantian, even if this merely implies that he regarded transcendental philosophy with interest.

In his 1905 lecture course, Hilbert described the "general idea" of the axiomatic method as follows:

> The general idea of this method unconsciously underlies all theoretical and practical thinking. One faces a body of factual material consisting of propositions, combinations of propositions, doubtful propositions, suppositions, etc. One picks out a certain number of these propositions and takes these as "principles" or "axioms," due to a certain triviality inherent in them [14].

Above all, it is remarkable that in this passage Hilbert upholds the traditional view that axioms are propositions, since back in 1899, in a famous controversy with Frege over the "Grundlagen der Geometrie," Hilbert already advanced the "modern" notion that the word "axiom" does not denote a proposition but rather a propositional function. Obviously, Hilbert was for many years unaware or unconcerned about this possible range of terminological distinctions. There is, however, still another point of interest in connection with this passage. I would like to compare it with a statement made by the Göttingen philosopher, Leonard Nelson, who received his doctoral degree shortly before attending Hilbert's 1905 lecture course. Already in 1904, Nelson published a paper entitled "Die kritische Methode und das Verhältnis der Psychologie zur Philosophie" [Nelson 1904], a sort of programme for the "neo-Friesian School," founded by Nelson. In this paper, Nelson elucidated the notion of "regressive method" as follows:

> If we pick out of the experiences of life such decisions and judgements, concerning which a consensus exists, we can dismember them, and so, by a regressive method, trace the common philosophical principles that were presupposed and applied in reaching these decisions and judgements [15].

A certain parallelism between the quoted passages from Hilbert and Nelson cannot be denied, and this parallelism is not at all accidental. Nelson's epistemology, which he borrowed from Jakob Friedrich Fries, a contemporary and critic of Hegel, served as the foundation for a scientific philosophy oriented towards the logical rigor of mathematics. The regressive method leads to a system of principles which are synthetic judgements *a priori*, and, therefore, should not be dogmatically stated but critically proved. Such an approach is possible with the help of Fries' method of deducing the principles from their underlying immediate cognitions. For Nelson, a deduction of the mathematical axioms is both possible and necessary:

> It must also be possible—apart from a justification through demonstration, which already provides evidence and which, therefore, is enough to satisfy the interests of the mathematician—to give a *critical deduction of the axioms of mathematics* in its full extension. This transfer of criticism to the axiomatic system of mathematics constitutes a separate scientific discipline: the philosophy of mathematics, or stated better: *critical mathematics* [16].

The conception of a critical mathematics goes back to Jakob Friedrich Fries' *Mathematische Naturphilosophie*, published in 1822. It was first brought into the modern discussion on foundations by Nelson's friend, Gerhard Hessenberg, in a lecture he delivered to the Mathematical Society of Berlin in November 1903 [Hessenberg 1904]. Although this conception cannot be discussed here in detail, it should be pointed out that Hilbert's axiomatization of geometry served as a model for the *mathematical* part of critical mathematics, which was to be supplemented, so to speak, with a "philosophical fundament" provided by subsequent *philosophical* efforts. Thus, for Nelson and his followers, it was a philosophical task to deduce the axioms of mathematics. Nelson identified mathematical judgements as synthetic *a priori* judgements in a Kantian sense, and their source of cognition as nonsensible, nonempirical, pure intuition.

This neo-Friesian conception of Hessenberg and Nelson proved highly attractive to the "working mathematician" due to the fact that it offered a philosophical foundation of mathematics without disturbing the everyday world of mathematicians, provided that they followed the axiomatic method in accordance with Hilbert's model.

Clearly, this philosophical orientation fits in beautifully with the programmatic character of Hilbert's early axiomatics. Indeed, Hilbert had formulated a programme which approached the various mathematical subdisciplines pragmatically. With his axiomatics of Euclidean geometry, he had redeemed only a part of it paradigmatically. As for possible solutions to other foundational questions, these were only outlined and then handed over to the respective experts. Therefore, the early axiomatic programme was a *research directive* designed to give researchers a guide for their own efforts. Hilbert did not need to present completely developed sets of axioms for set theory and logic, since he had found in Ernst Zermelo a collaborator in these fields. He also did not need to present a fully elaborated philosophy of mathematics, since he considered this domain to be in good hands: those of his Göttingen colleagues, Edmund Husserl and Leonard Nelson.

5. POLITICAL REALIZATION

I now turn to Hilbert's political activities in support of his ideas, activities which were quite essential to his overall programme and strategy. In this instance scientific politics meant, above all, personnel politics, i.e., Hilbert tried to use his influence to establish positions for researchers in Göttingen whom he expected to work on (and hopefully solve) the outstanding foundational problems. Since I have described these activities in other publications [Peckhaus 1990a; 1990b, 196–224], I shall only give a brief survey of them here.

Ernst Zermelo came to Göttingen in 1897 in order to work on his *Habilitation* in theoretical physics. From 1902 to 1907, he essentially lived on the proceeds of a *Privatdozenten* grant established by Friedrich Althoff, Head of the University Department of the Prussian Ministry of Education. This grant eventually terminated, and, after nine years as a *Privatdozent*, Zermelo still had yet to receive a permanent appointment. Thus, a precarious financial situation developed which carried with it the danger that Zermelo would have to suspend his academic teaching. In 1907, Hilbert managed to obtain another one-time grant from the Ministry; a stable financial solution, however, could only be reached through a change in Zermelo's academic status.

Therefore, on 25 July, 1907, the Directors of the Mathematical-Physical Seminar of the University of Göttingen made the following application to the Prussian Minister:

> Your Excellency should award a Lectureship in Mathematical Logic and related topics to the resident Professor Dr. Ernst Zermelo, and grant him a fixed annual reimbursement [17].

The application, which was drafted by Hilbert, proceeded to state that, subsequent to Augustus De Morgan and George Boole, "a domain midway between logic and mathematics" had come into being, called "logical calculus or algebra of logic." In this domain of knowledge, foreign countries had generally assumed the leading rôle, whereas among the younger German generation, "probably only Ernst Zermelo may be seen as a full representative of this direction." Hilbert was clearly playing for high stakes and with a rather weak hand, since at that time Zermelo had only published two brief papers on set theory [1902, 1904], one of them containing his famous proof of the well-ordering theorem [1904]. His only achievement in mathematical logic, in the narrower sense, was the *announcement* of a lecture course on this topic, a course which, however, never came to pass. Zermelo's work on the famous set-theoretic papers of 1908 [1908a, 1908b] was still in progress, their publication having been delayed because, among other things, Zermelo had been unsuccessfully searching for a consistency proof for his axioms of set theory [18]. The title of Zermelo's lectureship, "Mathematical Logic," was clearly related to Hilbert's revised axiomatic programme. With the discovery of the logical paradoxes, problems in set theory became problems in logic, and Hilbert chose Zermelo to solve them. In August 1907, less than four weeks after the filing of this application, the lectureship was granted. Zermelo's course, given in the summer term of 1908, was the first lecture course on mathematical logic in Germany on the basis of a ministerial commission.

Unlike the case of Zermelo, Hilbert's efforts on behalf of Edmund Husserl and Leonard Nelson led to lengthy disputes within the Göttingen Philosophical Faculty, and were one of the reasons for its division in 1922.

In 1901, on an initiative of the Ministry and against the resistance of the two Göttingen chair holders in philosophy, Julius Baumann and Georg Elias Müller, the philosopher and mathematician Edmund Husserl was appointed to an *Extraordinariat* in philosophy [19]. His advancement to a personal *Ordinariat* in 1905—with the strong backing of Hilbert—was also only brought about against the resistance of the Faculty. When plans were made to establish a (temporary) third chair of philosophy in Göttingen Hilbert tried to have Husserl appointed to it, thereby binding him more strongly to Göttingen. His efforts, however, failed, since Heinrich Maier, a historian of philosophy, was appointed. In this process, Hilbert gained only a formidable enemy, Georg Elias Müller [20].

This hostile confrontation had consequences for the quarrels concerning Leonard Nelson which began in 1906 as a result of Nelson's application to be admitted to the *Habilitation* [21]. As his *Habilitationsschrift* Nelson undiplomatically presented his already mentioned paper, "Die kritische Methode und das Verhältnis der Psychologie zur Philosophie" [1904], a work that had been rejected by Julius Baumann two years earlier, when Nelson submitted it as a doctoral thesis. In a detailed referee's report, Georg Elias Müller did not neglect to point this out. The majority of the commission assented to his negative vote. Although Felix Klein and David Hilbert demanded that the commission should convene and discuss the case orally, Nelson was advised by the Dean to withdraw his application. Klein and Hilbert argued for Nelson, referring to the considerable attention surrounding the kind of problems in the philosophy of mathematics that Nelson studied. Their arguments failed to prevail, however, and only in 1909, after a second attempt, Nelson was allowed to habilitate in Göttingen.

When Husserl left Göttingen in 1916, his personal *Ordinariat* was to be turned into an *Extraordinariat* [22]. The Faculty planned to appoint Georg Misch, again a historian of philosophy. A minority report, initiated by Hilbert, was submitted on March 3, 1917, the aim of which was the appointment of Nelson. It read as follows:

> With regard to the significance of the major philosophical problems—some of which [...] have been revealed by modern mathematics—it is in our opinion a *modest demand* that *at least here in Göttingen an Extraordinariat* should be created for a man who devotes all his working energy to that field of philosophy which in former times ruled alone, and who, because of the parallel development of mathematics, will nowhere find a better terrain than here for the realization of his efforts [23].

Epistemology was "that field of philosophy which in former times ruled alone," and, according to the report, it had not been developed further by philosophers since Kant. It goes without saying that this report evoked the opposition of the Göttingen philosophers, who triumphed in this little battle as well.

The quarrel surrounding Nelson was, in part, a power struggle between the two sections of the Philosophical Faculty over the allocation of chairs and positions in philosophy. This conflict flared up again when Heinrich Maier, who belonged to the Philological–Historical Section, was appointed to a chair at Heidelberg in July 1918 [24]. In searching for his successor, the Philological–Historical Section put together a list of candidates without consulting the Mathematical–Scientific Section. Again Hilbert initiated a vehement protest movement among the mathematicians and scientists, who claimed the right to voice their opinion in philosophical matters. They recommended the appointment of Misch to the Maier-*Ordinariat* and the transformation of the vacant extraordinary professorship into a "separate position for systematic philosophy of the exact sciences." In the course of this new quarrel, the majority overtly threatened to dissolve the Faculty, and pointed to the fact that during all three appointments since 1910 "Hilbert was on each occasion the soul of a repetitious opposition." The argument ended when the Ministry granted the Hilbert faction the desired position, which was finally filled by Nelson in June 1919.

These events dramatize Hilbert's persistence in trying to gain influence over philosophical teaching and research at Göttingen. The point is further illustrated by statements from Hilbert's papers and correspondence. In a private letter to the Ministerial Counsellor, Carl Heinrich Becker, dated 30 July, 1918 [25], Hilbert wrote that in his opinion mathematics, physics, and philosophy belong to a coherent scientific complex, and he wished "above all to propagate the connexions between mathematics and philosophy." Hilbert continued:

> I have always seen this as a part of my life's mission. Among those philosophers who were not predominantly historians or experimental psychologists, Husserl and Nelson, in my view, seem to be the most prominent personalities, and, to my mind, it is no accident that these two had appeared on the mathematical soil of Göttingen [26].

In the draft of a petition to the faculty which demanded the withdrawal of the philologists' list of candidates to succeed Maier and which proposed new negotiations, Hilbert referred to his own philosophical activities:

> I would like to know of anyone who claims that philosophy does not concern my own subject—in the past, present, and future. Half of my last term's lectures were filled with Aristotle's logic. In 1914 an international course of lectures on the principles of logic was to take place at the invitation of the Wolfskehl foundation. Even during the War the most eminent logician of England had promised to come here [27]. The aged Frege of Jena, a

European celebrity in the field of the principles of logic, will come in Autumn to lecture to us on (amongst other things) propositions, thought, truth, judgement, assertion, sense and reference, idealism, and realism [28]. Doctoral dissertation [of] Behmann [on] paradox of transfinite number [29]. Further on in our Mathematical Society lectures are being given all the time on all sorts of philosophical themes: e.g. Kant's antinomies (v. Kowalewski [sic!]) [30]—to say nothing of philosophy of physics, theories of space, time, and substance, which fill the world today, and which we practise constantly and from time to time exclusively [31].

Hilbert added that Nelson's appointment should help to make "Göttingen number one for scientific systematical philosophy through the establishment of a methodically working definitive school." This last quote makes clear that Hilbert intended to continue Felix Klein's policy of building a mathematical center in Göttingen which was also connected with the development of fields complementing mathematical research. Systematic philosophy could develop some sort of service function in founding mathematics. Hilbert was, in fact, an organizer of science who was not restricted, as is often assumed, to the narrow boundaries of his specialty. If one only considers Hilbert's research publications, then Paul Bernays was in fact right when he wrote in 1935 with regard to Hilbert's Heidelberg lecture of 1904 that "at this preliminary state Hilbert interrupted his investigations on the foundations of arithmetic for a long time" [Bernays 1935, 200]. This analysis, however, does not take into account Hilbert's activities in teaching—his lecture courses were research lectures—nor his engagement in university politics. This broader view shows how deeply Hilbert was involved with philosophical concerns even during periods when he fell silent in these areas of science. Hilbert set forth ideas which he hoped would be realized by younger researchers, and he saw as one of his main tasks to secure the institutional support necessary for their research. In light of these goals and Hilbert's growing insight into the philosophical problems inherent in his axiomatic research programme, one can better understand his rôle and motivation in the numerous conflicts within the Göttingen Philosophical Faculty.

Hilbert should have the last word. In a selection of arguments "for the Minister" he presented all of his personal perplexity as evoked by the quarrels of 1918:

For 15 years I have been fighting for philosophy. Althoff stamped an *Ordinariat* out of the ground [*Ordinariat* Maier]. Here only a small *Extraordinariat* according to my desire, while I am promising such great things: the number one center of philosophy. That can exist nowhere, or in Göttingen, so that no one has to go from here to Freiburg [like Husserl] or Heidelberg [like Maier]. A cultural question of the first order is at stake. Foreign countries! [...] Without Nelson I cannot realize an important part of my life's programme. Nelson is the leaven[,] he will represent here a definitive school, aiming for firm principles. His appointment is a cultural deed of the first rank: Reformation of the spirit of professorship. Without Nelson I'm nothing in the faculty [32].

ACKNOWLEDGMENTS

This paper gives some important results of a longer study which dealt with the interdisciplinary cooperation of mathematicians and philosophers in Göttingen between the turn of the century and World War I (Peckhaus *1990*). Its methodological conception is based on the results of a research project, "Case studies towards the foundation of a social history of formal logic," which was directed by Professor Christian Thiel in Erlangen and supported by a grant of the *Deutsche Forschungsgemeinschaft*. A first version of this paper was presented to the *Workshop on the History of Modern Mathematics* on 17 July, 1990, in Göttingen. I would like to thank Christian Thiel and Thony Christie for valuable comments and their aid in preparing the English version of this paper. In addition I am deeply indebted to David E. Rowe for his efforts to improve the final version.

NOTES

[1] Newson's English translation [Hilbert 1902]: "[...] and so to establish them upon a simple and complete system of axioms, [...]"
[2] Hilbert 1900b, 258 *f*., English transl. 1902, 442: "Wo immer von erkenntnistheoretischer Seite oder in der Geometrie oder aus den Theorien der Naturwissenschaft mathematische Begriffe auftauchen, erwächst der Mathematik die Aufgabe, die diesen Begriffen zu Grunde liegenden Principien zu erforschen und dieselben durch ein einfaches und vollständiges System von Axiomen derart festzulegen, daß die Schärfe der neuen Begriffe und ihre Verwendbarkeit zur Deduktion den alten arithmetischen Begriffen in keiner Hinsicht nachsteht."
[3] Hilbert 1900b, 264, English transl. 1902, 447: "Wenn es sich darum handelt, die Grundlagen einer Wissenschaft zu untersuchen, so hat man ein System von Axiomen aufzustellen, welche eine genaue und vollständige Beschreibung derjenigen Beziehungen enthalten, die zwischen den elementaren Begriffen jener Wissenschaft stattfinden."
[4] Newson's translation: "compatibility."

[5] Hilbert 1900b, 265, English transl. 1902, 448: "daß es gelingen muß, einen direkten Beweis für die Widerspruchslosigkeit der arithmetischen Axiome zu finden, wenn man die bekannten Schlußmethoden in der Theorie der Irrationalzahlen im Hinblick auf das bezeichnete Ziel genau durcharbeitet und in geeigneter Weise modificiert."

[6] "Der Satz vom Widerspruch die pièce de résistance." *Jahresbericht der Deutschen Mathematiker-Vereinigung* [*JDMV*] **12** (1903), 592.

[7] *JDMV* **14** (1905), 61.

[8] Hilbert 1905c, 122: "Es ist in der Entwicklungsgeschichte der Wissenschaft wohl immer so gewesen, dass man ohne viele Scrupel eine Disciplin zu bearbeiten begann und soweit vordrang wie möglich, dass man dabei aber, oft erst nach langer Zeit, auf Schwierigkeiten stiess, durch die man gezwungen wurde, umzukehren und sich auf die Grundlagen der Disciplin zu besinnen. Das Gebäude der Wissenschaft wird nicht aufgerichtet wie ein Wohnhaus, wo zuerst die Grundmauern fest fundamentiert werden und man dann erst zum Auf- und Ausbau der Wohnräume schreitet; die Wissenschaft zieht es vor, sich möglichst schnell wohnliche Räume zu verschaffen, in denen sie schalten kann, und erst nachträglich, wenn es sich zeigt, dass hier und da die locker gefügten Fundamente den Ausbau der Wohnräume nicht zu tragen vermögen, geht sie daran, dieselben zu stützen und zu befestigen. Das ist kein Mangel, sondern die richtige und gesunde Entwicklung."

[9] Hilbert 1905b, 195: "die noch wesentlich tiefer nach der theoretisch philosophischen Seite sich hinneigen, als die früher auftauchenden. Wie man stets in einer neuen Disciplin zuerst die altgewohnten, für selbstverständlich gehaltenen Gesetze anwendet, so nahm man in die Mengenlehre die herkömmliche aristotelische Logik mit und wandte ihre Begriffsbildungen skrupellos an."

[10] Hilbert 1905b, 216: "Ich möchte das hier bisher geleistete jedoch als zu sehr formal bezeichnen, da diese Logikkalküle keine weitergehenden principiellen Ziele und Aufgaben sich stellten, sondern ihnen nur an der formalen Ausbildung des Kalküls zur Darstellung der alten logischen Schlüsse lag."

[11] In a marginal note which he later added Hilbert wrote that one should simply use the sign of equality [1905b, 224].

[12] Hilbert 1905a, 178; cf. Moore 1987, 122.

[13] Hilbert 1905b, 219: "Ich habe die Fähigkeit, *Dinge* zu denken und sie durch einfache Zeichen $(a, b, \ldots, X, Y, \ldots)$ derart in vollkommen charakteristischer Weise zu bezeichnen, dass ich sie daran stets eindeutig wiedererkennen kann; mein Denken operiert mit diesen Dingen in dieser Bezeichnung in gewisser Weise nach bestimmten Gesetzen[,] und ich bin fähig, diese Gesetze durch Selbstbeobachtung zu erkennen und vollständig zu beschreiben." These considerations may have been motivated by Giuseppe Veronese's *Fondamenti di Geometria*, who starts his account of the fundamental principles which form abstract mathematics with the notion "Penso" ("I'm thinking") [Veronese

1891, 1], expressing the faculty and the act of thinking ("con ciò esprimo la facoltà e l'atto del pensare," *ibid.*, n. 1).

[14] Hilbert 1905c, 7 *f*.: "Die allgemeine Idee dieser Methode liegt wohl unbewusst allem theoretischen und practischen Denken zu Grunde. Man hat ein That-sachenmaterial vor sich, das in Sätzen, Combinationen von Sätzen, zweifel-haften Sätzen, Vermutungen etc. besteht. Aus diesen Sätzen greift man eine gewisse Anzahl heraus und nimmt diese als 'Grundsätze' oder 'Axiome' an, auf Grund einer gewissen Trivialität, die ihnen anhaften muss."

[15] Nelson 1904, 4 *f*.: "Greifen wir [. . .] aus den Erfahrungen des Lebens solche Urteile und Beurteilungen heraus, über die Einigkeit herrscht, so können wir diese zergliedern und so durch ein *regressives* Verfahren den philosophischen Principien nachspüren, die in den vorliegenden Urteilen und Beurteilungen zur Anwendung kommen und gemeinsam vorausgesetzt werden."

[16] Nelson 1904, 37: "Es muß—außer der die Evidenz schon mit sich führenden und darum dem Interesse des Mathematikers allein genügenden Begründung durch Demonstration—auch eine *kritische Deduktion der Axiome der Mathe-matik*, ihrem ganzen Umfange nach, möglich sein. Diese Übertragung der Kritik auf die Axiomsysteme der Mathematik konstituiert eine eigene wis-senschaftliche Disciplin: die Philosophie der Mathematik oder, nach besserer Bezeichnung, die *kritische Mathematik*."

[17] Geheimes Staatsarchiv Preußischer Kulturbesitz [GStA], Merseburg, Kultus-ministerium, Rep. 76 Va Sect. 6 Tit. IV No. 4 vol. 4, pp. 269−270; edited in Peckhaus 1990a, 51 *f*. "Eure Excellenz wolle dem hiesigen Professor Dr. Ernst Zermelo einen Lehrauftrage für mathematische Logik und verwandte Gegenstände erteilen und demselben dafür eine feste jährliche Remuneration bewilligen."

[18] Cf. Zermelo's letter to Hilbert, dated Arosa, March 25, 1907, Staats-und Universitäts-bibliothek [SUB] Göttingen, Cod. Ms. D. Hilbert 447, p. 5.

[19] For records concerning the appointment and the advancement of Husserl see GStA Merseburg, Rep. 76 Va Sect. 6 Tit. IV No. 1, vols. 17, 18, 22.

[20] See the draft of Hilbert's letter to Rudolf Otto, dated November 1918, SUB Göttingen, Cod. Ms. D. Hilbert 482, p. 17.

[21] Records: University Archive Göttingen, Phil. Fak., Nelson's personnel records.

[22] Records: GStA Merseburg, Rep. 76 Va Sect. 6 Tit. IV No. 1, vol. 25.

[23] *Ibid.*, pp. 159 *f*.: "Angesichts der Bedeutung der grossen philosophischen Probleme—deren einige, scharf zu nennende, die moderne Mathematik bloss-gelegt hat—ist es unserer Meinung nach eine *bescheidene Forderung*, dass *wenigstens hier in Göttingen ein Extraordinariat* einem Mann eingeräumt werde, der ausschliesslich dieser früher allein herrschenden Richtung in der Philoso-phie seine Arbeitskraft widmet und wegen der parallelen Entwickelung der Mathematik nirgends einen besseren Boden zur Verwirklichung seiner Bestre-bungen findet als hier."

[24] Records: *ibid.*

[25] *Ibid.*, pp. 315 *ff.*

[26] "Darin habe ich von jeher einen Teil meiner Lebensaufgabe erblickt. Unter den Philosophen, die nicht vorwiegend Historiker oder Experimentalpsychologen sind, erscheinen mir Husserl und Nelson als die markantesten Persönlichkeiten, und es ist für mich kein Zufall, dass sich diese beiden auf dem mathematischen Boden Göttingens eingefunden hatten."

[27] Hilbert refers to Bertrand Russell, who was also in discussion for a Göttingen visiting professorship.

[28] This lecture did not take place. In [Frege 1918] ("Der Gedanke") Frege discussed related topics.

[29] Of Behmann's doctoral dissertation of 1918, only an extract was published in [Behmann 1922].

[30] Hilbert refers to the lecture of Nelson's disciple Michael Kowalewsky "Über den Beweis des transzendentalen Idealismus" on June 3, 1913, obviously a report on the results of Kowalewsky's doctoral thesis [1914].

[31] SUB Göttingen, Cod. Ms. D. Hilbert 482, p. 11, appendix 1: "Ich möchte wohl jemand wissen, der behauptet, dass die Philosophie mein Fach nichts angeht —in Vergangenheit, Gegenwart u[nd] Zukunft. Die Hälfte meiner Vorlesungen des vorigen Semesters war mit Logik von Aristoteles erfüllt. Auf Einladung der Wolfskehlstiftung sollte 1914 ein internationaler Vortragscyk[lus] über die Principien der Logik stattfinden. Noch während des Krieges hat mir der hervorragendste Logiker Englands sein herkommen versprochen. Der greise Frege–Jena, europ[äische] Berühmtheit auf dem Gebiete der Principien der Logik[,] wird im Herbst kommen, um uns u.A. über Satz, Gedanke, Wahrheit, Urteil, Behaupt[ung,] Sinn u[nd] Bedeutung, Idealism[us], Realismus, vorzutragen. Dissertation Behmann Antinomie der transf[initen] Zahl. Ferner in unserer math[ematischen] Ges[ellschaft] beständig Vorträge über alle möglichen philos[ophischen] Themata: z.B. Kantische Antinomieen (v. Kowalewski [sic!])—ganz zu schweigen von Philos[ophie] der Physik, die Theorie von Raum, Zeit und Substanz, die heute die Welt erfüllt, u[nd] die wir ständig[,] zeitweise ausschliesslich betreiben."

[32] SUB Göttingen, Cod. Ms. D. Hilbert 482, p. 15: "Seit 15 Jahren kämpfe ich für die Philosophie. Althoff stampfte ein Ordinariat aus dem Boden [Ordinariat Maier]. Hier nur ein kleines Extraord[inariat] nach meinem Wunsch[,] wo ich so Grosses verspreche: erster Centralort für Philosophie. Das geht nirgends oder in Göttingen, so dass Niemand von hier nach Freiburg [wie Husserl] und Heidelberg [wie Maier] geht. Kulturfrage ersten Ranges steht auf dem Spiel. Ausland! [. . .] Ich kann ein[en] wichtigen Teil meines Lebensprogramms nicht durchführen ohne N[elson]. N[elson] ist der Sauerteig[,] er wird hier eine Ausschlag gebende, auf feste Principien gerichtete Schule vertreten: Seine Berufung ist Kulturtat 1sten Ranges: Reformation des Geistes des Professorentums. Ohne N[elson] bin ich Nichts in der Fakultät."

REFERENCES

Behmann, H. 1922. Die Antinomie der transfiniten Zahl und ihre Auflösung durch die Theorie von Russell und Whitehead. *Jahrbuch der Mathematisch-naturwissenschaftlichen Fakultät der Georg-August-Universität zu Göttingen* **23**, 55–64, extract of his doctoral thesis Göttingen 1918.

Bernays, P. 1935. Hilberts Untersuchungen über die Grundlagen der Arithmetik. In [Hilbert 1935, 186–216].

Blumenthal, O. 1935. Lebensgeschichte. In [Hilbert 1935, 388–429].

Frege, G. 1903. *Grundgesetze der Arithmetik, begriffsschriftlich abgeleitet.* Vol. 2, Jena: Hermann Pohle. Photo-reproduction Hildesheim: Olms, 1966.

—— 1918. Der Gedanke. Eine logische Untersuchung. *Beiträge zur Philosophie des deutschen Idealismus* **1** (1918/19), 58–77.

—— 1976. *Wissenschaftlicher Briefwechsel*, Gottfried Gabriel et al., eds. (= Frege, *Nachgelassene Schriften und Wissenschaftlicher Briefwechsel*; 2). Hamburg: Meiner.

Fries, J. F. 1822. *Die mathematische Naturphilosophie, nach philosophischer Methode bearbeitet. Ein Versuch.* Heidelberg: Winter. Photo-reproduction in J. F. Fries, *Sämmtliche Schriften*, G. König and L. Geldsetzer, Eds. Vol. 14. Aalen: Scientia, 1979.

Hessenberg, G. 1904. Über die kritische Mathematik. *Sitzungsberichte der Berliner Mathematischen Gesellschaft* **3**, 21–28, 20th meeting 25 November 1903, appendix of: *Archiv der Mathematik und Physik* (3) **7**.

Hilbert, D. 1899. Grundlagen der Geometrie. In *Festschrift zur Feier der Enthüllung des Gauss-Weber-Denkmals in Göttingen*, Fest-Comitee, Ed., 1–92 [separately page-numbered]. Leipzig: Teubner. Reprint (enlarged) as [Hilbert 1930].

—— 1900a. Über den Zahlbegriff. *Jahresbericht der Deutschen Mathematiker-Vereinigung* **8**, 180–184. Reprinted in [Hilbert 1930, 241–246]. All references are to the first printing.

—— 1900b. Mathematische Probleme. Vortrag, gehalten auf dem internationalen Mathematiker-Kongreß zu Paris 1900. *Nachrichten von der königl. Gesellschaft der Wissenschaften zu Göttingen. Mathematisch-physikalische Klasse aus dem Jahre 1900*, 253–297. Reprinted with additions by Hilbert in *Archiv für Mathematik und Physik* **1** (1901), 44–63, 213–237. This last text reprinted in [Hilbert 1935, 290–329]. All references are to the first printing.

—— 1902. Mathematical Problems. Lecture Delivered before the International Congress of Mathematicians at Paris in 1900, English translation by M. W. Newson. *Bulletin of the American Mathematical Society* **8**, 437–479.

—— 1905a. Über die Grundlagen der Logik und der Arithmetik. In *Verhandlungen des Dritten Internationalen Mathematiker-Kongresses in Heidelberg vom 8. bis 13. August 1904*, A. Krazer, Ed., 174–185. Leipzig: Teubner. Reprinted in [Hilbert 1930, 247–261]. All references are to the first printing.

—— 1905b. *Logische Principien des mathematischen Denkens*. Lecture course, summer term 1905, lecture notes by E. Hellinger (Library of the Mathematics Institute of the University of Göttingen).

—— 1905c. *Logische Principien des mathematischen Denkens*. Lecture course, summer term 1905, lecture notes by M. Born (SUB Göttingen, Cod. Ms. D. Hilbert 558a).

—— 1930. *Grundlagen der Geometrie*. 7th edition (of [Hilbert 1899], enlarged) Leipzig/Berlin: Teubner.

—— 1935. *Gesammelte Abhandlungen*. Vol. 3: *Analysis, Grundlagen der Mathematik, Physik, Verschiedenes, Lebensgeschichte*. Berlin/Heidelberg: Springer. Second edition Berlin/Heidelberg/New York: Springer, 1970.

Kitcher, P. 1976. Hilbert's Epistemology. *Philosophy of Science* **43**, 99–115.

Kowalewsky, M. 1914. Über die Antinomienlehre als Begründung des transzendentalen Idealismus. *Abhandlungen der Fries'schen Schule* N.F. **4**, no. 3, 695–764 (doctoral thesis Göttingen).

Moore, G. H. 1978. The Origins of Zermelo's Axiomatization of Set Theory. *Journal of Philosophical Logic* **7**, 307–329.

—— 1987. A House Divided Against Itself: The Emergence of First-Order Logic as the Basis for Mathematics. In: *Studies in the History of Mathematics*, Esther R. Phillips, ed. (= *MAA Studies in Mathematics*; 26), 98–136. Washington D.C.: MAA.

Moore, G. H. and Garciadiego, A. 1981. Burali–Forti's Paradox: A Reappraisal of Its Origins. *Historia Mathematica* **8**, 319–350.

Nelson, L. 1904. Die kritische Methode und das Verhältnis der Psychologie zur Philosophie. Ein Kapitel aus der Methodenlehre. *Abhandlungen der Fries'schen Schule* N.F. **1**, no. 1, 1–88. Reprinted in L. Nelson, *Gesammelte Schriften in neun Bänden*, P. Bernays et al., eds. Vol. 1. Hamburg: Meiner, 1970.

Peckhaus, V. 1990a. "Ich habe mich wohl gehütet, alle Patronen auf einmal zu verschießen." Ernst Zermelo in Göttingen. *History and Philosophy of Logic* **11**, 19–58.

—— 1990b. *Hilbertprogramm und Kritische Philosophie. Das Göttinger Modell interdisziplinärer Zusammenarbeit zwischen Mathematik und Philosophie* (= *Studien zur Wissenschafts-, Sozial- und Bildungsgeschichte der Mathematik*; 7). Göttingen: Vandenhoeck & Ruprecht.

Purkert, W. 1986. Georg Cantor und die Antinomien der Mengenlehre. *Bulletin de la Société Mathématique de Belgique*, Ser. A, **38** [Special issue *Mathematics. Topology, History, Philosophy. In Honor of Guy Hirsch*], 313–327.

Rang, B. and Thomas, W. 1981. Zermelo's Discovery of the "Russell Paradox." *Historia Mathematica* **8**, 15–22.

Russell, B. 1903. *The Principles of Mathematics*. Vol. 1. Cambridge: Cambridge University Press. Second edition London: Allen & Unwin, 1937.

Schüler, W. 1983. *Grundlegungen der Mathematik in transzendentaler Kritik. Frege und Hilbert* (= *Schriften zur Transzendentalphilosophie*, 3). Hamburg: Meiner.

Veronese, G. 1891. *Fondamenti di Geometria. A più dimensioni e a più specie di unità rettilinee. Esposti in forma elementare. Lezioni per la scuola di magistero in matematica*. Padova: Tipografia de seminario.

Zermelo, E. 1902. Ueber die Addition transfiniter Cardinalzahlen. *Nachrichten von der Königl. Gesellschaft der Wissenschaften zu Göttingen. Mathematisch-physikalische Klasse aus dem Jahre 1901* (Göttingen), 34–38.

—— 1904. Beweis, daß jede Menge wohlgeordnet werden kann. *Mathematische Annalen* **59**, 514–516.

—— 1908a. Neuer Beweis für die Möglichkeit einer Wohlordnung. *Mathematische Annalen* **65**, 107–128.

—— 1908b. Untersuchungen über die Grundlagen der Mengenlehre. I. *Mathematische Annalen* **65**, 261–281.

Ideas: Differential Geometry and Analysis

Rudolf Lipschitz's Work on Differential Geometry and Mechanics

Rossana Tazzioli

Dipartimento di Matematica, Università di Bologna, Italy

The concept of Riemannian manifold was fundamental not only in differential geometry, but also in mechanics. Indeed Lipschitz, Schering, Beltrami, and Killing tried to extend certain classical theories in mathematical physics —such as Hamiltonian mechanics and potential theory—to a Riemannian manifold with constant curvature. The author's aim is to show that their researches led to mathematical expressions containing Lorentz transformations as a particular case.

INTRODUCTION

The new concepts of 'manifold' and 'curvature' introduced by Riemann in 1854 in his *Habilitationsvortrag*, and the pertinent results achieved by him, besides launching a systematic study of differential geometry, constituted the starting point for new and fruitful developments in the field of mechanics. Through these researches it was possible to extend the principles of mechanics—like the motion equations of a point (whether or not subjected to external forces) and the Euler–Lagrange differential equations—to a Riemannian manifold provided with a nonzero curvature tensor. Thus, in order to be able to integrate and solve equations of motion in a Riemannian manifold it was necessary to generalize Hamiltonian mechanics and, in particular, the Hamilton-Jacobi method to the case of manifolds.

Attempts along these lines were made, above all, by Lipschitz, Schering, and Killing who, with the help of some results obtained by Beltrami [1868], formulated a potential theory in an n-dimensional space with constant curvature. These works, though little studied by historians, represent, in my opinion, a fundamental stage in the development of mechanics. In fact, by throwing light on profound relationships between mechanics and geometry, they make it possible to examine some questions in mathematical physics from a completely new viewpoint.

Moreover, the study of the laws of mechanics in a non-Euclidean space was done without making any hypothesis as to the nature of space or any dependence on the existence of a privileged coordinate system. Instead, it started from the analysis of a particular algebraic form, as in the case of Lipschitz, or, as in Killing's work, from a geometric conception of space (*geometrische Anschauung*). On the single hypothesis of existence in an n-dimensional manifold with constant curvature, Killing gave a transformation between a coordinate system moving with a body and another coordinate system thought to be fixed in space. H. A. Lorentz [1904] gave a particular case of the transformation law established by Killing. The first *physicist* to exhibit such Lorentz transformations (the coordinate transformations deduced by Lorentz) for a system of space–time coordinates was Woldemar Voigt [1887], who derived them using the invariance of Maxwell's equations.

1. GENERALIZATION OF RIEMANNIAN MANIFOLDS

In 1869, two articles were published in *Crelle's Journal*, one by Rudolf Lipschitz and the other by Elvin Bruno Christoffel, both with the common intention of solving a question left open by Riemann in his lecture on the principles of geometry delivered in 1854, but which appeared posthumously only in 1868. The question was to determine whether and under what conditions the metric tensor $g_{ij} \, dx^i \, dx^j$ associated with a Riemannian manifold could be transformed into constant coefficient form by means of a suitable change of coordinates. This problem, which represented the extension of Gauss' *theorema egregium* to the case of the manifolds, had already been solved by Riemann himself in a work of 1861. This paper [Riemann 1861], however, was destined to be published only in 1876 in the *Gesammelte Mathematische Werke* edited by H. Weber.

Though dealing with the same question, the articles by Lipschitz and Christoffel were profoundly different in content and purpose. While Christoffel was interested only in the analytical aspects of the problem, Lipschitz sought the mechanical meaning by linking the solution of the problem to the integration of a system of differential equations of motion. Moreover, Lipschitz analysed the question in the more general case where the line element $f(dx)$ was homogeneous in its differentials $dx_1, \ldots, dx_n$ of degree $p \geq 2$, positive-definite and with determinant

$$\left| \frac{\partial^2 f(dx)}{\partial \, dx_a \, \partial \, dx_b} \right| \neq 0.$$

In the case $p = 2$, he found the results obtained by Christoffel for quadratic forms. In this way the very concept of a Riemannian manifold could be considered as that particular case where the metric tensor was composed of a homogeneous form of second degree in its differentials.

The 'manifolds' introduced by Lipschitz are particular cases of what later came to be called *Finsler spaces*. In his Inaugural dissertation, Paul Finsler [1918] defined a new kind of space by means of a line element depending on the coordinates $x_1, \ldots, x_n$ and on the differentials $dx_1, \ldots, dx_n$, and supposed to be positively homogeneous of first degree with respect to dx_i. According to Elie Cartan, "les espaces de Finsler constituent la généralisation la plus naturelle des espaces riemanniens" [1934, 3].

The generalization proposed by Lipschitz had, in reality, been suggested by Riemann himself. Indeed, in his lecture of 1854, Riemann had observed that the next case in order of complexity would "probably" have included those manifolds whose line element was expressed by the fourth root of a fourth-degree differential expression. "Investigation of this more general class," Riemann wrote, "would actually require no essentially different principles, but it would be rather time-consuming and throw proportionally little new light on the study of space, especially since the results cannot be expressed geometrically; I consequently restrict myself to those manifolds where the line element can be expressed by the square root of a differential expression of the second degree" [1854, 310; trans. in Spivak 1970, 4A-10, 4A-11].

Lipschitz considered the transformation of coordinates:

$$y_1 = y_1(x_1, \ldots, x_n); \; y_2 = y_2(x_1, \ldots, x_n); \; \ldots; y_n = y_n(x_1, \ldots, x_n) \quad (1.1)$$

with nonzero Jacobian, and, by denoting the form $f(dx)$ as $g(dy)$ in the new coordinate system, he obtained the equations

$$\sum_a \left(\sum_b \frac{\partial^2 f(dx)}{\partial \, dx_a \, \partial \, dx_b} d^2 x_b + f_a(dx) \right) \delta x_a$$

$$= \sum_h \left(\sum_l \frac{\partial^2 g(dy)}{\partial \, dy_h \, \partial \, dy_l} d^2 y_l + g_h(dy) \right) \delta y_h, \quad (1.2)$$

where

$$f_a(dx) = \sum_c \frac{\partial^2 f(dx)}{\partial dx_a \, \partial dx_c} \, dx_c - \frac{\partial f(dx)}{\partial x_a},$$

$$g_h(dy) = \sum_m \frac{\partial^2 g(dy)}{\partial dy_h \, \partial dy_m} \, dy_m - \frac{\partial g(dy)}{\partial y_h}.$$

(1.3)

For $p = 1$, Lipschitz deduced from the previous relations integrability conditions for $f(dx)$. If the form $f(dx)$ was quadratic, Eqs. (1.3) became

$$f_a(dx) = \frac{1}{2} \sum_{sh} f_{ash} \, dx_s \, dx_h,$$

$$g_h(dy) = \frac{1}{2} \sum_{pq} g_{hpq} \, dy_p \, dy_q,$$

(1.4)

where f_{ash} and g_{hpq} were, respectively, the *Christoffel symbols* for $f(dx)$ $= \frac{1}{2}\sum_{ab} a_{ab} \, dx_a \, dx_b$ and $g(dy) = \frac{1}{2}\sum_{hl} e_{hl} \, dy_h \, dy_l$:

$$f_{ash} = \frac{\partial a_{as}}{dx_b} + \frac{\partial a_{ab}}{dx_s} - \frac{\partial a_{sh}}{\partial x_a},$$

$$g_{hpq} = \frac{\partial e_{hp}}{\partial y_q} + \frac{\partial e_{hq}}{\partial y_p} - \frac{\partial e_{pq}}{\partial y_h}.$$

(1.5)

Lipschitz extended the *Riemann curvature tensor* to handle quadratic forms with $p > 2$. Then each manifold with metric tensor equal to a homogeneous form of degree p in its differentials which is positive-definite with $\det(\partial^2 f(dx)/\partial dx_a \, \partial dx_b) \neq 0$ was provided with a 'curvature' generalizing the curvature tensor of a Riemannian manifold. If $p = 2$, Lipschitz proved that the two curvature tensors, associated with $f(dx)$ and $g(dy)$ coincided, under the transformation of coordinates (1.1), in which $f(dx) = g(dy)$.

In the case $n = 2$, that is, if a surface is considered, Lipschitz observed that the left-hand side of (1.2) was the Gaussian curvature corresponding to the form $f(dx)$, whereas the right-hand side represented the Gaussian curvature for the manifold defined by $g(dy)$. Thus, the form $f(dx)$ could be transformed into the form $g(dy)$ if and only if the two Gaussian curvatures of the surfaces associated with $f(dx)$ and $g(dy)$ were equal, thereby recovering, as Lipschitz noticed, the *theorema egregium*.

In order to give a mechanical interpretation of the differential equations (1.2) for any p and n, Lipschitz described the motion of a point not subjected to external forces with kinetic energy equal to $f(dx/dt)$. For $p = 2$, according to the principle of least action,

$$\delta \int_{t_0}^{t} T \, dt = 0, \tag{1.6}$$

where $T = \sum_{ik} m_{ik}(dx_i \, dx_k/dt^2)$ is the kinetic energy of the point (m_{ik} are functions of $x_i, \ldots, x_n$). If $f(dx) = \sum_{ik} m_{ik} \, dx_i \, dx_k$, then (1.6) becomes

$$\delta \int \sqrt{T} f(dx) = 0, \tag{1.7}$$

where $f(dx)$ represents the line element in the n-dimensional configuration space whose coordinates are $x_1, \ldots, x_n$, and the m_{ik} are the components of the metric tensor.

Lipschitz tried to generalize the principle of least action to his manifold with $p > 2$. To this end, he supposed that the path of the point having kinetic energy $f(dx/dt)$ causes the variation of the integral $\int f(x') \, dt$ to vanish. The trajectory is, therefore, a geodesic in the configuration space with which the metric tensor $f(dx)$ is associated. Thus Lipschitz found the 'generalized' Euler–Lagrange equations for $f(x')$:

$$\sum_b \frac{\partial^2 f(x')}{\partial x_a' \, \partial x_b'} x_b'' + f_a(x') = 0, \tag{1.8}$$

which coincided with the first term of (1.2) divided by $(dt)^p$. The second term divided by $(dt)^p$ equals

$$\sum_l \frac{\partial^2 g(y')}{\partial y_h' \, \partial y_l'} y_l'' + g_h(y') = 0, \tag{1.8'}$$

which was achieved by setting $\delta \int g(y') \, dt = 0$.

In this way, the system (1.2) was invested with a physical meaning: in fact it represented the 'inertial motions' of two points, and its integration was directly linked to the solution of the two systems of equations (1.8) and (1.8'). In particular, Lipschitz proved that if the change of coordinates (1.1) transformed $f(dx)$ into the form $g(dy)$ with constant coefficients, then it was possible to integrate the isoperimetric differential equations (1.8'). In that case, indeed, (1.8') would simply become

$$\sum_l \frac{\partial^2 g(y')}{\partial y_j' \, \partial y_l'} y_l'' = 0, \tag{1.9}$$

so that $y_i = y_i(0) + y_i'(0)(t - t_0)$ would integrate the system.

Lipschitz posed the following question: *"When is it possible that a form f(dx) will be transformed into a form g(dy) whose coefficients are constant?"* [1869, 91]. This *a priori* geometric problem now became central for mechanics in connection with the integration of the Euler–Lagrange differential equations (1.8). In the case $p = 2$, the necessary and sufficient condition for $f(dx)$ to be transformable into a form with constant coefficients was that the curvature tensor be identically zero. In order to attack this problem for the general case ($p > 2$), Lipschitz tacitly supposed it was possible to invert at least one of the coordinates $x_1, \ldots, x_n$ with respect to time, which enabled him to prove that the n quantities $x_1, \ldots, x_n$ could be expressed only as functions of $x_i(0)$ (0 was the initial point) and $x_i'(0)(t - t_0)$. He then considered the $x_i(0)$ to be fixed and the $x_i'(0)(t - t_0)$ to be variables, and further assumed that the partial derivatives of $x_1, \ldots, x_n$ with respect to $x_i'(0)(t - t_0)$ existed.

Based on these hypotheses, Lipschitz could prove the following theorem, which solved the general problem: *"The form f(δx) can, or cannot, be transformed into a form with constant coefficients if the equation*

$$\frac{d\varphi(\delta[(t - t_0)x'(0)])}{dt} = 0 \tag{1.10}$$

is, or is not, valid" [1869, 94], where $\varphi(\delta[(t - t_0)x'(0)])$ was $f(\delta x)$ in which the x_i were replaced by the variables $x_i'(0)(t - t_0)$.

2. MECHANICS IN A 'LIPSCHITZIAN MANIFOLD'

Lipschitz investigated the purely analytical aspects of the problems discussed above in a paper that appeared some months later in *Crelle's Journal* [1870c]. In another work [1872], he developed the mechanical implications of his researches by employing some results achieved by W. R. Hamilton in two papers entitled, "On a general method in dynamics" [1834 and 1835]. In his first essay, Hamilton had introduced the "characteristic function" for a moving point in space:

$$V = \int m(x_1' \, dx_1 + x_2' \, dx_2 + x_3' \, dx_3), \tag{2.1}$$

where m was the mass of the point. Since he considered only conservative systems, the Hamiltonian function $H = T - U$ (T kinetic energy, U the negative of the potential function) was constant in time, and he found after simple calculations

$$V = \int_0^t 2T \, dt. \qquad (2.2)$$

By varying the characteristic function, Hamilton [1834, 253] obtained the two differential equations:

$$\left[\left(\frac{\partial V}{\partial x_1}\right)^2 + \left(\frac{\partial V}{\partial x_2}\right)^2 + \left(\frac{\partial V}{\partial x_3}\right)^2\right] = 2m(U + H) \qquad (2.3)$$

$$\left[\left(\frac{\partial V}{\partial x_1}\right)_0^2 + \left(\frac{\partial V}{\partial x_2}\right)_0^2 + \left(\frac{\partial V}{\partial x_3}\right)_0^2\right] = 2m(U_0 + H), \qquad (2.3')$$

where (2.3') was given in terms of the initial point.

At the end of this paper [1834, 307], he introduced another 'auxiliary' function R linked to V by the relation

$$V = tH + R. \qquad (2.4)$$

It could also be expressed as

$$R = \int_0^t (T + U) \, dt. \qquad (2.5)$$

The advantage of this new "principal function," as Hamilton called it, over the old characteristic function was that in place of the variable energy H, the quantity R is expressed in terms of a new variable t.

Since $\partial R / \partial t = -H$ in the case when H is constant in time, Eqs. (2.3) and (2.3') were valid for the principal function too, that is

$$\left[\left(\frac{\partial R}{\partial x_1}\right)^2 + \left(\frac{\partial R}{\partial x_2}\right)^2 + \left(\frac{\partial R}{\partial x_3}\right)^2\right] = 2m(U + H) \qquad (2.6)$$

$$\left[\left(\frac{\partial R}{\partial x_1}\right)_0^2 + \left(\frac{\partial R}{\partial x_2}\right)_0^2 + \left(\frac{\partial R}{\partial x_3}\right)_0^2\right] = 2m(U_0 + H). \qquad (2.6')$$

Lipschitz intended to extend the Hamiltonian principal function to a manifold whose metric tensor was a homogeneous positive-definite form $f(dx)$ of degree $p \geq 2$ in its differentials and for which $\det(\partial^2 f(dx)/\partial dx_a \, \partial dx_b) \neq 0$. Indeed, Lipschitz [1872, 116–117] observed that "the latest speculations on the nature of space have shown that it is not necessary to assume that the element of a line through a point is expressible as the square root of an aggregate of squared differentials of appropriate coordinates for the given point."

He then set

$$R = \int_{t_0}^{t} \left[f\left(\frac{dx}{dt}\right) + U \right] dt. \tag{2.7}$$

The new principal function contained $f(dx/dt) = [f(dx)/dt^p]$ as kinetic energy [1], while U was the negative of the potential energy in the configuration space of the coordinates $x_1, \ldots, x_n$.

After introducing the function

$$F(dx) = \left(\frac{p(U + H)}{p - 1}\right)^{p-1} pf(dx), \tag{2.8}$$

the action integral (2.7) became simply

$$R = \int_{t_0}^{t} (F(x'))^{1/p} \, dt. \tag{2.9}$$

Then, by setting the first variation of the integral (2.9) equal to zero, Lipschitz obtained the differential equation of motion for $F(dx)$:

$$\frac{d\dfrac{\partial F(x')^{1/p}}{dx'_a}}{dt} - \frac{\partial F(x')^{1/p}}{dx_a} = 0. \tag{2.10}$$

When the metric tensor of the configuration space $f(dx)$ was a quadratic form in its differentials, the action expressed by (2.9) became

$$R = \int_{t_0}^{t} 2[(U + H)f(x')]^{1/2} \, dt. \tag{2.11}$$

This constituted a generalization of the integral of least action

$$\int \sqrt{2(U + H) \sum_{i=1}^{3} m_i \, ds_i^2}$$

given by C. G. J. Jacobi [1866, 45] in his *Vorlesungen*. Jacobi had confined himself to the case where the line element $f(dx)$ coincided with the constant coefficient form $\frac{1}{2} \Sigma_{i=1}^{3} m_i \, ds_i^2$.

After checking that his own theory agreed with the facts already known when the configuration space was provided with a Euclidean metric, Lipschitz proposed to solve the equations of motion (2.10) by the Hamilton–Jacobi method for $p > 2$ and arbitrary n. To this end, he considered the quantities $\partial R / \partial x_i$, which can be identified with the 'old momenta' p_i. The relations $p_i = \partial R / \partial x_i$ were susceptible to generalization by setting

$$\frac{\partial R}{\partial x_i} = \frac{\partial (F(x'))^{1/p}}{\partial x_i'}, \tag{2.12}$$

where $F(x')$, as introduced in (2.8), is the new Lagrangian function.

In the case where $f(dx) = \frac{1}{2}\sum_{ab} a_{ab}\, dx_a\, dx_b$, Lipschitz obtained from (2.12) the following two equations

$$\sum_{ab} \frac{A_{ab}}{\Delta} \frac{\partial R}{\partial x_a} \frac{\partial R}{\partial x_b} = 2(U + H), \tag{2.13}$$

$$\sum_{ab} \left(\frac{A_{ab}}{\Delta}\right)_0 \left(\frac{\partial R}{\partial x_a}\right)_0 \left(\frac{\partial R}{\partial x_b}\right)_0 = 2(U_0 + H), \tag{2.13'}$$

where $\Delta = \det(a_{ab})$, $A_{ab} = \partial \Delta / \partial a_{ab}$ and (2.13') is given for the initial point. The two equations were reduced to Hamilton's differential equations (2.6) and (2.6') whenever $f(dx)$ was a quadratic form with constant coefficients. Moreover, the left-hand side of (2.6) coincided with Lamé's first differential parameter $\Delta_1 R$. This quantity had been generalized by Beltrami [1868] to the case of n coordinates $x_1, \ldots, x_n$ not necessarily orthogonal with linear element $f(dx) = \frac{1}{2}\sum_{ab} a_{ab}\, dx_a\, dx_b$ by means of the formula

$$\Delta_1 R = \sum_{ab} \frac{A_{ab}}{\Delta} \frac{\partial R}{\partial x_a} \frac{\partial R}{\partial x_b}. \tag{2.14}$$

Since the previous expression for $\Delta_1 R$ is just the left-hand side of (2.13), the generalization of Hamilton's differential equations for R appears strictly connected with the 'geometrical' extension of the first differential parameter given by Beltrami.

For any p the solution of Eq. (2.12), which Lipschitz indicated with y_1, also solved the Hamilton–Jacobi equation:

$$H\left(x_i, \frac{\partial R}{\partial x_i}\right) = \alpha_1. \tag{2.15}$$

If $\alpha_2, \ldots, \alpha_n$ are the $(n - 1)$ integration constants of this equation representing the 'new momenta,' then $y_i = \partial R / \partial \alpha_i = \beta_i$ are the $(n - 1)$ integrals for the Euler–Lagrange differential equations (2.10).

Thanks to this generalization of the Hamilton–Jacobi method, Lipschitz obtained $(n - 1)$ integrals for Eqs. (2.10) in the case where the p-th degree form $F(dx)$ could be transformed into the dy_1^p form. To integrate a system of Euler–Lagrange equations was, therefore, a problem closely connected to the 'geometric' question of determining a system of coordinates $y_1, \ldots, y_n$ in which $F(dx)$ coincided with the constant coefficient form dy_1^p.

In the final part of his article [1872], Lipschitz attempted to give a dynamic meaning to Gauss' theorem according to which, if a set of equal length geodesics is orthogonal to a given curve, their final points formed a curve orthogonal to the geodesics themselves. This theorem was extended by Beltrami [1868] to n-dimensional spaces with line elements given by a quadratic positive-definite form.

To this end Lipschitz considered a solution P of (2.12) and the $(n - 1)$ integrals of system (2.10) determined by the Hamilton–Jacobi equation. He chose the values $x_i(0)$ in such a way that the function $P(0)$ was equal to a number A, and, consequently, obtained an $(n - 1)$-dimensional manifold. He then denoted by x_i the first-order manifold which, taking a point of $P(0) = A$ as initial datum, integrated the system (2.10) and was orthogonal to the manifold $P(0) = A$. Lipschitz considered these manifolds as P increased, and proved "that the equation $P = B$ (where B is a fixed value) is valid for all systems x_i of final points" [Lipschitz 1872, 141]. Moreover, he proved that the x_i were orthogonal to the $(n - 1)$-dimensional manifold $P = B$ and the action was $R = B - A$. This mechanical result corresponded to Gauss' theorem: "Taking the place of a geodesic on the given surface comes a solution of the mechanical problem, in place of the contour of the initial point is an equation for the initial values of the variables, in place of the condition that the geodesic and the contour of initial points be perpendicular comes a condition on the initial values of the variables, and in place of the final length of the geodesics a fixed value of the integral of least action" [Lipschitz 1872, 120].

3. CONNECTIONS BETWEEN GEOMETRY AND MECHANICS

His studies of the connections between differential geometry and mechanics provided Lipschitz with several noteworthy ideas: not only some of the geometrical statements enunciated by Gauss, Riemann, and Beltrami could be interpreted as mechanical laws, but these were also the very mechanical concepts that sometimes made it possible to deepen the corresponding geometrical relations. Indeed, wrote Lipschitz [1873a, 416], "through the progressive development of mechanics and geometry attention has been drawn to questions whose research is of equal interest to both disciplines and which, analytically expressed, depend on exactly the same algorithms."

In a work completed in October 1869 [1870a], Lipschitz studied the motion of a point not subjected to forces and forced to move under n constraints:

$$\Phi_1 = k_1, \ldots, \Phi_n = k_n \tag{3.1}$$

with kinetic energy $f(dx/dt)$, where $f(dx) = \frac{1}{2}\Sigma_{ab} a_{ab} dx_a dx_b$ was a quadratic positive-definite form. The author pointed out that if the pressure (*Druck*) taking place at every point of a path is inversely proportional to the curve's radius of curvature, the force can be used to define the radius of curvature.

Using the theory of Lagrange multipliers, Lipschitz went on to exhibit a quantity, which he denoted $D(\omega)$, that represented the determinant of a linear equation whose roots coincided with the multipliers λ_i up to a constant. If the point was then subjected to the single constraint $\Phi_1 = k_1$, Lipschitz was able to generalize the concept of radius of curvature ρ.

He set $-(1/\rho) = \lambda_1 \sqrt{(1, 1)} / 2f(dx)$, where $(1, 1) = \Sigma_{ab}(A_{ab}/\Delta) \times (\partial\Phi_1/\partial x_a)(\partial\Phi_1/\partial x_b)$, with $\Delta = \det(a_{ab})$ and $A_{ab} = \partial\Delta/\partial a_{ab}$.

By solving the equation $D[-(1/\rho)] = 0$, he obtained $(n - 1)$ 'principal curvature radii.' In three-dimensional Euclidean space, where $f(dx) = \frac{1}{2}\Sigma_{i=1}^3 dx_i^2$ with x_1, x_2, x_3 orthogonal Cartesian coordinates, the condition $D[-(1/\rho)] = 0$ led him to deduce the two principal curvature radii which, when multiplied together, yield the Gaussian curvature.

Based on these considerations, Lipschitz studied more general spaces and began to extend classical concepts like the theory of curvature [1876a], minimal surfaces [1874], and the radii of osculating circles [1876b] to these spaces. The mechanical implications of his deductions were highlighted in his paper, "Sätze aus dem Grenzgebiet der Mechanik und der Geometrie" [1873a], in which he examined the motion of a point having coordinates (x, y, z) and mass m which is not subjected to external forces but to a constraint $\Phi_1 = k_1$. If λ_1 was the Lagrange multiplier, which then was a pure function of time, Lipschitz observed that the expression $\lambda_1 \delta\Phi_1$ could be interpreted as the moment of force exercised by the motion of the point of mass m_1 on the surface $\Phi_1 = k_1$. Moreover, he deduced that quantity $G = \frac{1}{2}m[(x - a)^2 + (y - b)^2 + (z - c)^2]$, where (a, b, c) were the initial coordinates of the point considered, and the radius of curvature ρ of surface $\Phi_1 = k_1$ satisfied

$$2G = m\rho^2. \tag{3.2}$$

From (3.2), Lipschitz could deduce the well-known relation between the magnitude of the force and the radius of curvature:

$$-\frac{1}{\sqrt{2G}} = \frac{\lambda_1 \sqrt{\left(\frac{\partial \Phi_1}{\partial x}\right)^2 + \left(\frac{\partial \Phi_1}{\partial y}\right)^2 + \left(\frac{\partial \Phi_1}{\partial z}\right)^2}}{\sqrt{m^3}\left[\left(\frac{dx}{dt}\right)^2 + \left(\frac{dy}{dt}\right)^2 + \left(\frac{dz}{dt}\right)^2\right]}. \tag{3.3}$$

In the case of q points with masses $m_1, \ldots, m_q$ and coordinates (x_i, y_i, z_i), Lipschitz set

$$(1,1) = \frac{1}{m_i} \sum_i \left[\left(\frac{\partial \Phi_1}{\partial x_i}\right)^2 + \left(\frac{\partial \Phi_1}{\partial y_i}\right)^2 + \left(\frac{\partial \Phi_1}{\partial z_i}\right)^2\right],$$

$$G = \frac{1}{2} \sum_i m_i \left[(x_i - a_i)^2 + (y_i - b_i)^2 + (z_i - c_i)^2\right],$$

where (a_i, b_i, c_i) were the initial coordinates of the points. He then deduced the equation

$$-\frac{1}{\sqrt{2G}} = \frac{\lambda_1 \sqrt{(1,1)}}{\sum_i m_i \left[\left(\frac{dx_i}{dt}\right)^2 + \left(\frac{dy_i}{dt}\right)^2 + \left(\frac{dz_i}{dt}\right)^2\right]}. \tag{3.4}$$

"This subsumes Eq. (3.2) and characterizes the connection between the function $\sqrt{2G}$ and the function $\sqrt{\lambda_1(1,1)}$, of which the first represents a generalization of the radius of curvature and the second a generalization of the concept of pressure" [Lipschitz 1873a, 424].

Lipschitz supplied a further generalization of (3.3) by supposing that the q points moved in an n-dimensional non-Euclidean space, where the line element was expressed by the positive-definite form $f(dx) = \frac{1}{2}\sum_{ab} a_{ab} \, dx_a \, dx_b$, and where again $\Delta = \det(a_{ab})$, $A_{ab} = \partial \Delta / \partial a_{ab}$ are employed. If $\varphi(du)$ denoted the function obtained from $f(dx)$ by introducing 'normal variables' $u_i = x_i'(0)(t - t_0)$, he showed that, by setting

$$(1,1) = \sum_{ab} \frac{A_{ab}}{\Delta} \frac{\partial \Phi_1}{\partial x_a} \frac{\partial \Phi_1}{\partial x_b},$$

the expression $\lambda_1(1,1)$ was an invariant quantity for realizing a transformation of coordinates from $x_1, \ldots, x_n$ to $u_1, \ldots, u_n$. Moreover, under appropriate hypotheses, the following relation was valid:

$$\frac{-1}{\sqrt{2f_0(u)}} = \frac{\sqrt{\lambda_1(1,1)}}{2\varphi(u')}, \tag{3.5}$$

where $f_0(u)$ was the $f(u)$ form in which the coefficients were calculated at the initial time. It contained (3.3) as a particular case.

Whenever $n = 3$ and $f(dx) = \frac{1}{2}\Sigma_{i=1}^3 dx_i^2$, the normal variables u_i coincide with $x_i - x_{0i}$ (where x_{0i} are the coordinates of the point at the instant t_0), the 'normal type' $\varphi(du) = G$, and $\lambda_1\sqrt{(1,1)}$ is exactly the quantity Kronecker [1869b] denoted ρ (radius of curvature) in his work on functions of several variables [2]. Thus the two concepts of radius of curvature and pressure could be generalized to non-Euclidean spaces with constant curvature, and their measurements were deducible from the same algorithms.

The basic procedure of the research method elaborated by Lipschitz was thus to seek in any expressions both the mechanical and the geometric meaning of various analytic expressions. Moreover, this approach often uncovered profound relationships "aus dem Grenzgebiet der Mechanik und der Geometrie." In a paper on the principle of least action [1877], Lipschitz gave a mechanical interpretation of some conditions that Riemann had introduced in the *Commentatio* [Riemann 1861] in a purely geometric vein. After observing that the form Ψ he had introduced earlier [Lipschitz 1869] coincided with the Riemann curvature tensor, Lipschitz showed how this quantity equalled the sum of the two expressions

$$\frac{1}{2}dd\sum_{ab} a_{ab}\,\delta x_a\,\delta x_b - d\delta\sum_{ab} a_{ab}\,dx_a\,\delta x_b + \frac{1}{2}\delta\delta\sum_{ab} a_{ab}\,dx_a\,dx_b \tag{3.6}$$

and

$$\sum_{bc} \frac{A_{bc}}{\Delta}[\Psi_b(dx,dx)\Psi_c(\delta x,\delta x) - \Psi_b(dx,\delta x)\Psi_c(dx,\delta x)], \tag{3.7}$$

where

$$\Psi = \sum_{absh}\left(\frac{\partial f_{abs}}{\partial x_h} - \frac{\partial f_{abh}}{\partial x_s} + \frac{1}{2}\sum_{cb}\frac{A_{cb}}{\Delta}(f_{cas}f_{cbh} - f_{cah}f_{bhs})\right)du_a\,\delta u_b\,dx_s\,\delta x_h,$$

and f_{abc} are the Christoffel symbols for $f(dx) = \frac{1}{2}\Sigma_{ab} a_{ab}\,dx_a\,dx_b$.

The expression (3.7) had to equal zero by virtue of certain conditions imposed by Riemann [3]. Moreover, Lipschitz noted that "the last covariant [(3.7)] is in its essential parts the same covariant that must be made minimal by applying the principle of least action, and thus characterizes the subject of the present work" [Lipschitz 1877, 317].

By making precise what Gauss [1829, 26] had expressed "only in words" about the principle of least action, Lipschitz introduced the n Cartesian coordinates $z_1, \ldots, z_n$, the forces $Z_1, \ldots, Z_n$ and the n constraints $\Phi_1 = k_1, \ldots, \Phi_r = k_n$, in order to determine the quantities $d^2 z_i / dt^2$ which minimized

$$\sum m_a \left(\frac{d^2 z_a}{dt^2} - \frac{Z_a}{m_a} \right)^2. \tag{3.8}$$

By introducing any n variables $x_1, \ldots, x_n$ and setting $f(dx) = \frac{1}{2} \sum_{ab} a_{ab} \, dx_a \, dx_b$, Lipschitz deduced the following formula:

$$\sum_{ab} \frac{A_{ab}}{\Delta} \left[\Psi \left(\frac{dx}{dt'} \frac{dx}{dt} \right) - X_a \right] \left[\Psi_b \left(\frac{dx}{dt'} \frac{dx}{dt} \right) - X_b \right] \tag{3.9}$$

(where $X_1, \ldots, X_n$ are the forces in the new coordinates $x_1, \ldots, x_n$). This represented a covariant form with respect to $f(dx)$ and was equivalent to the vanishing of (3.8). Thus, Lipschitz [1877, 330] observed that the quantity to be minimized by means of the principle of least action is expressed by the covariant (3.9) in the mechanical problem. The conditions 'analytically' deduced by Riemann in the *Commentatio* (see [3]) by means of which (3.7) vanished extended the principle of least action to all Riemannian manifolds and, consequently, carried profound physical meaning. After all, observed Lipschitz [1873a, 416], "because mechanics deals with bodies in motion it must take into account geometry to set down its foundations, and [geometry] has always proven worthy for attaining its goals."

4. THE "PLANET–PROBLEM" AND POTENTIAL THEORY IN NON-EUCLIDEAN SPACE

Lipschitz applied the general results obtained in his 1869 and 1872 papers on the integration of the Euler–Lagrange differential equations to "the problem of an attraction according to the Newtonian law, or the planet–problem" [1873b, 351]. He intended to extend this problem to an n-dimensional space whose metric tensor was expressed by a positive-definite form of the pth degree in its differentials. In his article with the explanatory title, "Extension of the planet–problem to a space of n dimensions and constant integral curvature" [1873b], whose introductory note had been drafted by Cayley, he proposed to generalize the mechanical problem to a 'Lipschitzian manifold,' where a point was attracted to a fixed point by a force depending only on the distance. In order to do this, Lipschitz once again returned to the variation of the integral

$$R = \int_{t_1}^{t} [f(x') + U] \, dt, \qquad (4.1)$$

where $f(dx)$ was a positive-definite form of pth degree in its differentials, with $\det(\partial^2 f(dx)/\partial dx_a \, \partial dx_b) \neq 0$ and U a function of $x_1, \ldots, x_n$ alone.

When $U = 0$, that is when no external forces acted, the inertial motion of the point followed a path which was represented by a geodesic in the configuration space. This geodesic was expressed by the function $r = \sqrt[p]{p f_0(u)}$, where $f_0(u)$ denoted the function $f(dx)$ whose coefficients were determined at the initial instant, and the differentials were replaced by the normal variables $u_1, \ldots, u_n$. If $p = 2$, then $r = \sqrt{2 f_0(u)}$ and if $U = U(r)$, then, as Lipschitz had already proved in a previous work [1870c, 25], the normal type $\varphi(du)$ and the form $f_0(u)$ were linked by the relation

$$\varphi(du) = \left[d\sqrt{\{2 f_0(u)\}} \right]^2 + \frac{m^2}{2 f_0(u)} \left[f_0(du) - \left\{ d\sqrt{f_0(u)} \right\}^2 \right] \quad (4.2)$$

in which m was a function of r.

He considered the two cases where $m = r$ and $m = c(\sin r/c)$ (c a constant) which corresponded to the two curvatures $k = 0$ and $k = 1/c^2$ respectively, since $k = -(1/m) \, d^2 m/dr^2$ was the Riemannian curvature of the manifold. He concluded "that [for $k = 1/c^2$] the planet–problem is extended to a space of n dimensions of constant integral curvature" [Lipschitz 1873b, 352].

The already well-known results were found in a three-dimensional Euclidean space [1873b, 352], whereas if the space had nonzero curvature, that is if $m = (1/c)\sin(r/c)$, Lipschitz showed that the integral of least action was

$$\int_{t_1}^{t} \left[\frac{1}{2} \left\{ \left(\frac{dr}{dt} \right)^2 + c^2 \sin^2 \frac{r}{c} \left(\frac{d\varphi}{dt} \right)^2 \right\} + U \right] dt. \qquad (4.3)$$

By setting the variation of (4.3) equal to zero, he deduced the equations

$$\frac{d^2 r}{dt^2} - c \sin \frac{r}{c} \cos \frac{r}{c} \left(\frac{d\varphi}{dt} \right)^2 = \frac{dU}{dr}$$

$$\frac{d}{dt} \left(c^2 \sin^2 \frac{r}{c} \frac{d\varphi}{dt} \right) = 0. \qquad (4.4)$$

From (4.4), he determined the differential equations of motion. The integration of these equations made use of the theory of elliptic functions and was dealt with in the volume by Weierstrass and Schellbach, *Lehre von den elliptischen Integralen und den Theta-functionen* (*theory of elliptic integrals and theta functions*) [1864].

Besides explicitly writing down the equations of motion of the planet, it was also important to define the potential function U in a space with nonzero curvature tensor. As Lipschitz [1872, 117] reported in a note, some twenty years earlier Dirichlet had already expressed an interest in "how the theory of attraction according to the Newtonian law appears if the Gaussian theory of imaginary space is taken as the foundation." In fact, Lipschitz himself had already carried out some researches on potential functions in non-Euclidean spaces with constant curvature in a paper of 1869 [Lipschitz 1870c]. Here, on the basis of ideas Beltrami [1868] had set forth in the theory of differential parameters, Lamé's second differential parameter was generalized to any coordinates by the formula

$$\Delta_2(U) = \Delta^{-1/2} \sum_{ab} \frac{\partial \left(\Delta^{-1/2} A_{ab} \dfrac{\partial U}{\partial x_b} \right)}{\partial x_a}. \qquad (4.5)$$

In the case where $x_1, \ldots, x_n$ are orthogonal Cartesian coordinates, (4.5) coincides with the usual Laplacian $(\partial^2 U/\partial x_1^2) + (\partial^2 U/\partial x_2^2) + (\partial^2 U/\partial x_3^2) = 0$. Already in 1847, Jacobi had studied the transformation properties of the Laplace equation for transformations of coordinates that preserved orthogonality between coordinates [Jacobi 1847]. This condition was dropped by Eugenio Beltrami, who observed that a transformation of the second differential parameter "requires nothing beyond knowledge of the form which the line element assumes in the new system of variables" [Beltrami 1868, 75].

For $m = r$, Lipschitz obtained as a solution of (4.5), $U = \text{const } t/r^{n-2}$, whereas if $m = c \sin(r/c)$, $U = \text{const} \int m^{-(n-1)} dr$. The latter was also found by Ernst Schering [1870, 160] in the case of a three-dimensional space with negative curvature. In a subsequent paper, "Die Schwerkraft im Gaussischen Raume" [Schering 1873a], he described the potential function W due to the attraction between two particles, m and μ, in a homogeneous n-dimensional space with constant curvature. It had the form $W = m\mu\omega_n(r)$, where $\omega_n(r)$ depended on the distance r between the two particles and was expressed as a function of "Gauss' Π-Function" defined by $\Pi(s) = \int t^{s-1} e^{-t} dt$ [Schering 1873a, 153–154].

Some properties of potential functions in spaces of constant curvature analogous to those of the logarithmic and Newtonian potential functions had been deduced not only by Schering but also by Beltrami [1868]. Later on, Alberto Tonelli [1882, 295] found that potential functions in n-dimensional spaces of constant curvature were completely determined by the boundary values, and (up to any arbitrary constant) by the values of the derivative with respect to the normals on the boundary. From this theorem Tonelli deduced the following formula for two functions U and V such that $\Delta_2 U = 0$ and $\Delta_2 V = 0$:

$$\int_{S_{n-1}} U \frac{dV}{dp} \, dS_{n-1} = \int_{S_{n-1}} V \frac{dU}{dp} \, dS_{n-1}. \tag{4.6}$$

Here S_{n-1} is the boundary of an n-dimensional volume S_n and p is the normal to S_{n-1} pointing inwards. Expression (4.6) and the equation

$$\int_{S_{n-1}} \frac{dV}{dp} \, dS_{n-1} = 0. \tag{4.7}$$

"represent an extension to the potentials in spaces of n dimensions of the well-known properties of the logarithmic and Newtonian potentials" [1882, 295]. Simpler and more complete results were deduced if the space was a flat manifold.

The mathematical relations which hold for the potential functions in n-dimensional non-Euclidean spaces and the theorems deduced from them had no confirmation in the physical world where two particles interacted according to the Newtonian attractive force. However, if the entire universe were conceived as filled with a homogeneous ether and provided with a metric with positive constant curvature, the particles making up the ether would have to attract each other in such a way as to assure equilibrium of the whole ethereal mass. Beltrami [1876a] solved this problem by applying certain general considerations on potential theory to the case where the 'elementary potential function' (the law by which two elementary particles attract each other) was $\varphi(r) = \exp(-\mu r^2)$, with μ a positive constant and r the distance between the two particles. Using this hypothesis, Beltrami was able to prove the following property: "*matter uniformly distributed throughout space and acting on itself with the potential law* $\exp(-\mu r^2)$ *is in equilibrium at all its points*" [1876a, 34]. Then, he assumed the density of the "infinite sphere" was no longer constant but now variable according to the law $h(r) = K/r$, K constant, and deduced that "*the action thus exercised by the matter thus condensed on a point of space, very rapidly tends, with the moving away of this point from the condensation centre, to follow the Newtonian law*" [1876a, 36]. In this case the potential function $\varphi(r)$ is compatible with what happens physically.

Beltrami then went on to consider the potential function

$$V = \int h \exp(-\mu r^2) \, dS \tag{4.8}$$

of a quantity of matter distributed within a bounded space S, where h was the density which was assumed to be independent of μ. If A was an absolute constant and t a parameter (which here signifies "the time passed since the initial instant"), he found that the function

$$T = \frac{1}{2A\sqrt{\pi}\,t^{3/2}} \int h \exp(-r^2/4A^2 t) \, dS \tag{4.9}$$

satisfied the equation describing heat motion in isotropic bodies: $\partial T/\partial t = A^2 \, \Delta T$. Then, Beltrami [1876a, 37] observed, the temperature of an isotropic body may be considered as the potential function of actions at a distance between points of the body according to the potential law

$$\frac{\exp(-r^2/4A^2 t)}{2A\sqrt{\pi}\,t^{3/2}}. \tag{4.10}$$

The function h, which here acts as the density represents, in heat theory, the condition of initial temperature. If the function $U = \int T \, dt$ is then introduced, it is easily found that U also satisfied the equation $u = \int (h \, dS/r)$. So the function U was 'mathematically' the Newtonian potential function of matter having density h and occupying the portion of space S. If ether particles interacted in accordance with this potential law, they were able to determine macroscopic phenomena like heat. "What is the physical significance of this function U?" Beltrami [1876a, 38] wondered. "On this and other questions suggested by these considerations we hope to be able to return later." But he did not deal with these subjects in any of his subsequent works.

5.   SOME REMARKS ON KILLING'S PAPER [1885b]

A different way to frame the study of mechanics and potential theory in non-Euclidean spaces was proposed by Wilhelm Killing in his article "Die Mechanik in den Nicht-*Euklid*ischen Raumformen" [1885b]. In this paper, whose content represents an application to mechanics of some results set forth by Killing in his book *Die Nicht-Euklidischen Raumformen in analytischer Behandlung* (*Non-Euclidean space forms in an analytical treatment*) [1885a], the author explicitly referred back to the works of Lipschitz, Beltrami, and others concerning questions of geometry and mathematical physics. However, Killing's approach was very different. He made use of

"Weierstrass Coordinaten" in an n-dimensional space having constant curvature $1/k^2$ (k a complex number), coordinate systems which Killing himself had already employed in his book [1885a, 73–74], and which were introduced for the first time by Weierstrass in 1872 [4].

In order to give the Weierstrassian coordinates of a point P in a Cartesian coordinate system with centre O, Killing called a_i the orthogonal lines from P to the coordinate planes. He posed:

$$x_0 = \cos \frac{OP}{k}, \qquad x_i = k \sin \frac{a_i}{k}, \qquad i = 1, \ldots, n, \qquad (5.1)$$

where $x_0, x_1, \ldots, x_n$ were linked by the relation

$$k^2 x_0^2 + x_1^2 + \cdots + x_n^2 = k^2. \qquad (5.2)$$

"We denote this coordinate system as a Weierstrassian system," he wrote [1885a, 71], and for $k = \infty$ obtained the usual orthogonal Cartesian coordinates.

Imagining (x, y, z) as coordinates of a point not subject to external forces in a three-dimensional space of constant curvature, Killing deduced from the expression $k^2 p^2 + x^2 + y^2 + z^2 = k^2$ the inertial equations of motion in a non-Euclidean space:

$$\frac{d^2 p}{dt^2} = -\frac{v^2}{k^2} p; \quad \frac{d^2 x}{dt^2} = -\frac{v^2}{k^2} x; \quad \frac{d^2 y}{dt^2} = -\frac{v^2}{k^2} y; \quad \frac{d^2 z}{dt^2} = -\frac{v^2}{k^2} z,$$

$$(5.3)$$

(v denotes constant velocity). After studying the kinematics of a point in a space of constant curvature, a subject which Lipschitz [1872] and, above all, Beltrami [1876b] had explicitly dealt with, Killing examined the motion of a point subject to forces. In particular, he considered a central force and, as a relevant example, the problem of planetary motion. Implicitly critical of the method of investigation used by Lipschitz, Killing [1885b, 7] wrote: "In order to carry over the Newtonian law of gravitation to a non-Euclidean space form, one cannot begin with the algebraic form but rather one must derive the corresponding analytic expression from the geometric conceptions of space (*Anschauungen*) that are fundamental to the law." He went on to consider a point moving under the action of a central force with intensity proportional to the quantity $1/\sin^2(r/k)$, where r represented the distance of the point from the centre of attraction. He then deduced an extension to non-Euclidean spaces of Kepler's three laws and the Gaussian flux theorem [1813, 9]. According to Gauss, if S is a closed surface, P a fixed point of the surface element dS, M the point of attraction, r the distance MP, and u the angle between MP and the normal to dS pointing inwards, the value of the integral

$$\int \frac{dS \cos u}{r^2}$$

over the surface S, was 0 if M lies outside of S, 2π if M lies on S, and 4π if M lies inside S, respectively.

Killing generalized this theorem to an n-dimensional space with constant curvature k by considering the following integral extended to an $(n-1)$-dimensional closed space S, where u, r, and dS corresponded to the quantities defined by Gauss:

$$\int \frac{dS \cos u}{k^{n-1} \sin^{n-1} \dfrac{r}{k}} = 0.$$

Killing further proved that the previous integral was equal to 0, or to ω, or to $\frac{1}{2}\omega$, depending on whether M lies outside, inside, or on S, where ω denotes the area of a spherical surface with radius 1.

After finding the equations that describe the motion of a point, the next step was to integrate them. To accomplish this, Killing [1885b, 16] observed "that the methods of Hamilton and Jacobi can be carried over to non-Euclidean space forms" and, as Lipschitz [1872] and Schering [1873b, c] had already tried to do, he extended these methods to manifolds of constant curvature. In order to explicitly describe the motion of a point, Killing found a suitable transformation between two coordinate systems, one 'fixed' and the other moving with the body. Denoting by x_i the coordinates of the point with respect to the first system and by ξ_i those with respect to the second system, he showed that the following equations are valid:

$$k^2 \xi_0 = k^2 a_{00} x_0 + a_{01} x_1 + \cdots + a_{0n} x_n,$$
$$\xi_i = a_{i0} x_0 + a_{i1} x_1 + \cdots + a_{in} x_n, \qquad i = 1, \ldots r \tag{5.4}$$

under the "well-known conditions"

$$k^2 a_{00}^2 + \sum_{i=1}^{n} a_{i0}^2 = k^2 a_{00}^2 + \sum_{i=1}^{n} a_{0i}^2 = k^2,$$

$$\sum_{i=0}^{n} a_{i0} a_{ij} = \sum_{i=0}^{n} a_{0i} a_{ji} = 0, \qquad j = 1, \ldots, n$$

$$\frac{a_{0i} a_{01}}{k^2} + \sum_{i=1}^{n} a_{ij} a_{il} = \frac{a_{i0} a_{l0}}{k^2} + \sum_{i=1}^{n} a_{ji} a_{li} = \delta_{jl}, \qquad j, l = 1, \ldots, n. \tag{5.5}$$

Killing had found Eqs. (5.5) in his book [1885a, 73–74] in a purely geometrical setting, that is by imposing (5.2) on a system of Weierstrassian coordinates.

If

$$
\eta = \begin{bmatrix} k^2 & 0 & \cdots & 0 & 0 \\ 0 & 1 & \cdots & 0 & 0 \\ \vdots & \vdots & \vdots & \vdots & \vdots \\ 0 & 0 & \cdots & 1 & 0 \\ 0 & 0 & \cdots & 0 & 1 \end{bmatrix}, \quad
\lambda = \begin{bmatrix} a_{00} & \dfrac{a_{01}}{k^2} & \cdots & \dfrac{a_{0n}}{k^2} \\ a_{10} & a_{11} & \cdots & a_{1n} \\ \vdots & \vdots & \vdots & \vdots \\ a_{n0} & a_{n1} & \cdots & a_{nn} \end{bmatrix},
$$

then relations (5.4) and (5.5) can be rewritten as

$$
\begin{bmatrix} \xi_0 \\ \xi_1 \\ \vdots \\ \xi_n \end{bmatrix} = \lambda \begin{bmatrix} x_0 \\ x_1 \\ \vdots \\ x_n \end{bmatrix},
\tag{5.4'}
$$

under the condition

$$
\lambda^T \eta \lambda = \eta.
\tag{5.5'}
$$

In the particular case where $k^2 = -1$ and $n = 4$, expression (5.5') is the 'Lorentz condition' (as it is known today). Whenever $a_{00} > 0$, Eqs. (5.4') represents a 'Lorentz transformation.'

One sees how natural the development was arising from the researches carried out by Lipschitz, Schering, and Beltrami which led Killing to give the transformations of coordinates that later became the basis of the special theory of relativity. His deduction did not require any assumptions concerning the physical properties of space or the nature of reference sets, but unfolded starting from geometric considerations alone and from the unique hypothesis of existence in a 'curved' space.

NOTES

[1] Even Somoff [1869] considered a positive-definite, quadratic form of the type $\sum_{rs} a_{rs} x'_r x'_s$ in order to extend Lamé's first and second differential parameters to arbitrary coordinates. This expression is evidently found by setting $p = 2$ in the definition given by Lipschitz: $T = f(dx)/dt^p$.

[2] Concerning this work, read by Kronecker in August 1869 to the Berlin Academy of Science, Lipschitz [1870a, 274] wrote: "Herr Kronecker hat an den von ihm

gegründeten Begriff der Charakteristik eines Functionensystems eine Ausdehnung der Theorie der Krümmung von Flächen auf Functionen mehrerer
Variabeln angeschlossen. Die betreffenden Resultate, welche in dem Monats-
bericht der Berliner Akademie vom August 1869 mitgetheilt sind, stehen in
einer gewissen Verwandtschaft mit den Ergebnissen einer Arbeit über die
ganzen homogenen Functionen von n Differentialen [Lipschitz 1869], die ich
ausgeführt habe, ehe ich zu derjenigen Auffassung dieses Gegenstandes kam,
welche in der bezüglichen publicirten Abhandlung festgehalten ist."

[3] Riemann [1861, 402–403] had found that the curvature tensor was given by the
following relation:

$$\delta\delta \sum a_{ab}\, dx_a\, dx_b - 2d\delta \sum a_{ab}\, dx_a\, \delta x_b + dd \sum a_{ab}\, \delta x_a\, \delta x_b$$

subject to the three conditions:

$$\delta' \sum a_{ab}\, dx_a\, dx_b - \delta \sum a_{ab}\, dx_a\, \delta'x_b - d\sum a_{ab}\, \delta x_a\, \delta'x_b = 0$$

$$\delta' \sum a_{ab}\, dx_a\, dx_b - 2d\sum a_{ab}\, dx_a\, \delta'x_b = 0$$

$$\delta' \sum a_{ab}\, \delta x_a\, \delta x_b - 2\delta \sum a_{ab}\, \delta x_a\, \delta'x_b = 0.$$

[4] "Diese Coordinaten verdanke ich meinem hochverehrten Lehrer, Herrn Weierstrass, der dieselben bei einigen Vorträgen im mathematischen Seminar 1872
benutzte," Killing [1879, 74] indeed observed in a note to a previous paper of
his.

REFERENCES

Arnold, V. I. 1978. *Mathematical methods of classical mechanics.* New York:
Springer. [Translation from Russian of edition printed 1974. Moscow: Nauka.]

Beltrami, E. 1868. Sulla teorica generale dei parametri differenziali. *Memorie
dell'Accademia delle Scienze dell'Istituto di Bologna* **8**, II serie, 551–590; all page
references are to [Beltrami 1902–1920 II, 74–118].

Beltrami, E. 1876a. Considerazioni sopra una legge potenziale. *Rendiconti del R.
Istituto Lombardo* **9**, II serie, 725–733; all page references are to [Beltrami
1902–1920 III, 30–38].

Beltrami, E. Formules fondamentales de cinématique dans les espaces de courbure
constante, in [Beltrami 1902–1920 III, 23–29]. [Abstract of lecture read by
Beltrami in 1876 to Accademia Reale dei Lincei in Rome.]

Beltrami, E. 1902–1920 *Opere matematiche* (4 vol.), Facoltà di Scienze della R.
Università di Roma, Ed. Milano: Hoepli.

Cartan, E. 1934. *Les espaces de Finsler.* Paris: Hermann & Cie.

Cayley, A. 1873. Note in illustration of certain general theorems obtained by dr.
Lipschitz. *Quarterly Journal of Pure and Applied Mathematics* **12**, 346–349.

Reprinted in *The Collected Papers of Arthur Cayley*, A. R. Forsyth, ed., IX, 110–112. Cambridge: University Press, 1896.

Christoffel, E. B. 1869. Über die Transformation der homogenen Differentialausdrücke zweiten Grades. *Journal für die reine und angewandte Mathematik* **70**, 46–69.

Dugas, R. 1950. *Histoire de la mécanique.* Neuchâtel: Editions du Griffon.

Finsler, P. 1918. *Ueber Kurven und Flächen in allgemeinen Raümen.* Dissertation. Göttingen.

Gauss, C. F. 1813. Theoria attractionis corporum sphaeroidicorum ellipticorum homogeneorum methodo nova tractata. *Commentationes societatis regiae scientiarum Gottingensis recentiores* **2**, 1–24; all page references to [Gauss 1863–1933 V, 1–22].

Gauss, C. F. 1828. Disquisitiones generales circa superficies curvas. *Commentationes societatis regiae scientiarum Gottingensis recentiores* **6**, 99–146; all page references are to [Gauss 1863–1933 IV, 217–258].

Gauss, C. F. 1829. Über ein neues allgemeines Grundgesetz der Mechanik. *Journal für die reine und angewandte Mathematik* **4**, 232–235; all page references are to [Gauss 1863–1933 V, 23–28].

Gauss, C. F. 1863–1933. *Werke* (12 vol.), Königliche Gesellschaft der Wissenschaften Göttingen, Ed. Göttingen.

Goldstein, H. 1951. *Classical Mechanics.* Cambridge: Addison-Wesley.

Hamilton, W. R. 1834. On a general method in dynamics; by which the study of the motion of all free systems of attracting or repelling points is reduced to the search and differentiation of one central relation, or characteristic function. *Philosophical Transactions of the Royal Society of London* 1834, 247–308. Reprinted in [Hamilton 1940, 103–161].

Hamilton, W. R. 1835. Second essay on a general method in dynamics. *Philosophical Transactions of the Royal Society of London* 1835, 95–144. Reprinted in [Hamilton 1940, 162–211].

Hamilton, W. R. 1940. *The Mathematical Papers of Sir William Rowan Hamilton*, A. W. Conway and A. J. Connell, eds., II. Cambridge: University Press.

Hankins, T. L. 1980. *Sir William Rowan Hamilton.* Baltimore: Johns Hopkins Univ. Press.

Jacobi, C. G. J. 1847. Eine particuläre Lösung der partiellen Differentialgleichung $(\partial^2 V/\partial x^2) + (\partial^2 V/\partial y^2) + (\partial^2 V/\partial z^2) = 0$, *Journal für die reine und angewandte Mathematik* **36**, 113–134.

Jacobi, C. G. J. 1866. *Vorlesungen über Dynamik.* Berlin: Reimer.

Jammer, M. 1960. *Concepts of space.* Cambridge: Harvard University Press.

Killing, W. 1879. Über zwei Raumformen mit constanter positiver Krümmung. *Journal für die reine und angewandte Mathematik* **86**, 72–83.

Killing, W. 1880. Die Rechnung in den Nicht-*Euklid*ischen Raumformen. *Journal für die reine und angewandte Mathematik* **89**, 265–287.

Killing, W. 1885a. *Die Nicht-Euklidischen Raumformen in analytischer Behandlung.* Leipzig: Teubner.

Killing, W. 1885b. Die Mechanik in den Nicht-*Euklid*ischen Raumformen. *Journal für die reine und angewandte Mathematik* **98**, 1–48.

Klein, F. 1926–1927. *Vorlesungen über die Entwickelung der Mathematik im 19. Jahrhundert* (2 vol.). Berlin: Springer.

Kline, M. 1972. *Mathematical Thought from Ancient to Modern Times.* New York: Oxford University Press.

Kronecker, L. 1868a. Über Systeme von Functionen mehrerer Variabeln. *Monatsberichte der Königlichen Preussischen Akademie der Wissenschaften* 1869, 159–193. Reprinted in [Kronecker 1895, 175–212].

Kronecker, L. 1869b. Über Systeme von Functionen mehrerer Variabeln (Zweite Abhandlung). *Monatsberichte der Königlichen Preussischen Akademie der Wissenschaften* 1869, 688–698. Reprinted in [Kronecker 1895, 213–226].

Kronecker, L. 1895. *Leopold Kronecker's Werke*, K. Hensel, ed., I. Leipzig: Teubner. [Repr. New York: Chelsea, 1968.]

Lipschitz, R. 1869. Untersuchungen in Betreff der ganzen homogenen Functionen von n Differentialen. *Journal für die reine und angewandte Mathematik* **70**, 71–102.

Lipschitz, R. 1870a. Entwickelung einiger Eigenschaften der quadratischen Formen von n Differentialen (Erste Mittheilung). *Journal für die reine und angewandte Mathematik* **71**, 274–287.

Lipschitz, R. 1870b. Entwickelung einiger Eigenschaften der quadratischen Formen von n Differentialen (Zweite Mittheilung). *Journal für die reine und angewandte Mathematik* **71**, 288–295.

Lipschitz, R. 1870c. Fortgesetzte Untersuchungen in Betreff der ganzen homogenen Functionen von n Differentialen. *Journal für die reine und angewandte Mathematik* **72**, 1–56.

Lipschitz, R. 1872. Untersuchung eines Problems der Variationsrechnung, in welchem das Problem der Mechanik enthalten ist. *Journal für die reine und angewandte Mathematik* **74**, 116–149.

Lipschitz, R. 1873a. Sätze aus dem Grenzgebiet der Mechanik und der Geometrie. *Mathematische Annalen* **6**, 416–435.

Lipschitz, R. 1873b. Extension of the planet–problem to a space of n dimensions and constant integral curvature. *Quarterly Journal of Pure and Applied Mathematics* **12**, 349–370.

Lipschitz, R. 1874. Ausdehnung der Theorie der Minimalflächen. *Journal für die reine und angewandte Mathematik* **78**, 1–45.

Lipschitz, R. 1876a. Beitrag zu der Theorie der Krümmung. *Journal für die reine und angewandte Mathematik* **81**, 230–242.

Lipschitz, R. 1876b. Généralisation de la théorie du rayon osculateur d'une surface. *Journal für die reine und angewandte Mathematik* **81**, 295–300.

Lipschitz, R. 1877. Bemerkungen zu dem Princip des kleinsten Zwanges. *Journal für die reine und angewandte Mathematik* **82**, 316−342.

Lorentz, H. A. 1904. Electromagnetische verschijnselen in een stelsel dat zich met willekeurige snelheid, kleiner dan die van het licht, beweegt. *Koninklijke Akademie van Wetenschappen te Amsterdam. Wis-en Natuurkundige Afdeling. Verslagen van de Gewoone Vergaderingen* **12**, 986−1009. [Translation: Electromagnetic phenomena in a system moving with any velocity smaller than that of light. *Koninglijke Akademie van Wetenschappen te Amsterdam. Section of sciences. Proceedings* (1904) **6**, 809−831.]

Riemann, B. 1854. Über die Hypothesen, welche der Geometrie zu Grunde liegen. *Habilitationsschrift*, Göttingen. First printed in *Abhandlungen der Königlichen Gesellschaft der Wissenschaften zu Göttingen* (1868) **13**, 133−152. All page references are in [Riemann 1990, 304−319].

Riemann, B. 1861. Commentatio mathematica, qua respondere tentatur quaestioni ab Ill.ma Academia Parisiense propositae: "Trouver quel doit être l'état calorifique d'un corps solide homogène indéfini pour qu'un système de courbes isothermes, à un instant donné, restent isothermes après un temps quelconque, de telle sorte que la température d'un point puisse s'exprimer en fonction du temps et de deux autres variables indépendantes." In [Riemann 1990, 423−436].

Riemann, B. 1990. *Bernhard Riemann gesammelte mathematische Werke, wissenschaftlicher Nachlass und Nachträge*, R. Narasimhan, ed. Leipzig: Teubner. [Includes *Bernhard Riemann gesammelte mathematische Werke und wissenschaftlicher Nachlass*, H. Weber and R. Dedekind, eds. Leipzig: Teubner, 1876 (repr. 1892) and supplement prepared by M. Noether and W. Wirtinger in 1902. Third repr. New York: Dover, 1953.]

Rosenfeld, B. A. 1988. *A history of non-Euclidean geometry. Evolution of the concept of a geometrical space*. New York: Springer. [Translation from Russian of edition printed 1976. Moscow: Nauka.]

Schering, E. 1870. Die Schwerkraft im Gaussischen Raume. *Nachrichten von der Königlichen Gesellschaft der Wissenschaften und der Georg-Augusts-Universität zu Göttingen* 1870, 311−321. Reprinted in [Schering 1902, 155−162].

Schering, E. 1873a. Die Schwerkraft in mehrfach ausgedehnten Gaussischen und Riemannschen Räumen. *Nachrichten von der Königlichen Gesellschaften und der Georg-Augusts-Universität zu Göttingen* 1873, 149−159. Reprinted in [Schering 1902, 177−184].

Schering, E. 1873b. Hamilton-Jacobische Theorie für Kräfte, deren Maas von der Bewegung der Körper abhängt (Erste Abtheilung). *Nachrichten von der Königlichen Gesellschaft der Wissenschaften und der Georg-Augusts-Universität zu Göttingen* 1873, 744−753. Reprinted in [Schering 1902, 185−192].

Schering, E. 1873c. Hamilton-Jacobische Theorie für Kräfte, deren Maas von der Bewegung der Körper abhängt (Zweite Abtheilung), *Abhandlungen der Königlichen Gesellschaft der Wissenschaften zu Göttingen* **18**, 3−54. Reprinted in [Schering 1902, 193−245].

Schering, E. 1902. *Gesammelte mathematische Werke von Ernst Schering*, R. Haussner and K. Schering, eds., I. Berlin: Mayer and Müller.

Sommerfeld, A. 1919. Klein, Riemann und die mathematische Physik. *Die Naturwissenschaften* **7**, 300–303.

Somoff, J. 1865. Moyen d'exprimer directement en coordonnées curvilignes quelconques, orthogonales ou obliques, les paramètres différentielles du premier et du second ordres et la courbure d'une surface. *Mémoires de l'Académie de St. Petersbourg* **8**, série VII, 1–45.

Spivak, M. 1970. *A Comprehensive Introduction to Differential Geometry II*. Boston: Publish or Perish.

Stäckel, P. 1903. Bericht über die Mechanik mehrfacher Mannigfaltigkeiten. *Jahresbericht der deutschen Mathematiker-Vereinigung* **12**, 469–481.

Stäckel, P. 1893. Über die Bewegung eines Punktes in einer n-fachen Mannigfaltigkeit. *Mathematische Annalen* **42**, 537–563.

Tonelli, A. 1882. Sopra la funzione potenziale in uno spazio di n dimensioni. *Annali di matematica pura ed applicata* **10**, 291–321.

Voigt, W. 1887. Über das Doppler'sche Princip. *Nachrichten von der Königlichen Gesellschaft der Wissenschaften zu Göttingen* 1887, 44–51.

Weierstrass, K. and Schellbach, K. H. 1864. *Lehre von den elliptischen Integralen und den Theta-functionen*. Berlin: Reimer.

Karl Weierstrass (1815–1897)
Portrait from The David Eugene Smith Collection,
Rare Book and Manuscript Library,
Columbia University. Reproduced by permission.

Karl Weierstrass (1815–1897)
Courtesy of Dr. Reinhard Bölling

The Proof of the Laurent Expansion by Weierstrass*

Peter Ullrich

Westfälische Wilhelms-Universität, Mathematisches Institut, Münster, Germany

In a manuscript that was written in the year 1841 Weierstrass proved the existence of the Laurent expansion for functions on an annulus. In this proof, which is independent of the one by Laurent and at most indirectly influenced by the works of Cauchy, he used complex integration in a decisive way. Since later on during his teaching activities at Berlin he intended to give an integral-free foundation of the theory of analytic functions, it therefore seems plausible that there were "ideological" reasons that made him publish this work of his youth only over half a century later. This simple picture, however, has to be differentiated in the light of the attitude of Weierstrass towards his foundational studies, especially towards the methods used therein and the position of these studies in relation to his other mathematical works.

A hundred and fifty years ago Karl Weierstrass (1815–1897) completed a manuscript entitled "Darstellung einer analytischen Function einer complexen Veränderlichen, deren absoluter Betrag zwischen zwei gegebenen Grenzen liegt" (= "Representation of an analytic function of a complex variable whose absolute value lies between two given bounds") [Weierstrass 1841a]. In this paper the twenty-six year old student at Münster, who had passed his teaching examination and had begun his preparation year at the Gymnasium Paulinum in Münster the same year, proved the existence of the Laurent expansion, two years before the note [Cauchy 1843] by Augustin Louis Cauchy (1789–1857) made public the statement of this theorem by Pierre Alphonse Laurent (1813–1854).

---

*Enlarged and translated version of a talk given on September 21, 1990 at the section "Elementarmathematik, Didaktik und Geschichte der Mathematik" of the Jahrestagung of the DMV at Bremen.

With respect to the later image of Weierstrass as the great master of power series, this manuscript seems to be a sin of his youth: He uses integrals as the essential means in his proof.

Weierstrass himself kept this work concealed for not less than fifty-three years, obviously gave no hints as to its existence both to his colleagues and to his students, and did not publish the paper until the edition of the first volume of his collected works [Weierstrass 1894–1927] in 1894. So Alfred Pringsheim (1850–1941) writes in 1895 [Pringsheim 1896, 123]:

"Herr Weierstrass hat in seiner fundamentalen Abhandlung: *"Zur Theorie der eindeutigen analytischen Functionen"* und, soweit ich feststellen konnte, auch in seinen Vorlesungen den Laurent'schen Satz weder explicite bewiesen noch direct angewendet." (= "In his fundamental treatise *"Zur Theorie der eindeutigen analytischen Functionen"* and, as far as I could ascertain, also in his lectures, Mr. Weierstrass has neither explicitly proven nor directly used Laurent's theorem.")

This reserve on the side of Weierstrass is reflected by the accounts of his mathematics in the literature: Gustav Mittag-Leffler (1846–1927) gives a discussion of [Weierstrass 1841a] in [Mittag-Leffler 1923, 27–36] which says much more about the theory of complex functions in the works of Cauchy, Carl Friedrich Gauss (1777–1855) and Bernhard Riemann (1826–1866) than about the manuscript of Weierstrass. Modern presentations mostly restrict to mentioning that Weierstrass had Laurent's theorem already in 1841 but did not publish it until 1894, e.g. [Bottazzini 1986, 178], [Dugac 1973, 48], [Neuenschwander 1981, 228–229]. Only [Manning 1975, 360–362, esp. 361] and [Remmert 1991, 174–175, 239, and 351] comment on mathematical details; Remmert even raises the question why Weierstrass delayed the publication of his result [ibid. 351].

1.   [WEIERSTRASS 1841a]

In [Weierstrass 1841a] Weierstrass wishes to expand functions $F(x)$ of a complex variable x on an annulus around 0 with radii A and B, $A < B$, into series of the form

$$\sum_{\nu=-\infty}^{+\infty} A_\nu x^\nu.$$

Besides the "harmless" conditions on the function $F(x)$—that it is univalent, finite, and continuous—he assumes the following crucial one:

For x in the annulus and h a complex number, together with k the term

$$\frac{F(x + k) - F(x)}{k} - \frac{F(x + hk) - F(x)}{hk}$$

will become "unendlich klein" (= "infinitely small"), i.e., will tend to 0.

[ibid., 51]. (For F real differentiable this implies that all directional derivatives must be equal, so that the Cauchy–Riemann differential equations must hold.) If one analyses the formulations of Weierstrass more thoroughly, one recognizes that he demands the above limit process to be uniform in h as long as the absolute value of h remains under a fixed bound, and uniform in x on the annulus. But later on [ibid., 56] he needs the uniform convergence with respect to x just on compact sets. Therefore, one easily sees that complex continuously differentiable functions fulfil the above assumptions. Hence the proof by Weierstrass gives the existence of the Laurent expansion for this kind of functions. This remark, however, is not made in [Weierstrass 1841a]. It was not until 1899 that Edouard Goursat (1858–1936) could get rid of the condition that the derivative is continuous [Goursat 1900].

As has been pointed out in the beginning, the proof of Weierstrass uses integrals in an essential way. It should be noted first that Weierstrass always parametrizes the circular line around 0 with radius 1 by

$$w = \frac{1 + \lambda i}{1 - \lambda i} \quad \text{with} \quad -\infty < \lambda < +\infty$$

(cf. [ibid., 52]), hence substitutes the integral along a finite path in the complex domain by an improper integral along the real axis. For the sake of simplicity and clearness, however, this parametrization will not be written out in the sequel.

For F a function that fulfills the above conditions, Weierstrass first shows that

$$\int_{|w|=1} \frac{F(x_0 w)}{w} \, dw$$

has the same value for all x_0 with same absolute value between A and B [ibid., 53–55]. The proof is just a re-parametrization of the integral, which, indeed, is a little bit tedious because of the special parametrization [ibid., 54]. The condition on the difference quotients is not yet needed for this, but for the next step, namely to show that, in fact, the above integral does not depend on the absolute value of x_0, i.e., the Cauchy integral theorem for annuli [ibid., 56–57].

To this end Weierstrass writes

$$x_1 = x_0 \frac{1 + \varepsilon i}{1 - \varepsilon i} = x_0 + x_0 h \varepsilon$$

for $\varepsilon > 0$. Then one has $|x_1| = |x_0|$ and $|h| \le 2$. With this h and

$$x = x_0 w, \qquad k = x_0 \varepsilon w,$$

the earlier assumption gives that, for ε sufficiently small, the absolute value of

$$\frac{F(x_0 w + x_0 \varepsilon w) - F(x_0 w)}{x_0 \varepsilon w} - \frac{F(x_0 w + x_0 h \varepsilon w) - F(x_0 w)}{x_0 h \varepsilon w}$$

is less than or equal to a given bound $E > 0$ for all w of absolute value 1 and all x_0 whose absolute value lies between two given radii a and b, $A < a < b < B$. The numerator of the second fraction equals $F(x_1 w) - F(x_0 w)$, so the integral of the second fraction along the boundary of the unit circle vanishes by what has been proven before. Therefore, Weierstrass has

$$\left| \int_{|w|=1} \frac{F(x_0 w + x_0 \varepsilon w) - F(x_0 w)}{x_0 \varepsilon w} \, dw \right| \le 2\pi E.$$

Since ε appears in the denominator of the integrand, one can substitute $n\varepsilon$ for ε everywhere in this estimate without changing the bound on the right-hand side. Writing the distance of the two given radii a and b as $n x_0 \varepsilon$ with n sufficiently great and x_0 real, one gets by this device

$$\left| \int_{|w|=1} \frac{F(bw) - F(aw)}{(b - a)w} \, dw \right| \le 2\pi E.$$

Since $E > 0$ is arbitrary, the Cauchy integral theorem for annuli is proven.

Now Weierstrass defines the coefficients of the Laurent expansion of F by

$$A_n := \frac{1}{2\pi i} \int_{|w|=1} (x_0 w)^{-n} \frac{F(x_0 w)}{w} \, dw,$$

where, by what has just been shown, the integral is independent of x_0 in the annulus since $x^{-n} \cdot F(x)$ fulfills the above conditions if $F(x)$ does so [ibid., 57–58].

While motivating that the coefficients must look like this, he slips into an error: He just assumes that the Laurent expansions converge absolutely [ibid., 51], but when he substitutes the Laurent series for $F(x)$ in the above integral in order to obtain the coefficient A_n, he interchanges integration and summation without any remark on uniform convergence [ibid., 58]. One could ascribe this to juvenile impetuosity or to the fact that he wants to give a motivation rather than a proof. But one should, of course, be aware of the foundational problems connected with interchanging limit processes and uniform convergence in the first half of the nineteenth century; (cf., e.g., [Giusti 1984], [Grabiner 1981], [Grattan-Guinness 1970], [Guitard 1986], and [Laugwitz 1989] and the bibliography cited therein).

On the other hand, Weierstrass correctly shows that the Laurent series defined by these coefficients converges on the annulus at all. For this he uses the Cauchy estimates for Laurent coefficients, which immediately result from the integral representation [ibid., 59]. Yet one should not be surprised that, contrary to his lectures in later years (cf. [Weierstrass 1988, 91–92]), he does not verify that a continuous function on a circular line is bounded.

The concluding proof that this Laurent series really coincides with $F(x)$ [Weierstrass 1841a, 60–63] is an essentially canonical manipulation of the expressions defined before. The only critical point is the interchanging of integration and summation.

It was already mentioned that Weierstrass seems to have certain difficulties with respect to this point. Indeed, he does not choose the direct way for the interchanging but introduces an auxiliary parameter $\varepsilon < 1$ and considers

$$\sum_{\nu=1}^{\infty} A_\nu \varepsilon^\nu x^\nu,$$

analogously for the principal part of the series. For these series he can, in fact, interchange without encountering any problems when he substitutes integral expressions for the A_n; however, he does not justify the admissibility of this procedure [ibid., 60]. In order to prove that the limit process "ε to 1" commutes with the summation, i.e., that one has

$$\lim_{\varepsilon \to 1} \sum_{\nu=1}^{\infty} A_\nu \varepsilon^\nu x^\nu = \sum_{\nu=1}^{\infty} A_\nu x^\nu,$$

he refers to the original paper of Niels Henrik Abel (1802–1829) with the continuity theorem of Abel ([Abel 1826, 314] resp. [Abel 1881 1, 223]). Such auxiliary parameters had already been used by Siméon Denis Poisson

(1781–1840) in order to interchange integration and summation, e.g. [Poisson 1820, 422–424]; cf. [Poisson 1835, 187–191]. Poisson, however, had not realized that he had to prove something like Abel's theorem (cf. [Grattan-Guinness 1970, 88–90]; also [Laugwitz 1989, 225]).

So it only remains to study the interchanging of "ε to 1" with the integration. Here Weierstrass says something about the uncertainties, in fact, he is forced to do so since for $\varepsilon = 1$ the integrand

$$F(xw)\left\{\frac{1}{w - \varepsilon} + \frac{\varepsilon}{1 - \varepsilon w}\right\}$$

exhibits difficulties on the path of integration, and he treats the problem by *ad hoc* considerations [Weierstrass 1841a, 60–62].

Christoph Gudermann (1798–1852), the academic teacher of Weierstrass at Münster, had called attention to the importance of uniform convergence in an article [Gudermann 1838, 251–252] in 1838. In fact, this is the first place where this concept appears explicitly (cf. [Bottazzini 1986, 204], [Dugac 1973, 47], [Remmert 1991, 97]), while the well-known considerations of Abel [1826] use it only implicitly. Though Weierstrass will have learnt this concept from Gudermann, probably both through the article and through the lectures in the years 1838–39, he is, nevertheless, relatively uncertain about this phenomenon in [Weierstrass 1841a], and instead retreats to explicit calculations rather than relying on theoretical arguments. In the manuscript "Zur Theorie der Potenzreihen" (= "On the theory of power series") [Weierstrass 1841b] that was completed in the autumn of the same year, Weierstrass is somewhat more advanced, at least as far as the interchanging of two infinite summations is concerned: He proves the double series theorem that was subsequently named in his honor, resorting solely to the Cauchy estimates of the coefficients, which are deduced without the use of integrals [ibid., 70–73], but under the superfluous additional assumption that the series converge "unbedingt" (= "unconditionally"), i.e., absolutely ([ibid., 70]; cf. [Manning 1975, 363–366], also [Ullrich 1990, 96]).

In any event, in the paper [Weierstrass 1841a] Weierstrass completes the proof of the Laurent expansion in a correct way. As an application he proves the Riemann Removable Singularity Theorem by a further use of the Cauchy estimates of the coefficients ([ibid., 63]; see [Ullrich 1994] for a discussion of the history of this theorem). Furthermore, he gives formulas for the derivatives of the function F [Weierstrass 1841a, 64–66].

2. CAUCHY, LAURENT AND WEIERSTRASS

In order to evaluate the performance of Weierstrass correctly in the historical context, one should point out that it was not until 1843, two years after the manuscript had been completed by Weierstrass, that Laurent published his result [Laurent 1843] and Cauchy read a note on it to the academy at Paris [Cauchy 1843] which made the theorem public. Therefore, simply on the basis of chronology, the proof cannot have influenced Weierstrass. Nor could Weierstrass have inspired Laurent since, as mentioned before, the paper of Weierstrass was not published until 1894. Remarkably enough, a detailed proof of the Laurent theorem by Laurent's own hand was published very late, too, namely posthumously in 1863 [Laurent 1863, esp. 121–125]. (For a discussion of Laurent's proof cf., [Bottazzini 1986, 162–163] or [Remmert 1991, 350–351].)

Of course, the expansion theorem of Laurent can easily be deduced from the Cauchy integral theorem for annuli in a way analogous to the usual Cauchy theory. This was already noted by Cauchy in his *Comptes Rendus* note [Cauchy 1843, 116]. However, he explicitly conceded the status of a "nouveau théorème' (= "new theorem") to Laurent's result [ibid., 117]. So Cauchy himself did not go as far as Kronecker, who wrote in [Kronecker 1894, 177] on the Laurent expansion: "Diese Entwickelung wird manchmal als Laurent'scher Satz bezeichnet; aber da sie eine unmittelbare Folge des Cauchy'schen Integrals ist, so ist es unnütz, einen besonderen Urheber zu nennen." (= "This expansion is sometimes called Laurent's theorem, but as it is an immediate consequence of Cauchy's integral, it is pointless to ascribe it to a specific author.")

A look at Cauchy's proof of the integral theorem for annuli [Cauchy 1840] is of interest, though. Like the treatise of Weierstrass, which surprisingly shows Weierstrass at work with integrals and not with power series, this work of Cauchy shakes the standard image of the master of integrals. Here he does not consider integrals but mean values of a function $\bar{\omega}$ [ibid., 335, 337], to be precise, for θ a primitive n-th root of unity he studies the behavior of

$$\Pi(r) := \frac{1}{n} \sum_{\nu=0}^{n-1} \bar{\omega}(\theta^{\nu} r)$$

for n to infinity, where, incidentally, he assumes $\bar{\omega}$ to be continuously differentiable [ibid., 333, 337]. Naturally, the proof works without any integrals, and it would rather fit into the so-called Weierstrassian power series approach. (For example, in his attempt to build up a function theory which is "elementary" in the Weierstrassian sense, Pringsheim explicitly referred to this article of Cauchy [Pringsheim 1896, 125].)

One may ask whether this 1840 paper of Cauchy had an influence on [Weierstrass 1841a]. The study [Manning 1975] shows how the early mathematics of Weierstrass is rooted in the works of Abel, Gudermann, and Carl Gustav Jacob Jacobi (1804–1851). These are also the names Weierstrass himself refers to when describing the fundaments of his research, for example in his inaugural address to the Berlin academy [Weierstrass 1857]. Especially, to Cauchy, Mittag-Leffler transmits statements by Weierstrass to the effect that he (Weierstrass) learned of the works of Cauchy "erst in einer späten Periode seines Lebens" (= "only in a late period of his life") [Mittag-Leffler 1897, 81] resp., more precisely, "erst 1842" (= "only in 1842") [Mittag-Leffler 1923, 35–36].

Yet, with explicit reference to the first quote of Mittag-Leffler from the year 1897, Émile Borel (1871–1956) set up the hypothesis a year later in a footnote to his "Leçons sur la théorie des fonctions" [Borel 1898, 80] that during the time of the mathematical studies of Weierstrass Cauchy's influence was already so great that he must have influenced Weierstrass, at least indirectly. For example, [Manning 1975, 316] constructs a possible path by which Cauchy's theory of convergent series could have reached Weierstrass via Abel and Gudermann.

But supposing that Weierstrass was not cut off from the diffusion process of mathematical results does not necessarily imply he would have learned of Cauchy's result within one year after its publication. Even if by some chance this had taken place, the performance of Weierstrass with respect to the proof of Laurent's theorem has to be evaluated by no means less than that of Laurent.

3. A CONCEALED PAPER

So it is, to say the least, surprising that this paper was just published in 1894 in the first volume of the collected works [Weierstrass 1894–1927] of Weierstrass. In the first volume of the collected works there are five papers that had not been published previously, whether in a mathematical journal or in the prospectus of a school. Of these papers, which date from the years 1840 to 1844, his thesis for the teaching examination "Über die Entwicklung der Modular-Functionen" (= "On the expansion of the modular functions") [Weierstrass 1840] was published, at least in extracts, in

1856 in the article, "Theorie der Abel'schen Functionen" (= "Theory of the Abelian functions") [Weierstrass 1856, §§9–13, 346–379]. The papers "Zur Theorie der Potenzreihen" (= "On the theory of power series") [Weierstrass 1841b] and "Definition analytischer Functionen einer Veränderlichen vermittelst algebraischer Differentialgleichungen" (= "Definition of analytic functions of one variable by means of algebraic differential equations") [Weierstrass 1842] were admittedly published only in the collected works, too, but already decades before their contents obtained entry in the lectures of Weierstrass on the theory of analytic functions (cf. [Manning 1975, 363–373]; also [Dugac 1973, 48–49], [Neuenschwander 1981, 228–229], or [Ullrich 1989, 158]). So the paper on the Laurent expansion shares the fate of having been kept totally hidden with only a treatise five printed pages in length, entitled "Reduction eines bestimmten dreifachen Integrals" (= "Reduction of a definite threefold integral"), [Weierstrass 1844], essentially a work of mere computation.

For a modern mathematician is is even more incomprehensible that Weierstrass did not use the paper on the Laurent expansion, since it is proven in that paper that, in particular, a complex continuously differentiable function can locally be expanded into a power series, a fact he did not display anywhere in his introductory lectures on the theory of analytic functions (cf. [Weierstrass 1988, esp. xvi–xx]).

However, Weierstrass had no need to do so because of his notion of function: The functions he wanted to consider in analysis are not abstract correspondences, but have to be described by the four basic operations of arithmetic and locally uniform limit processes ([Weierstrass 1925, esp. 1], [Weierstrass 1988, 49], cf. also [Bottazzini 1986, 287–288], [Dugac 1973, 70–72]). In the complex case each such function locally has a power series expansion as he actually proves in his lectures [Weierstrass 1988, 65, 67–68, and 111–116, esp. 111]. So it makes little sense for him to show that a complex differentiable function can be expanded locally into a power series, since each function he admits to complex analysis has this property.

By these considerations it becomes comprehensible why in a series of notes taken by Adolf Kneser (1862–1930) during the winter term 1880/81 from a lecture of Weierstrass the definition of a differentiable function is found to be defective in that

"den differenzierbaren Variabeln ein funktionentheoretisch nicht zu rechtfertigender Vorzug vor den undifferenzierbaren eingeräumt wird." (= "the differentiable variables are given a preference over the nondifferentiable ones that cannot be justified function-theoretically.")

(for a longer quote cf. [Ullrich 1989, 153]). Of course, all his functions in complex analysis are differentiable, but this is only a result *a posteriori* [Weierstrass 1988, 49]. (In fact, in the real case the condition of differentiability is really restrictive, as the example [Weierstrass 1872] of a continuous nowhere differentiable function shows, since it is the locally uniform limit of polynomials.)

Yet, even if one follows Weierstrass and regards the analytic functions —i.e., those which locally have a power series expansion—as the "right ones" in the complex case, the existence of the Laurent expansion and the Riemann Removable Singularity Theorem nevertheless represent topics that can be formulated and are of interest within the setting of the theory of analytic functions. Indeed, there are proofs for these theorems which use only power series, but these proofs were not found until 1884, 43 years after the treatise of Weierstrass (cf. [Remmert 1991, 352–355] and [Ullrich 1990, 101–108]).

Though Mittag-Leffler writes [Mittag-Leffler 1923, 35]:

> "In der WEIERSTRASSschen Funktionentheorie hat der LAURENTsche Satz nicht den elementaren Platz und spielt nicht die grundlegende Rolle wie in der CAUCHYschen." (= "In the function theory of Weierstrass Laurent's theorem does not have the elementary place and does not play the fundamental rôle as in the one of Cauchy.")

there are, in fact, situations where Weierstrass himself could not avoid Laurent expansions. Then he proved their existence by methods that were developed *ad hoc*. For example, in his proof of the Jacobian triple product identity,

$$\prod_{n=1}^{\infty} (1 - h^{2n})(1 + h^{2n-1}z^2)(1 + h^{2n-1}z^{-2}) = \sum_{\nu=-\infty}^{\infty} h^{\nu^2}z^{2\nu},$$

in the fifth volume of his works, the "Vorlesungen über die Theorie der elliptischen Funktionen" (= "Lectures on the theory of elliptic functions"), he made an indefinite setup for the Laurent expansion and then determined the coefficients from the behavior of the product under transformations [Weierstrass 1894–1927 5, 165–169]. That this setup is possible at all, i.e., that a Laurent expansion exists, is not established by the argument that the above product is analytic on an annulus, but apparently—the text is somewhat vague here and uses iterated references—by decomposing the product into three infinite partial products, where, for example, $\prod_{n=1}^{\infty}(1 + h^{2n-1}z^2)$ can be written as a series of positive powers of z^2 [ibid., 165]. Similar auxiliary constructions are used by Weierstrass for the

Laurent expansion of logarithmic derivatives (cf. [Weierstrass 1988, 108 and 154–156]).

One knows from his "Glaubensbekenntnis" (= "confession of faith") [Weierstrass 1894–1927 2, 235] that Weierstrass thought "dass es...nicht der richtige Weg ist, wenn...zur Begründung einfacher und fundamentaler algebraischer Sätze das »Transcendente«...in Anspruch genommen wird" (= "that it is not the right way if one indents upon the »transcendental« for the establishment of simple and fundamental algebraic theorems"). So it is quite obvious to assume—as has [Remmert 1991, 351]—that Weierstrass avoided his general proof of the Laurent expansion because of his aversion to integrals.

Remmert even uses the word "phobia" in this context, [ibid., 239 and 450(!)]. Weierstrass, however, seems to have seen the question, "to integrate or not to integrate," more as matter of taste than of substance or even of ideology. In a talk in 1884 to the Mathematisches Seminar at Berlin he said [Weierstrass 1925, 8]:

> "So grosse Wichtigkeit nun auch der Integralbegriff für die ganze Analysis hat, möchte ich doch die Funktionentheorie begründen blos mit Hilfe der elementaren Sätze über die Grundoperationen.... Ich sage nicht, dass man in jedem Fall einen solchen direkten Beweis [d.h., nur mittels der Grundoperationen] führen soll [sic!] oder führen kann; diese Frage lasse ich unentschieden. Aber ich suche die direkten Beweise soweit durchzuführen, als es eben geht und möchte mich bei der Begründung der Funktionentheorie dieser Methode vorzugsweise bedienen." (= "However great the importance of the notion of the integral for all the analysis, I nevertheless wish to found the theory of functions solely with the help of the elementary theorems of the basic operations.... I do not say that one should give [sic!] or can give such a direct proof [i.e. by the mere use of the basic operations] in any case; I leave this question undecided. But I try to give direct proofs as far as possible, and I want to use this method preferably with the foundation of the theory of functions.")

Furthermore, since Weierstrass took part in the edition of the first volume of his collected works, in fact, even signed its preface [Weierstrass 1894–1927 1, V–VI], the paper [Weierstrass 1841a] would not have been included if he had thought that something was wrong with the proof. This consideration is supported by the following quote from Mittag-Leffler, who indeed personally knew Weierstrass, [Mittag-Leffler 1923, 35]:

> "Wenn WEIERSTRASS bei der strengen Kritik, die er an seinen eigenen Arbeiten übte, seinen mehr als 40 Jahre früher abgefaßten Beweis für CAUCHYS und LAURENTS Sätze in seinen Mathematischen Werken...ohne irgend eine erklärende Anmerkung...veröffentlichte, so kann der Grund

kein anderer gewesen sein, als daß er seine Beweismethode in ihrer scharfen und knappen Art auch fernerhin für zwingend hielt." (= "If, in view of his sharp criticism of his own works, Weierstrass published his proof of the theorems of Cauchy and of Laurent, which had been written over 40 years before, in his collected mathematical works without any explanatory comment, this can only mean that he believed his method of proof, in its sharp and concise manner, to be henceforth as conclusive.")

Perhaps we also find the "publication policy" of Weierstrass more than surprising, since from the present point of view we are mainly interested in his contributions to the foundations of the theory of functions. He himself, however, regarded these contributions only as prerequisites to his real aim, which was a theory of the Abelian integrals and Abelian functions, as one learns from his inaugural address to the academy at Berlin [Weierstrass 1857, esp. 224]. Also his disciples and contemporaries shared this valuation of his work. Emil Lampe (1840–1918), for example, expressed this pointedly in 1897 in his obituary of Weierstrass [Lampe 1899, 34]:

"In dem Centrum aller Arbeiten von Weierstraß stehen die Abel'schen Functionen; man könnte sogar sagen, daß alle allgemeinen functionentheoretischen Untersuchungen von ihm nur zu dem Zwecke unternommen sind, um das Problem in Vollständigkeit und Klarheit zu lösen, das durch die Forderung der Darstellung der Abel'schen Functionen jener Zeit gestellt war." (= "The Abelian functions stand at the center of all the works of Weierstrass; one could even say that all of his general function-theoretic investigations were undertaken by him solely for the purpose of solving the problem posed that time by the need to represent Abelian functions in a complete and clearcut fashion.")

A similar statement was given by Mittag-Leffler 26 years later [Mittag-Leffler 1923, 49].

So the question regarding the Laurent theorem, which is of significance for the development of modern function theory, was a question of lesser importance for Weierstrass. Heinrich Behnke (1898–1979) has made this paradox of the history of mathematics, which, up to a certain extent, is also a tragic feature of Weierstrass, poignant [Behnke 1966, 24]:

"Geblieben ist gerade das Gerüst, das Weierstraß zur Sicherung seiner Theorie konstruierte. Wenig Interesse aber finden die speziellen Funktionen, die er behandelte." (= "What remains is the scaffolding that Weierstrass constructed in order to secure his theory. Little interest is found, however, for the special functions that he dealt with.")

REFERENCES

Abel, N. H. 1826. Untersuchungen über die Reihe: $1 + (m/1)x + [m \cdot (m - 1)/1 \cdot 2] \cdot x^2 + [m \cdot (m - 1) \cdot (m - 2)/1 \cdot 2 \cdot 3] \cdot x^3 + \ldots$ u.s.w., *Journal für die reine und angewandte Mathematik (Crelle's Journal)* **1**, 311–339; also in: [Abel 1881 1, 219–250].

—— 1881. *Œuvres complètes de Niels Henrik Abel. Nouvelle édition publiée aux frais de l'état norvégien par MM. L. Sylow et S. Lie*, 2 volumes, Christiania: Grøndahl & Søn.

Behnke, H. 1966. Karl Weierstraß und seine Schule, in: *Festschrift zur Gedächtnisfeier für Karl Weierstraß 1815–1965*, 13–40, H. Behnke and K. Kopfermann, eds. Wissenschaftliche Abhandlungen der Arbeitsgemeinschaft für Forschung des Landes Nordrhein-Westfalen 33. Köln, Opladen: Westdeutscher Verlag.

Borel, É. 1898. *Leçons sur la théorie des fonctions*, Paris: Gauthier-Villars.

Bottazzini, U. 1986. *The Higher Calculus: A history of real and complex analysis from Euler to Weierstrass*. New York: Springer.

Cauchy, A. L. 1840. Considérations nouvelles sur la théorie des suites et sur les lois de leur convergence, in: *Exercices d'analyse et de physique mathématique*, Paris: Bachelier; here in: [Cauchy 1882–1974, 2. ser. XI, 331–353].

—— 1843. Rapport sur un Mémoire de M. Laurent, qui a pour titre: Extension du théorème de M. Cauchy relatif à la convergence du développement d'une fonction suivant les puissances ascendantes de la variable x, *Comptes Rendus hebdomadaires des séances de l'Academie des Sciences (Paris)* **17**, 938–942 (30 octobre 1843), here in: [Cauchy 1882–1974, 1. ser. VIII, 115–117].

—— 1882–1974. *Œuvres complètes d'Augustin Cauchy*. 27 volumes. Paris: Gauthier-Villars.

Dugac, P. 1973. Eléments d'analyse de Karl Weierstrass, *Archive for History of Exact Sciences* **10**, 41–176.

Giusti, E. 1984. Gli "errori" di Cauchy e i fondamenti dell'analisi, *Bollettino di Storia delle Scienze Matematiche* **4**, 24–54.

Goursat, E. 1900. Sur la définition générale des fonctions analytiques, d'après Cauchy, *Transactions of the American Mathematical Society* **1**, 14–16.

Grabiner, J. V. 1981. *The origins of Cauchy's Rigorous Calculus*. Cambridge, Mass.: MIT Press.

Grattan-Guinness, I. 1970. *The Development of the Foundations of Mathematical Analysis from Euler to Riemann*. Cambridge, Mass.: MIT Press.

Gudermann, Chr. 1838. Theorie der Modular-Functionen und der Modular-Integrale, 3. part, *Journal für die reine und angewandte Mathematik (Crelle's Journal)* **18**, 220–258.

Guitard, T. 1986. La querelle des infiniment petits à l'École polytechnique au XIX siècle. *Historia Scientiarium* **30**, 1–61.

Kronecker, L. 1894. *Vorlesungen über die Theorie der einfachen und der vielfachen Integrale*, E. Netto, ed. Leipzig: Teubner.

Lampe, E. 1899. Karl Weierstraß, *Jahresbericht der Deutschen Mathematiker-Vereinigung* **6**, 27–44.

Laugwitz, D. 1989. Definite values of infinite sums: Aspects of the foundations of infinitesimal analysis around 1820, *Archive for History of Exact Sciences* **39**, 195–245.

Laurent, P. A. 1843. Extension du théorème de M. Cauchy relatif à la convergence du développement d'une fonction suivant les puissances ascendantes de la variable x, *Comptes Rendus hebdomadaires des séances de l'Académie des Sciences* (*Paris*) **17**, 348–349 (21 août 1843).

—— 1863. Mémoire sur la théorie des imaginaires, sur l'équilibre des températures et sur l'équilibre d'élasticité, *Journal de l'école impériale polytechnique* **23**, 75–204.

Manning, K. R. 1975. The Emergence of the Weierstrassian Approach to Complex Analysis, *Archive for History of Exact Sciences* **14**, 297–383.

Mittag-Leffler, G. 1897. Weierstrass, *Acta Mathematica* **21**, 79–82.

—— 1923. Die ersten 40 Jahre des Lebens von Weierstraß, *Acta Mathematica* **39**, 1–57.

Neuenschwander, E. 1981. Über die Wechselwirkungen zwischen der französischen Schule, Riemann und Weierstraß. Eine Übersicht mit zwei Quellenstudien, *Archive for History of Exact Sciences* **24**, 221–255.

Poisson, S. D. 1820. Mémoire sur la manière d'exprimer les fonctions par des séries de quantités périodiques et sur l'usage de cette transformation dans la résolution de différens problèmes, *Journal de l'École royale polytechnique* **11** cah. 18, 417–489.

—— 1835. *Théorie mathématique de la chaleur*. Paris: Bachelier.

Pringsheim, A. 1896. Ueber Vereinfachungen in der elementaren Theorie der analytischen Functionen, *Mathematische Annalen* **47**, 121–154.

Remmert, R. 1991. *Theory of Complex Functions*. Readings in Mathematics. Graduate Texts in Mathematics 122. New York: Springer.

Ullrich, P. 1989. Weierstraß' Vorlesung zur "Einleitung in die Theorie der analytischen Funktionen," *Archive for History of Exact Sciences* **40**, 143–172.

—— 1990. Wie man beim Weierstraßschen Aufbau der Funktionentheorie das Cauchysche Integral vermeidet, *Jahresbericht der Deutschen Mathematiker-Vereinigung* **92**, 89–110.

—— 1994. The Riemann Removable Singularity Theorem from 1841 onwards, in: *The History of Modern Mathematics* 3, 155–178. E. Knobloch and D. E. Rowe, eds. (Special volume of *Historia Mathematica*). Boston, San Diego: Academic Press.

Weierstrass, K. 1840. *Über die Entwicklung der Modular-Functionen*. Westernkotten. in: [Weierstrass 1894–1927 1, 1–49].

—— 1841a. *Darstellung einer analytischen Function einer complexen Veränderlichen, deren absoluter Betrag zwischen zwei gegebenen Grenzen liegt.* Münster. in: [Weierstrass 1894–1927 1, 51–66].

—— 1841b. *Zur Theorie der Potenzreihen.* Münster. in: [Weierstrass 1894–1927 1, 67–74].

—— 1842. *Definition analytischer Functionen einer Veränderlichen vermittelst algebraischer Differentialgleichungen.* Münster. in: [Weierstrass 1894–1927 1, 75–84].

—— 1844. *Reduction eines bestimmten dreifachen Integrals.* Deutsch-Crone. in: [Weierstrass 1894–1927 1, 105–109].

—— 1856. Theorie der Abel'schen Functionen, *Journal für die reine und angewandte Mathematik (Crelle's Journal)* **52**, 285–380.

—— 1857. *Akademische Antrittsrede.* in: [Weierstrass 1894–1927 1, 223–226].

—— 1872. *Über continuirliche Functionen eines reellen Arguments, die für keinen Werth des letzteren einen bestimmten Differentialquotienten besitzen. Gelesen in der Königl. Akademie der Wissenschaften am 18. Juli 1872.* in: [Weierstrass 1894–1927 2, 71–74].

—— 1894–1927. *Mathematische Werke*, 7 volumes. Berlin: Mayer & Müller; Reprint: Hildesheim: Georg Olms and New York: Johnson Reprint.

—— 1925. Zur Funktionentheorie, *Acta Mathematica* **45**, 1–10.

—— 1988. *Einleitung in die Theorie der analytischen Funktionen, Vorlesung Berlin 1878, in einer Mitschrift von Adolf Hurwitz, bearbeitet von Peter Ullrich.* Dokumente zur Geschichte der Mathematik 4. Braunschweig/Wiesbaden: Vieweg.

Research Communities and International Collaboration

Georg Friedrich Bernhard Riemann (1826–1866)
Courtesy of the Niedersächsische
Staats-und Universitätsbibliothek Göttingen

The Riemann Removable Singularity Theorem from 1841 Onwards*

Peter Ullrich

Westfälische Wilhelms-Universität, Mathematisches Institut, Münster, Germany

The theorem on the characterization of removable singularities of functions of a complex variable has had an odd fate. It was first discovered by Weierstrass in 1841, but he did not publish it until 1894. Then it was independently rediscovered by Riemann in 1851, and this time some of his disciples did harm to the spreading of this result. But once difficulties with the single variable case had been settled, the theorem was relatively quickly generalized to several complex variables and has influenced research on complex analysis of several variables up to the present.

Since Bernhard Riemann (1826–1866) was just 15 years old in 1841, the above title implies that the history of the "Riemann" Removable Singularity Theorem does not begin with Riemann, yet another example illustrating the phenomenon in which the name associated with a mathematical theorem turns out not to be the name of the person who stated or proved the result for the first time.

But the story of this theorem is even stranger. Karl Weierstrass (1815–1897) discovered and proved it in 1841, but kept it hidden for 53 years. Riemann found the theorem anew and gave correct proofs of it. Some of his disciples, however, published incorrect proofs or did not even mention it in their textbooks on Riemann's theory.

So this theorem is an instructive example that, contrary to the often expressed belief, knowledge in mathematics is not monotonically increasing.

*Expanded version of a talk given on July 16, 1990 at the Workshop on the History of Modern Mathematics at Göttingen.

1. STATEMENT OF THEOREM

In a modern textbook on complex analysis of one variable one finds the Riemann Removable Singularity Theorem in the following formulation [Remmert 1991, 212].

Riemann Removable Singularity Theorem. *Let $D \subset \mathbb{C}$ be open, $c \in D$, and $f : D - \{c\} \to \mathbb{C}$ be holomorphic (that is, analytic, i.e., complex differentiable). Then the following are equivalent*:

   i) There is a holomorphic $g : D \to \mathbb{C}$ with $g|_{D-\{c\}} = f$ (i.e., f has a holomorphic extension to c, i.e., c is a removable singularity of f).
   ii) There is a continuous $g : D \to \mathbb{C}$ with $g|_{D-\{c\}} = f$ (i.e., f has a continuous extension to c).
   iii) f remains bounded in a neighborhood of c.
   iv) $\lim_{z \to c} (z - c)f(z) = 0$.

The implications from i) to ii), ii) to iii), and iii) to iv) are clear, of course. The hard part is to show that iv) implies i). In the following historical analysis, however, not much emphasis will be placed on distinguishing between some author's proof of the implication from iv) to i) versus a proof "solely" of the implication from iii) to i). Namely, if one has proven that iii) implies i) and has a function $f(z)$ which fulfills iv), one can apply the proven result to the function $(z - c)f(z)$ and look at its power series expansion around c. This kind of argument was quite common in the time under consideration, in fact, it was used in proofs of the Removable Singularity Theorem, sometimes, however, in a fallacious way, as will be seen in Section 2.3.

Perhaps another glance at this theorem from the mathematical point of view would be valuable. Up to a linear fractional transformation, the Removable Singularity Theorem, to be precise, the implication from iii) to i), is equivalent to the Casorati–Weierstrass theorem that in any neighborhood of an essential singularity a holomorphic function attains values that are arbitrarily close to any given complex number. The implication from the Casorati–Weierstrass theorem to the Removable Singularity Theorem is evident. On the other hand, for c an essential singularity of f and b an arbitrary complex number one gets from the Removable Singularity Theorem that $1/[f(z) - b]$ cannot be bounded in any neighborhood of c, hence the Casorati–Weierstrass theorem.

This argument was indeed used by Weierstrass in 1868 [Weierstrass 1986, 76-79] and 1876 [Weierstrass 1876, 122-124] and Felice Casorati (1835-1890) in 1868 [Casorati 1868, 435] in their respective proofs of the Casorati-Weierstrass theorem and also in the proof of Otto Hölder (1859-1937) in 1882 [Hölder 1882, 141]. These proofs will be analysed in greater detail later on. (For the history of the Casorati-Weierstrass theorem cf. [Neuenschwander 1978a]; also [Ullrich 1989, 163-164].)

But first the "1841" in the title of the present paper shall be justified.

2. ONE VARIABLE

2.1. Weierstrass

As already discussed in [Ullrich 1994], Weierstrass proved in [Weierstrass 1841a] the existence of the Laurent expansion

$$F(x) = \sum_{\nu=-\infty}^{+\infty} A_\nu x^\nu$$

for continuously differentiable functions $F(x)$ and the Cauchy estimates for the coefficients A_ν—that $|F(x)| \le M$ for all x with $|x| = a$ implies $|A_\nu| \le Ma^{-\nu}$ for all ν—but kept his results hidden for over half a century. In this 1841 paper he also wrote [ibid., 63]:

> "Wenn die Veränderliche x alle Werthe haben kann, deren absoluter Betrag eine bestimmte Grenze nicht überschreitet, und wenn $F(x)$ für keinen Werth von x in der Nähe von 0 unendlich gross wird, so kann man aus dem in §3 bewiesenen Umstande [d.h. den Cauchyschen Abschätzungen]...folgern, dass
>
> $$A_\nu = 0$$
>
> ist für jeden negativen Werth von ν. Denn man kann unter den gemachten Voraussetzungen a beliebig klein annehmen, während M eine bestimmte Grösse nicht überschreitet. In diesem Falle läßt sich daher $F(x)$ in eine Reihe von der Form
>
> $$A_0 + A_1 x + A_2 x^2 + \cdots + A_\nu x^\nu + \cdots,$$
>
> die nur ganze positive Potenzen enthält, entwickeln." (= "If the variable x can attain all values whose absolute value does not exceed a given bound, and if $F(x)$ does not become infinitely great for any value of x near 0, then one can conclude from the fact proven in art. 3 [i.e., the Cauchy estimates]...that
>
> $$A_\nu = 0$$

holds for any negative value of ν. Namely, under the present assumptions one can assume a to be arbitrarily small, whereas M does not exceed a certain value. Hence in this case $F(x)$ can be expanded in a series of the form

$$A_0 + A_1 x + A_2 x^2 + \cdots + A_\nu x^\nu + \cdots$$

that contains only positive powers.")

One can interpret this, of course, just in the way that Weierstrass wants to state that his theorem specializes to the power series expansion if one considers a disc instead of an annulus. But if 0 would be a point where $F(x)$ is defined *a priori* then, according to the assumptions stated in the beginning of the paper [ibid., 51], $F(x)$ has to fulfill a condition on the difference quotients and, especially, has to be finite and continuous around 0, so the assumption "$F(x)$ does not become infinitely great in a neighborhood of 0" would be superfluous. This argument is enforced by the fact that Weierstrass analogously discusses the case of a function bounded around infinity without explaining how to interpret the condition on the difference quotients for the point infinity [ibid., 63–64]. Hence, indeed, he states and proves the implication from iii) to i) of the Removable Singularity Theorem. (This fact was already noted in [Osgood 1901, 18, footnote 26]; modern references are [Neuenschwander 1978a, 158] and [Remmert 1991, 213 and 358]. It is, however, mentioned neither in the discussion of [Weierstrass 1841a] by Mittag-Leffler [1923, 27–36] nor in the one by Manning [1975, 360–362].)

If one supposes that a dislike of integrals was the reason why Weierstrass delayed publication of [Weierstrass 1841a], then Laurent's theorem in fact is the critical point of the proof of the Removable Singularity Theorem, since for the Cauchy estimates he had already given an integral-free proof in the autumn of the very year 1841 in his paper "Zur Theorie der Potenzreihen" (= "On the theory of power series") [Weierstrass 1841b].

As to the Removable Singularity Theorem, the behavior of Weierstrass seems even odder, since in 1876 he considered it "bekannten Satze" (= "well-known theorem") [Weierstrass 1876, 77], whereas two years later he seems to have forgotten this theorem [Weierstrass 1988, 127].

Furthermore, as mentioned before, Weierstrass knew how to deduce the Casorati–Weierstrass theorem from results of this type. In his introductory lectures on the theory of analytic functions in 1868 [Weierstrass 1986, 76–79] he used Liouville's theorem—which can be interpreted as the Removable Singularity Theorem for the singularity infinity of an entire function—in order to prove the Casorati–Weierstrass theorem for entire functions.

In his article [Weierstrass 1876] he considered functions that are analytic on the complex plane with the exception of arbitrarily many poles and finitely many essential singularities. For these functions he proved the Removable Singularity Theorem without explicitly calling this result a "lemma" [ibid., 122–124] and concluded from it the Casorati–Weierstrass theorem for these functions [ibid., 124]. (In fact, this is the best Casorati–Weierstrass type result for one variable that can be found in the works of Weierstrass.) One should note that these results are just presented as corollaries of the main part of [Weierstrass 1876] which is concerned with explicit expressions for functions of the above type. So one has a further illustration of the fact that Weierstrass placed emphasis in his mathematical work in a way that differs from our present approach.

To finish this section on Weierstrass, one may note that, in contrast to the tensions in his relation to Riemann, which are noted by Neuenschwander [1981, 229–235], Weierstrass waited for over half a century before he let anybody know that he had the Removable Singularity Theorem ten years before Riemann.

2.2. *Riemann*

Returning to the chronological order, the next proof after the one by Weierstrass in 1841 is the one in Riemann's doctoral dissertation [Riemann 1851] in 1851. Though Weierstrass is prior to Riemann with his result, it was somewhat of a dead-end, since Weierstrass made no use of it. Riemann's thesis, however, was at least read and studied.

Even the notation, the German word "hebbar" for "removable," seems to have appeared here for the first time as "hebbare Unstetigkeitsstelle" (="removable place of non-continuity") [ibid., art. 10, 21, and art. 12, 23]. In the twelfth article [ibid., 23–24] Riemann proves the following "Lehrsatz" (="theorem"):

"Wenn eine Function w von z eine Unterbrechung der Stetigkeit jedenfalls nicht längs einer Linie erleidet und ferner für jeden beliebigen Punkt O' der Fläche, wo $z = z'$ sei, $w(z - z')$ mit unendlicher Annäherung des Punktes O unendlich klein wird, so ist sie nothwendig nebst allen ihren Differentialquotienten in allen Punkten im Innern der Fläche endlich und stetig." (="If a function w of z does at least not become discontinuous along a line and if, furthermore, for any arbitrary point O' of the surface where $z = z'$, $w(z - z')$ becomes infinitely small for infinite approach of the point O, then it [= the function w] and all its differential quotients are necessarily finite and continuous at all points in the interior of the surface.")

One should add some remarks on the notions in this statement. First, as is well-known, for Riemann "function" means differentiable function. Then, his condition that w does "not become discontinuous along a line" has to be interpreted in the light of his considerations in article 10 [ibid., 19] in such a way that the possible points of discontinuity can be treated one by one, which can be translated into modern language to state "that the set of these points is discrete." So Riemann formulates and proves the crucial implication from iv) to i) of the Removable Singularity Theorem. Even more, he considers a discrete set of possible singularities instead of a single one, but this is only an easy generalization.

As to Riemann's proof, we writes $w = u + iv$ with real-valued functions u, v and considers the contour integral

$$\int \left(u \frac{\partial x}{\partial s} - v \frac{\partial y}{\partial s} \right) ds.$$

He has shown before in part II of article 9 [ibid., 15] that only the singularities of $w = u + iv$ can contribute to this integral, since w fulfills the Cauchy–Riemann differential equations elsewhere. But the condition on the singularities—that $(z - z')w$ tends to zero as z tends to z'—implies that also the contributions from the singularities vanish according to part III of article 9 [ibid., 16]. So $[u\partial x/\partial s - v\partial y/\partial s] ds$ is integrable throughout the domain in question and hence has a primitive function U with $u = \partial U/\partial x$ and $v = -\partial U/\partial y$, as he has shown in part IV of article 9 [ibid., 16–18].

Now, this function U is harmonic with the possible exception of the singularities of w. But Riemann has shown in article 10 [ibid., 20–21] that, in fact, in this situation U and all its derivatives are "finite and continuous" throughout the whole domain under consideration. The proof of that theorem works by the same arguments as given before and by the analytic representation

$$\int \left(\log r \frac{\partial U}{\partial p} - U \frac{\partial \log r}{\partial p} \right) ds$$

for a harmonic function U, which is the equivalent of Cauchy's integral formula. Now, since $w = \partial U/\partial x - i\partial U/\partial y$, the Removable Singularity Theorem is proven.

Riemann generalizes the Removable Singularity Theorem in the following article 13 to functions w with the property that w times a higher power of $(z - z')$ tends to zero as z tends to the singular place z' [ibid., 24–25]. From the Laurent expansion with finite principal part which he gets in this situation, he deduces the Puiseux expansion

$$\sum_{\nu=-m}^{\infty} a_\nu (z - z')^{\nu/n}$$

of a "Fläche," nowadays called a "Riemann surface," at a branching point [ibid., art. 14, 25–27].

The paper [Puiseux 1850] of Victor Puiseux (1820–1883), where the Puiseux expansion is proved for algebraic functions, was published one year before Riemann finished his thesis. So, simply as a question of chronology, the paper of Puiseux could well have exerted an influence on Riemann's work. Furthermore, Riemann is known to have studied the current French literature; in a draft for the defense of his doctoral thesis he even refers to a report of Cauchy "über eine Arbeit von Puiseux" (= "on a work by Puiseux") [Neuenschwander 1981, 226]. But, as Bottazzini has already pointed out, this does not mean that Riemann's thesis is "a simple translation of Puiseux's results into geometrical terminology" [Bottazzini 1986, 252, also 221–223]. Indeed, Puiseux arrived at the expansion in a way very different from Riemann's approach.

Puiseux does not consider general functions of a complex variable but restricts discussion *a priori* to functions defined by algebraic equations [Puiseux 1850, 365 and 375, footnote]. His proof works via an explicit study of these equations and the corresponding Galois group at a branching point [ibid., 386–397]. Especially, there is no analogue or preliminary version of the Removable Singularity Theorem in his article.

Furthermore, one should keep in mind that Puiseux-type series expansions had been used long before Puiseux by Isaac Newton (1643–1727), for example, but also by Jean le Rond d'Alembert (1717–1783) in his proof of the fundamental theorem of algebra [d'Alembert 1746]—d'Alembert assumed without proof that the inverse of a polynomial can locally be written as a Puiseux series, from which he deduced that one can decrease the absolute value of a polynomial so long as a root has not yet been reached. Puiseux, however, had to suppose the validity of the fundamental theorem of algebra for his proof of the Puiseux expansion. (For a discussion of d'Alembert's proof and its recovery by Jean Robert Argand (1768–1822) see [Remmert 1988, 83, 88, and 90–92].) And Carl Friedrich Gauss (1777–1855), the doctoral father of Riemann, had analysed d'Alembert's proof in his dissertation ([Gauss 1799, 7–11] resp. [Gauss 1890, 8–12]).

Also, the fact that Riemann's proof of the Removable Singularity Theorem uses real analysis, to be precise, the theory of the logarithmic potential, hints rather at an influence of Gauss and his work on potential theory than any influence from the French school, even if the only two works of Gauss that Riemann quoted in his dissertation are on conformal mapping [Gauss 1825] and on differential geometry [Gauss 1828], respectively.

From "Bernhard Riemann's Lebenslauf" by Richard Dedekind (1831–1916) [Riemann 1876, 507–526] one knows the story of Riemann visiting Gauss after finishing his thesis [ibid., 513]. Gauss informed Riemann that he had a paper in preparation that would treat the same subject as Riemann's thesis but would go beyond it. Sketches of this paper have been identified in the hand-written notes left by Gauss (see [Schlesinger 1912] in the collected works of Gauss [ibid., 209]). The article of Gauss would bear the title "(Bestimmung der) Convergenz der Reihen, in welche die periodischen Functionen einer veränderlichen Grösse entwickelt werden" (= "(Determination of the) convergence of the series in which the periodic functions of one variable quantity are developed"), and one complete version of sketches and parts of another are printed in the collected works of Gauss [1863–1933 10.1, 400–419]. In fact, they contain the idea of something like Riemann's "Zweige" (= "branches") of an analytic function, Gauss called them "Schichten" (= "layers") [ibid., 407–412], but he had obviously not gotten the right idea as to how to treat branching points [ibid., 414]. Especially, one does not find any reference to the Removable Singularity Theorem in these notes.

So one can conclude that there are no predecessors of Riemann's proof of the Removable Singularity Theorem in 1851 either in the French school or in Gauss' works.

Ten years later, in his lectures on function theory in Göttingen, Riemann also mentioned this theorem and sketched a proof (see the lecture notes from "Sommersemester" (= "summer term") 1861 edited by Neuenschwander [1987, 46–47]). Here Riemann shows that a function $f(t)$ that is complex differentiable on a disc with the possible exception of singularities a_k can be expressed as a sum of an integral of $f(z)/(z - t)$ along the boundary of the original disc and of integrals of $f(z)/(z - t)$ along the boundaries of discs around each of the a_k. He expands the first integral into a power series and the other integrals into Laurent series around a_k which only have a principal part. Then he notes that if condition iv) of the Removable Singularity Theorem is fulfilled for a_k in place of c, the value of the integral around a_k is zero, since the radius of the disc can be taken arbitrarily small and the value does not depend on the radius.

2.3. The "Heirs"

In autumn 1858 Enrico Betti (1823–1892), Francesco Brioschi (1824–1897), and Casorati visited Riemann at Göttingen, which resulted in Riemann exerting a profound influence on the mathematical works of Betti and Casorati (see [Bottazzini 1977] and [Weil 1979]). In particular, one and a half years before Riemann gave the lectures mentioned earlier, the Removable Singularity Theorem appeared as Teorema 2 in Section 7 of the Betti lectures for the winter term 1859/60 on elliptic functions [Betti 1860/61, 248]. In the beginning of his paper Betti refers to Riemann's thesis [ibid., 229, footnote 1] and his immediate application of the theorem [ibid., 248, Teorema 3] is the same Laurent expansion that Riemann discusses in article 13 of his thesis [Riemann 1851, 24–25]. Betti's proof, however, is written in the language of the integration theory of Augustin Louis Cauchy (1789–1857) rather than in that of potential theory. (This is in accord with the opinion expressed by Bottazzini [1977, 31] that "BETTI [bewahrte] RIEMANN gegenüber eine eigene Originalität der Forschungsmethoden" (= "in relation to Riemann Betti maintained his originality in the methods of research which he made use of").)

Furthermore, contrary to Riemann, Betti does not impose conditions on the position of the possible singularities, so that the modern reader might think of two different branches of the logarithm meeting at the negative real axis as an obvious counterexample to Betti's formulation of the theorem. But one should be aware that his terminology is different from the terminology used today—and even from that of Riemann. For example, Betti's condition that the function under consideration should be "monodroma" [Betti 1860/61, 248] obviously means more than just that it be univalent, as one would conclude from simply reading the definition [ibid., 228]. Namely, Betti's criterion for a function to be "monodroma" can only be applied to continuous functions, so it is not fulfilled in the above counterexample. (It should also be mentioned that, while Riemann concludes in the Removable Singularity Theorem that both the function under consideration and all its differential quotients are finite and continuous [Riemann 1851, 23], Betti only states that the function itself is "finita e continua," (= "finite and continuous") [Betti 1860/61, 248], though he has proven the same result as Riemann.)

As to the arguments in Riemann's lectures in 1861, they can also be found in the first edition of a book by H. Durège, "Elemente der Theorie der Functionen einer complexen veränderlichen Grösse. Mit besonderer

Berücksichtigung der Schöpfungen Riemanns" (= "Elements of the theory of functions of a complex variable, with special reference to Riemann's constructions") [Durège 1864] that was published in 1864. According to the preface [ibid., III–IV], Durège did not attend Riemann's lectures but used, among other sources, two sets of lecture notes by students of Riemann. In article 26 [ibid., 104–106] he reproduces Riemann's proof that in the situation just described, the function can be expressed as the sum of a power series and of Laurent series. But then he suddenly starts a new section and, in article 27 [ibid., 107–108], gives a proof of the implication from iv) to i) of the Removable Singularity Theorem which simplifies the discussion of the previous article in the present situation.

If φ, as he denotes the function, is analytic on the domain T with the possible exception of a where $\lim_{z \to a}(z - a)\varphi(z) = 0$ holds, one has

$$\varphi(t) = \frac{1}{2\pi i} \int_{\partial T} \frac{\varphi(z)}{z - t}\, dz - \frac{1}{2\pi i} \int_{\kappa_r(a)} \frac{\varphi(z)}{z - t}\, dz,$$

where ∂T is the boundary of T and $\kappa_r(a)$ is the boundary of a disc around a with radius r. Durège parameterizes the second integral and shows that it becomes arbitrarily small as r tends to zero.

The argument of the vanishing second integral and the $1/[f(z) - b]$ trick (cf. section 1) were used by Casorati in his proof of the Casorati–Weierstrass theorem in 1868 [Casorati 1868, 434–435]. As mentioned earlier, Casorati was deeply influenced by the mathematics of Riemann, in fact, he "was one of the most convincing advocates of Riemann's ideas in Italy" ([Bottazzini 1986, 221], cf. also [Bottazzini 1977, 33–35]). For example, pages 118–136 of his book [Casorati 1868] give an account of Riemann's doctoral thesis, quoting the Riemann Removable Singularity Theorem on page 130. Furthermore, he had been advised by Emil Friedrich Prym (1841–1915) to study the book of Durège (cf. [Neuenschwander 1978b, 40]) and he had obviously followed this advice, as pages 141–142 of [Casorati 1868] summarize the contents of [Durège 1864]. So there is strong evidence that the Removable Singularity Theorem had influenced the proof of the Casorati–Weierstrass theorem given by Casorati. Surprisingly, however, he does not mention the Removable Singularity Theorem or Riemann's name in connection with his proof nor when he uses the implication from ii) to i) in his discussion of the poles [Casorati 1868, 421].

Back to Durège. So far we have analysed only the first edition of his book. But the second edition [Durège 1873] from 1873 is also worth an intense look. In fact, article 26 remains the same [ibid., 106–109], but in article 27a [ibid., 112–113] Durège gives a defective proof of the Removable Singularity Theorem. Again he considers a function φ which has a singular point a and fulfills condition iv). From this he concludes that $(z - a)\varphi(z)$ is continuous at a and infers from this with a reference to formula (14) in article 25 that $(z - a)\varphi(z)$ can be expanded into a power series,

$$(z - a)\varphi(z) = c_0 + c_1(z - a) + c_2(z - a)^2 + c_3(z - a)^3 + \cdots,$$

even claiming this holds if only the finiteness of $\lim_{z \to a}(z - a)\varphi(z)$ is assumed. This guaranteed, one can, of course, get the claim since, by condition iv), the left-hand side tends to zero as z tends to a.

To make things clear, in article 25 [ibid., 102–103] Durège has proven the power series expansion only for functions that are "functions"—which implies "differentiable" [ibid., 26–27]—throughout the whole domain under consideration and not the implication from ii) to i) of the Removable Singularity Theorem, which he would have to use to make the above argument correct. He even mentions an example $\sin(1/z)$ [ibid., 99] that represents a counterexample to his argument if one restricts it to the real numbers.

At least in the fifth edition [Durège and Maurer 1906] of Durège's book, which was edited by Ludwig Maurer (1859–1927) in 1906, one finds a correct proof of the implication from iii) to i) by Laurent's theorem and the Cauchy estimates [ibid., 131–132].

Of course, Durège was not the only mathematician to publish a book inspired by Riemann's ideas. There is also the book by Carl Neumann (1832–1925) "Vorlesungen über Riemann's Theorie der Abel'schen Integrale" (= "Lectures on Riemann's theory of Abelian integrals") [Neumann 1865] from 1865. Like Durège, Neumann had not attended Riemann's original lectures but referred to Riemann's published papers [ibid., VI], especially his thesis, which he mentioned in the first sentence of the preface [ibid., III].

But, what does he write on isolated singularities of functions [ibid., 94–96]? He considers two categories of points of discontinuity, the first one where the function makes a finite jump and the second one where it jumps to infinity. (He calls a point c a pole of f [ibid., 94] if f is not continuous at c, whereas $1/f$ is continuous at c, and concludes that a pole must belong to the second category. This, in fact, seems to be the place where the notation "pole" has been introduced to the literature.) Of course, everything Neumann says is correct, but yet it is surprising that somebody who explicitly refers to Riemann does not even allude to the "Lehrsatz" in Riemann's thesis which implies that the first category is void. Also in the second edition [Neumann 1884] of his book, Neumann again mentions the finite jumps, however his discussion is shorter [ibid., 38].

From this one gets the impression that the disciples of Riemann presented a greater danger to him than his supposed adversary Weierstrass, at least as far as the Removable Singularity Theorem is concerned. Even more, a correct proof of the implication iv) to i) of this theorem, which strongly resembles Riemann's argument in his 1861 lecture, can be found on pages 114–116 of a book [Koenigsberger 1874] by Leo Koenigsberger (1837–1921), a disciple of Weierstrass (!), where may be found a discussion of "die von Riemann gegebenen Grundlagen der Functionentheorie" (= "the foundations of function theory as given by Riemann") [ibid., III].

However, there is also a book by Axel Harnack (1851–1888), entitled "Die Elemente der Differential- und Integralrechnung" (= "The elements of differential and integral calculus") [Harnack 1881] that does not refer to Riemann, but reproduces the defective proof from the second edition of Durège's book [ibid., 369]. It is not the intention of the author to give a complete list of fallacious proofs of the Removable Singularity Theorem, as this list would be a long one and only cause grief. (Some further "proofs" are discussed in an article [Osgood 1896] by William Fogg Osgood (1864–1943); cf. [ibid., 300].) In fact, Harnack's book is mentioned here mainly because it contains a modification of the Removable Singularity Theorem that—contrary to [Osgood 1928, 330] and [Remmert 1991, 213]—is not contained in Riemann's doctoral dissertation. Written in modern language it reads as follows:

Let $D \subset \mathbb{C}$ be open, $C \subset D$ a smooth real 1-dimensional curve and $f : D \to \mathbb{C}$ continuous with $f|_{D-C}$ holomorphic. Then $f : D \to \mathbb{C}$ is holomorphic. (I.e., a continuous extension across a real 1-dimensional curve is automatically holomorphic).

Contrary to his proof of the original Removable Singularity Theorem, Harnack is very careful with respect to this theorem. He requests that f is *a priori* not only continuous but real differentiable, which means that only the Cauchy–Riemann differential equations might fail to hold. At least, his proof for this restricted case is correct and even works for the general case [Harnack 1881, 348–349 and 369].

Perhaps Harnack was interested in the version of the Removable Singularity Theorem just given because he dealt with similar problems of exceptional sets in connection with his studies on Fourier series (e.g., [Harnack 1880] and [Harnack 1882], published one year before, respectively, one year after, the textbook). Though it is not impossible that this version had implicitly been used before, he seems to be the first to explicitly state it. For example, his book—and not a work of Riemann—is the quote that Osgood gave for the above result in his "Encyklopädie" article [Osgood 1901, 18, footnote 26]. Some years later, however, when Osgood himself worked on this problem and generalized this version of the Removable Singularity Theorem to several variables (cf. Section 3) he erroneously ascribed the result to Riemann ([Osgood 1914, 164]; also [Osgood 1928, 330]).

The above version of the Removable Singularity Theorem became a simple corollary when in 1886 Giacinto Morera (1856–1909) proved his theorem that a locally integrable continuous function is analytic [Morera 1886, 305].

2.4. *Modern Proofs*

But the situation got better for the original version as well.

In 1882 Hölder gave a proof of the Casorati–Weierstrass theorem [Hölder 1882] and, as a "Hilfssatz" (= "auxiliary theorem"), proved the implication from iii) to i) of the Removable Singularity Theorem [ibid., 138–140]. That is, he integrates the function in question twice, finds that he gets a complex differentiable, hence analytic function, looks at its power series expansion around the possible singularity and finds that its first two terms vanish. His analysis is very careful, for example, he notes that it would not suffice to integrate only once since, *a priori*, the function one gets then is just continuous and not necessarily differentiable at the singularity [ibid., 139].

Four years later, in 1886, Morera proved his aforementioned theorem that a locally integrable function is analytic [Morera 1886, 305]. This directly gives the implication from ii) to i) of the Removable Singularity Theorem, but, using auxiliary functions, can also be used to prove the implications from iii) and iv), respectively.

Another ten years later, in 1896, Osgood gave a critical historical account of several proofs of the Removable Singularity Theorem [Osgood 1896] and proved the theorem in a way similar to the method of Hölder. Osgood applied Morera's theorem to the once integrated function. In fact, he had turned Morera's theorem into the definition of an analytic function [ibid., 297]. However, he did not mention Morera and, according to a footnote, had just learned of Hölder's proof and found his own proof independently [ibid., 302].

Still another ten years later, in 1906, Edmund Landau (1877–1938), David Raymond Curtiss (1878–1953), and Maxime Bôcher (1867–1918) each published a paper ([Landau 1906], [Curtiss 1906], and [Bôcher 1906]) that contained other proofs using auxiliary functions $[(z - c)f(z)$ or $(z - c)^2 f(z)]$ instead of a double integration. Bôcher even quoted Morera's theorem [ibid., 164].

So, only sixty-five years after its first appearance and correct proof the Removable Singularity Theorem had become fully established in the theory of functions of one complex variable.

3. SEVERAL VARIABLES

In the case of several complex variables, again Weierstrass' name is the first to appear. In his 1879 paper on the theory of analytic functions of several variables [Weierstrass 1879] he considered functions that can be expressed as the quotient of two power series in n variables. He showed [ibid., 146] that if there is a bound on the value of this quotient which holds whenever the denominator is not zero, the quotient has an analytic extension to the set where the denominator vanishes.

One should note that one is no longer dealing with singular points—or at most singular lines of real dimension 1—but with singular sets of considerable dimension. The zero set of the denominator in the result of Weierstrass has complex dimension $n - 1$ or, what is a more convenient formulation in our situation, complex codimension 1 in the space $\mathbb{C}^n$ of n complex variables. With this notion, the implication from iii) to i) of the Removable Singularity Theorem generalizes to several variables as follows:

Any analytic function has an analytic extension across an analytic set (of possible singularities) if the set has at least complex codimension 1 (i.e., real codimension 2) and the function remains (locally) bounded around it.

In [Osgood 1914, 164] and [Osgood 1929, 188, footnote 1] Osgood gave credit to the nowadays forgotten mathematician Hugo Kistler (1880–19??) for having explicitly stated and proven the above Removable Singularity Theorem for the first time in his Göttingen doctoral dissertation [Kistler

1905, 34] in 1905. On the other hand, Osgood remarked "This theorem...was probably used before Kistler's enunciation and proof of it." [Osgood 1914, 164, footnote *]. In fact, one easily has the idea to get this theorem via reduction to the case of one variable, i.e., one considers any variable, fixes the values of the other variables, and studies the extension of the resulting function of one variable. This kind of proof had already become possible seven years before Kistler when in 1898 Osgood himself showed [Osgood 1899] that a (locally) bounded function of n variables is analytic if (and only if) all these n functions of one variable are analytic.

There is also a generalization of the pseudo-Riemann version of the Removable Singularity Theorem which has been discussed in connection with Harnack's textbook. Osgood noted in 1913 that continuous extension implies analytic extension if the exceptional analytic set has at least real codimension 1 [Osgood 1914, 164]. Since, as seen above, Osgood was very careful in giving credit, but did not name an author in this case, one may conclude that this result is due to him. In fact, it was attributed to him by Ludwig Bieberbach (1886–1982) in his "Encyklopädie" report [Bieberbach 1921, 526].

Besides these more or less straightforward translations to the case of several variables there are also new phenomena which do not have a single variable counterpart. For example, the old single variable situation of an isolated singularity does not exist any more. As Adolf Hurwitz (1859–1919) was the first to explicitly note in a lecture at the first International Congress of Mathematicians in 1897 [Hurwitz 1898, 104], functions of more than one complex variable cannot have isolated singularities, to be precise, each isolated singularity is a removable one, without any condition —boundedness or whatever—on the function. Hurwitz only gave an idea of the proof by referring to the generalized theorem of Laurent.

This remark of Hurwitz admits a strong generalization, which is sometimes called the Second Removable Singularity Theorem:

Any analytic function has an analytic extension across an analytic set if the set has at least complex codimension 2 (*i.e., real codimension* 4).

Although there are some ideas of this theorem in the aforementioned article by Weierstrass [1879, 158], again, according to Osgood ([1913, 164] and [1929, 188, footnote 1]), Kistler's statement and proof of the theorem in [Kistler 1905, 31–33] was the first one in the literature.

One should not be too surprised that nowadays Kistler is forgotten since already in the first decade of this century his thesis had only been listed without being reviewed in "Jahrbuch über die Fortschritte der Mathematik" (see Volume 37 (1906), 447).

One reason for this might have been that Kistler was interested in the Second Removable Singularity Theorem for analytic functions only as an auxiliary result for the analogous statement for meromorphic functions, i.e., functions with poles [Kistler 1905, 3–4 and 30–31]. And in the reduction step from the meromorphic to the analytic case he used a result of Pierre Cousin (1867–1933) on the expression of meromorphic functions as quotients of analytic ones that Cousin had stated only for connected domains of the form $D_1 \times \cdots \times D_n \subset \mathbb{C}^n$ with D_ν a connected domain in $\mathbb{C}$ [Cousin 1895, 60, Théorème XIV], whereas Kistler tacitly assumed it to be true for any connected domain in $\mathbb{C}^n$ [Kistler 1905, 30–31]. This fault in Kistler's argument had been discussed by Fritz Hartogs (1874–1943) in a lecture at the annual meeting of the Deutsche Mathematiker-Vereinigung in 1906 [Hartogs 1907, 237–239], but the situation was even worse. As Thomas H. Gronwall (1877–1932) remarked in 1913 [Gronwall 1917, 53], Cousin's result holds only under even more restrictive conditions, namely, at least $n - 1$ of the D_ν have to be simply connected.

On the other hand, in 1903, two years before Kistler's thesis, Hartogs proved his "Kontinuitätssatz" for two complex variables in his doctoral thesis [Hartogs 1904, 66], which can canonically be generalized to several variables [Hartogs 1906, 239]:

Let $(a_1, a_2, \ldots, a_n)$ be fixed and $f(w, z_2, \ldots, z_n)$ be analytic for all points $(w, a_2, \ldots, a_n)$ with $|w - a_1| = r$, however singular in $(a_1, a_2, \ldots, a_n)$. Then there is a $\delta > 0$ such that for any $(n - 1)$-tuple $b_2, \ldots, b_n$ with $|b_\nu - a_\nu| < \delta$ for $\nu = 2, \ldots, n$ there is a w with $|w - a_1| < r$ such that $(w, b_2, \ldots, b_n)$ is a singularity of f.

And from this theorem one can easily infer the Second Removable Singularity Theorem not only for complex codimension 2 (i.e., real codimension 4), but even for real codimension 3. Hartogs did not state this very fact explicitly in his papers [Hartogs 1904], [1906], and [1907]. But he definitely considered his theorem as a generalization of the nonexistence of isolated singularities in several complex variables (cf. [Hartogs 1904, 63], [Hartogs 1907, 236]).

So, compared to the story of the single variable case, the translation of the Riemann Removable Singularity Theorem from one to several variables took place without any greater difficulties. It laid the foundations for a variety of results on the extension of analytic objects, which are central in the modern research on complex analysis, e.g., the results of Thullen [1935] and Remmert and Stein [1953] on the extension of analytic sets. A complete account of the descendants of the Removable Singularity Theorem is beyond the scope of the present article, and for a report on results through 1968 the reader is referred to [Behnke and Thullen 1970, 73–89, esp. 73–75 and 88–89]. However, one should be aware that in the case of several variables too, the Removable Singularity Theorem and related results can lead to confusion, as the concluding example illustrates.

In the case of one variable each connected domain is a domain of holomorphy of an analytic function. Weierstrass made a cryptic remark on this in 1880 [Weierstrass 1880b, 223], Gustav Mittag-Leffler (1846–1927) discussed and showed this fact implicitly in 1884 [Mittag-Leffler 1884, esp. 66], while Carl Runge (1856–1927) gave an explicit statement and proof in 1885 [Runge 1885, 229 and 239–244]. Runge even remarked that his proof yields that each connected domain in $\mathbb{C}$ is a domain of meromorphy [ibid., 229].

But, as the result of Hurwitz already showed, for $n > 1$ there are connected domains in $\mathbb{C}^n$ that are not domains of holomorphy, for example, the whole space minus a point. Even more restrictive conditions on domains of holomorphy follow from results that Hartogs proved, e.g., his "Kontinuitätssatz" and, especially, his famous "Kugelsatz" ([Hartogs 1904, 67–68], [Hartogs 1906, 231, 239]): If D is a bounded connected domain in $\mathbb{C}^n$, $n \geq 2$, with connected boundary ∂D, then each function that is holomorphic on a neighborhood of ∂D has a holomorphic continuation on the whole of D.

Contrary to domains of holomorphy, Hartogs was very uncertain as to domains of meromorphy even in 1907. It was mentioned before that he correctly criticized Kistler's proof of the generalization of the Second Removable Singularity Theorem to meromorphic functions [Hartogs 1907, 237–239], but he definitely abstained from conjecturing whether at least the statement of that generalization was true or whether Weierstrass was right in claiming in [Weierstrass 1880a, 129] that each (connected) domain in $\mathbb{C}^n$ is a domain of meromorphy [Hartogs 1907, 239, footnote 1].

It was left to Eugenio Elia Levi (1883–1917) to translate both the "Kontinuitätssatz" [Levi 1910, 66] and the "Kugelsatz" [ibid., 71] of Hartogs to meromorphic functions in 1909 and, by this result, to demonstrate the generalization of the Second Removable Singularity Theorem Kistler had failed to prove. (For a careful treatment of the involved mathematical problems see [Kneser 1932a] and [Kneser 1932b] by Hellmuth Kneser (1898–1973).) Especially, as Levi pointed out [Levi 1910, 61–62 and 71], the claim of Weierstrass on domains of meromorphy turned out to be erroneous.

In fact, [Levi 1910] gave rise to the study of the so-called "Levi problem" to characterize domains of holomorphy (respectively, meromorphy) in terms of their boundary. Namely, one calls a (relatively compact, connected) domain D in $\mathbb{C}^n$ "strictly pseudoconvex in the sense of Levi" if for each point x_0 of the boundary of D there is a neighborhood U of x_0 and a strictly plurisubharmonic function $p : U \to \mathbb{R}$ with $D \cap U = \{x \in U : p(x) < p(x_0)\}$, where "strictly plurisubharmonic" means that the Levi-form

$$\sum_{\mu, \nu = 1}^{n} \frac{\partial^2 p}{\partial z_\mu \, \partial \overline{z_\nu}} \zeta_\mu \overline{\zeta_\nu}$$

is positive-definite. Levi observed [ibid., 80–81] that each domain of meromorphy (respectively, of holomorphy) is strictly pseudoconvex in the sense of Levi and asked whether the converse is also true [ibid., 81]. It first seemed that the answer would be in the negative since in 1912 Otto Blumenthal (1876–1944) published in [Blumenthal 1912] what he thought to be a counterexample. However, 14 years later Heinrich Behnke (1898–1979) showed that Blumenthal's argument was defective [Behnke 1926, esp. 347–348 and 363–365] so that the Levi problem was still open. (In fact, Blumenthal had realized his mistake long before [ibid., 365, footnote 20].)

Levi himself had already been able to show that each point of the boundary of a domain D that is strictly pseudoconvex in the sense of Levi has a neighborhood U such that $D \cap U$ is a domain of meromorphy (respectively, of holomorphy) ([Levi 1910, 83–86] and [Levi 1911]). This property of being "locally" a domain of holomorphy was studied by Henri Cartan and is referred to as the property of being "pseudoconvex in the sense of H. Cartan" ([Cartan 1931, esp. 48] respectively [Cartan 1979 I, 303–326, esp. 305]). In 1942 Kiyoshi Oka (1901–1978) succeeded in solving the Levi problem for the case of two variables by showing that each domain that is pseudoconvex in the sense of H. Cartan is globally a domain of holomorphy ([Oka 1942, 37] respectively [Oka 1984, 65]). (For the contents of [Oka 1942] cf. also commentaries by Cartan in [Oka 1984, 77–79]. Note, for example, that Oka's general notion of pseudoconvexity is the one "in the sense of Hartogs(-Kneser)," which turns the property which each domain of holomorphy must have according to the "Kontinuitätssatz" into a definition.)

As to the general case of several complex variables, the Levi problem was solved independently and nearly simultaneously by Oka ([1953, 146] respectively, [1984, 185]), Hans J. Bremermann [1954, 68] and François Norguet [1954, 137]. A consequence of this result is, for example, that a connected domain in $\mathbb{C}^n$ is a domain of holomorphy if and only if it is a domain of meromorphy (cf. [Behnke and Thullen 1970, 81–87, esp. 85–86]).

REFERENCES

d'Alembert, J. 1746. Recherches sur le Calcul intégral, first part. *Histoire de l'Académie Royale des Sciences et Belles Lettres, Année MDCCXLVI* (Berlin 1748), 182–224.

Behnke, H. 1926. Über analytische Funktionen mehrerer Veränderlicher. I. Teil. Die Kanten singulärer Mannigfaltigkeiten. *Abhandlungen aus dem Mathematischen Seminar der Hamburgischen Universität* **4**, 347–365.

—— and Thullen, P. 1970. *Theorie der Funktionen mehrerer komplexer Veränderlichen.* Ergebnisse der Mathematik und ihrer Grenzgebiete, 2. ser., vol. 51. Berlin: Springer, second edition.

Betti, E. 1860/61. La teorica delle funzioni ellittiche. *Annali di matematica pura ed applicata*, first series **3** (1860), 65–159 and 298–310, and **4** (1861), 26–45, 57–70 and 297–336; here in: Betti, E. *Opere matematiche*, vol. 1, Milano: Hoepli 1903, 228–412.

Bieberbach, L. 1921. Neuere Untersuchungen über Funktionen von komplexen Variablen, in: *Encyklopädie der mathematischen Wissenschaften mit Einschluß ihrer Anwendungen*, second vol.: *Analysis*, third part, first moiety. Leipzig, Teubner 1909–1921, 379–532.

Blumenthal, O. 1912. Bemerkungen über die Singularitäten analytischer Funktionen mehrerer Veränderlichen, in: *Festschrift Heinrich Weber zu seinem siebzigsten Geburtstag am 5. März 1912 gewidmet von Freunden und Schülern*, 11–22. Leipzig: Teubner.

Bôcher, M. 1906. Another proof of the theorem concerning artificial singularities. *Annals of Mathematics* 2. ser. **7**, 163–164.

Bottazzini, U. 1977. Riemanns Einfluß auf E. Betti und F. Casorati. *Archive for History of Exact Sciences* **18**, 27–37.

—— 1986. *The Higher Calculus: A history of real and complex analysis from Euler to Weierstrass.* New York: Springer.

Bremermann, H. J. 1954. Über die Äquivalenz der pseudokonvexen Gebiete und der Holomorphiegebiete im Raum von n komplexen Veränderlichen. *Mathematische Annalen* **128**, 63–91.

Cartan, H. 1931. Sur les domaines d'existence des fonctions de plusieurs variables complexes. *Bulletin de la Société mathématique de France* **59**, 46–69, also in: [Cartan 1979 I, 303–326].

—— 1979. *Œuvres. Collected Works*. 3 volumes. R. Remmert and J-P. Serre, eds. Berlin, Heidelberg: Springer.

Casorati, F. 1868. *Teorica delle funzioni di variabili complesse*. Pavia: Fratelli Fusi.

Cousin, P. 1895. Sur les fonctions de *n* variables complexes. *Acta Mathematica* **19**, 1–61.

Curtiss, D. R. 1906. A proof of the theorem concerning artificial singularities. *Annals of Mathematics* 2. ser. **7**, 161–162.

Durège, H. 1864. *Elemente der Theorie der Functionen einer complexen veränderlichen Grösse. Mit besonderer Berücksichtigung der Schöpfungen Riemanns*. Leipzig: Teubner.

—— 1873. *Elemente der Theorie der Functionen einer complexen veränderlichen Grösse. Mit besonderer Berücksichtigung der Schöpfungen Riemanns*. Leipzig: Teubner, second edition of [Durège 1864].

—— and Maurer, L. 1906. *Elemente der Theorie der Funktionen einer komplexen veränderlichen Größe*. Leipzig: Teubner, fifth edition of [Durège 1864].

Gauss, C. F. 1799. *Demonstratio nova theorematis omnem functionem algebraicam rationalem integram unius variabilis in factores reales primi vel secundi gradus resolvi posse*. Dissertation Helmstedt, in: [Gauss 1863–1933 3, 1–31].

—— 1825. *Allgemeine Auflösung der Aufgabe: die Theile einer gegebenen Fläche auf einer andern gegebenen Fläche so abzubilden, dass die Abbildung dem Abgebildeten in den kleinsten Theilen ähnlich wird (Als Beantwortung der von der Königlichen Societät der Wissenschaften in Copenhagen für MDCCCXXII aufgegebenen Preisfrage.)* printed in: Astronomische Abhandlungen, H. C. Schumacher, ed. third number, Altona 1825; also in: [Gauss 1863–1933 4, 189–216].

—— 1828. Disquisitiones generales circa superficies curvas, *Commentationes societatis regiae scientiarum Gottingensis recentiores* **VI** (ad a. 1823–1827), Commentationes classis mathematicae, 99–146; also in: [Gauss 1863–1933 4, 217–258].

—— 1890. *Die vier Gauss'schen Beweise für die Zerlegung ganzer algebraischer Funktionen in reelle Factoren ersten und zweiten Grades (1799–1849)*. E. Netto, ed. Ostwald's Klassiker der exakten Wissenschaften 14. Leipzig: Engelmann.

—— 1863–1933. *Werke*. 12 volumes. Göttingen: (Königliche) Gesellschaft der Wissenschaften.

Gronwall, T. H. 1917. On the expressibility of a uniform function of several complex variables as the quotient of two functions of entire character. *Transactions of the American Mathematical Society* **18**, 50–64.

Harnack, A. 1880. Ueber die trigonometrische Reihe und die Darstellung willkürlicher Functionen. *Mathematische Annalen* **17**, 123–132.

—— 1881. *Die Elemente der Differential- und Integralrechnung*. Leipzig: Teubner.

—— 1882. Vereinfachung der Beweise in der Theorie der Fourier'schen Reihe. *Mathematische Annalen* **19**, 235–279, 524–528.

Hartogs, F. 1904. *Beiträge zur elementaren Theorie der Potenzreihen und der eindeutigen analytischen Funktionen zweier Veränderlicher.* Dissertation München. Leipzig: Teubner.

—— 1906. Einige Folgerungen aus der Cauchyschen Integralformel bei Funktionen mehrerer Veränderlichen, *Sitzungsberichte der mathematisch-physikalischen Klasse der Königlich Bayerischen Akademie der Wissenschaften zu München* **36**, 223–242.

—— 1907. Über neuere Untersuchungen auf dem Gebiete der analytischen Funktionen mehrerer Variablen. *Jahresbericht der Deutschen Mathematiker-Vereinigung* **16**, 223–240.

Hölder, O. 1882. Beweis des Satzes, dass eine eindeutige analytische Function in unendlicher Nähe einer wesentlich singulären Stelle jedem Werth beliebig nahe kommt, *Mathematische Annalen* **20**, 138–143.

Hurwitz, A. 1898. Über die Entwickelung der allgemeinen Theorie der analytischen Funktionen in neuerer Zeit, in: *Verhandlungen des ersten internationalen Mathematiker-Kongresses in Zürich vom 9. bis 11. August 1897.* F. Rudio, ed. Leipzig: Teubner, 91–112.

Kistler, H. 1905. *Über Funktionen von mehreren komplexen Veränderlichen.* Dissertation Göttingen. Basel: Basler Berichthaus.

Kneser, H. 1932a. Der Satz von dem Fortbestehen der wesentlichen Singularitäten einer analytischen Funktion zweier Veränderlichen. *Jahresbericht der Deutschen Mathematiker-Vereinigung* **41**, 164–168.

—— 1932b. Ein Satz über die Meromorphiebereiche analytischer Funktionen von mehreren Veränderlichen. *Mathematische Annalen* **106**, 648–655.

Koenigsberger, L. 1874. *Vorlesungen über die Theorie der elliptischen Functionen nebst einer Einleitung in die allgemeine Functionenlehre.* part 1. Leipzig: Teubner.

Landau, E. 1906. On a familiar theorem of the theory of functions. *Bulletin of the American Mathematical Society* **12**, 155–156.

Levi, E. E. 1910. Studii sui punti singolari essenziali delle funzioni analitiche di due o più variabili complesse. *Annali di Matematica pura ed applicata*, 3. ser. **17**, 61–87.

—— 1911. Sulle ipersuperficie dello spazio a 4 dimensioni che possono essere frontiera del campo di esistenza di una funzione analitica di due variabili complesse. *Annali di Matematica pura ed applicata*, 3. ser. **18**, 69–79.

Manning, K. R. 1975. The Emergence of the Weierstrassian Approach to Complex Analysis. *Archive for History of Exact Sciences* **14**, 297–383.

Mittag-Leffler, G. 1884. Sur la représentation analytique des fonctions monogènes uniformes d'une variable indépendante. *Acta Mathematica* **4**, 1–79.

—— 1923. Die ersten 40 Jahre des Lebens von Weierstraß. *Acta Mathematica* **39**, 1–57.

Morera, G. 1886. Un teorema fondamentale nella teorica delle funzioni di una variabile complessa. *Rendiconti Reale Istituto Lombardo di scienze e lettere*, 2. ser. **19**, 304–307.

Neuenschwander, E. 1978a. The Casorati–Weierstrass Theorem (Studies in the History of Complex Function Theory I). *Historia Mathematica* **5**, 139–166.

—— 1978b. Der Nachlaß von Casorati (1835–1890) in Pavia. *Archive for History of Exact Sciences* **19**, 1–89.

—— 1981. Über die Wechselwirkungen zwischen der französischen Schule, Riemann und Weierstraß. Eine Übersicht mit zwei Quellenstudien. *Archive for History of Exact Sciences* **24**, 221–255.

—— 1987. *Riemanns Vorlesungen zur Funktionentheorie. Allgemeiner Teil.* Preprint Nr. 1086, September 1987, Technische Hochschule Darmstadt, Fachbereich Mathematik.

Neumann, C. 1865. *Vorlesungen über Riemann's Theorie der Abel'schen Integrale.* Leipzig: Teubner.

—— 1884. *Vorlesungen über Riemann's Theorie der Abel'schen Integrale.* Leipzig: Teubner, second edition of [Neumann 1865].

Norguet, F. 1954. Sur les domaines d'holomorphie des fonctions uniformes de plusieurs variables complexes. (Passage du local au global). *Bulletin de la Société Mathématique de France* **82**, 137–159.

Oka, K. 1942. Domaines pseudoconvexes. *Tôhoku Mathematical Journal* **49**, 15–52, English translation in: [Oka 1984, 48–77].

—— 1953. Domaines finis sans point critique intérieur. *Japanese Journal of Mathematics* **27**, 97–155, English translation in: [Oka 1984, 135–194].

—— 1984. *Collected papers. Translated from the French by R. Narasimhan. With Commentaries by H. Cartan. Edited by R. Remmert.* Berlin, Heidelberg: Springer.

Osgood, W. F. 1896. Some points in the elements of the theory of functions. *Bulletin of the American Mathematical Society* **2**, 296–302.

—— 1899. Note über analytische Functionen mehrerer Veränderlichen. *Mathematische Annalen* **52**, 462–464.

—— 1901. Allgemeine Theorie der analytischen Funktionen a) einer und b) mehrerer komplexen Grössen, in: *Encyklopädie der mathematischen Wissenschaften mit Einschluß ihrer Anwendungen*, second vol.: *Analysis*, second part. Leipzig: Teubner 1901–1921, 1–114.

—— 1914. Topics in the theory of functions of several complex variables, in: *The Madison Colloquium 1913.* American Mathematical Society, Colloquium Lectures, vol. IV. New York: American Mathematical Society, 111–230.

—— 1928. *Lehrbuch der Funktionentheorie*, vol. 1. Leipzig and Berlin: Teubner, fifth edition.

—— 1929. *Lehrbuch der Funktionentheorie*, vol. 2, part 1. Leipzig and Berlin: Teubner, second edition.

Puiseux, V. 1850. Recherches sur les fonctions algébriques. *Journal de mathématiques pures et appliquées*, 1. ser. **15**, 365–480.

Remmert, R. 1988. Fundamentalsatz der Algebra, in: H.-D. Ebbinghaus et al. *Zahlen*, 79–99. Grundwissen Mathematik 1. Berlin, Heidelberg: Springer, second edition.

—— 1991. *Theory of Complex Functions*. Readings in Mathematics. Graduate Texts in Mathematics 122. New York: Springer.

—— and Stein, K. 1953. Über die wesentlichen Singularitäten analytischer Mengen, *Mathematische Annalen* **126**, 263–306.

Riemann, G. F. B. 1851. *Grundlagen für eine allgemeine Theorie der Functionen einer veränderlichen complexen Grösse*. Dissertation Göttingen. Here in: [Riemann 1876, 3–45].

—— 1876. *Bernhard Riemann's gesammelte mathematische Werke und wissenschaftlicher Nachlass*. H. Weber and R. Dedekind, eds. Leipzig: Teubner.

Runge, C. 1885. Zur Theorie der eindeutigen analytischen Functionen, *Acta Mathematica* **6**, 229–244.

Schlesinger, L. 1912. Über Gauss' Arbeiten zur Funktionentheorie, *Nachrichten der Königlichen Gesellschaft der Wissenschaften zu Göttingen. Mathematisch-physikalische Klasse*. Here in: [Gauss 1863–1933 10.2, Abh. 2].

Thullen, P. 1935. Über die wesentlichen Singularitäten analytischer Funktionen und Flächen im Raume von *n* komplexen Veränderlichen. *Mathematische Annalen* **111**, 137–157.

Ullrich, P. 1989. Weierstraß' Vorlesung zur "Einleitung in die Theorie der analytischen Funktionen." *Archive for History of Exact Sciences* **40**, 143–172.

—— 1994. The proof of the Laurent expansion by Weierstraß, in: *The History of Modern Mathematics 3*, 139–153. E. Knobloch and D. E. Rowe, eds. (Special volume of *Historia Mathematica*). Boston, San Diego: Academic Press.

Weierstrass, K. 1841a. *Darstellung einer analytischen Function einer complexen Veränderlichen, deren absoluter Betrag zwischen zwei gegebenen Grenzen liegt*. Münster. In: [Weierstraß 1894–1927 1, 51–66].

—— 1841b. *Zur Theorie der Potenzreihen*. Münster. In: [Weierstraß 1894–1927 1, 67–74].

—— 1876. Zur Theorie der eindeutigen analytischen Functionen, *Mathematische Abhandlungen der Akademie der Wissenschaften zu Berlin* 1876, 11–60. Here in: [Weierstrass 1894–1927 2, 77–124].

—— 1879. Einige auf die Theorie der analytischen Functionen mehrerer Veränderlichen sich beziehenden Sätze. Berlin. In: [Weierstrass 1894–1927 2, 135–188].

—— 1880a. Untersuchungen über die 2*r*-fach periodischen Functionen von *r* Veränderlichen, *Journal für die reine und angewandte Mathematik* (*Crelle's Journal*) **89**, 1–8. Here in: [Weierstrass 1894–1927 2, 125–133].

—— 1880b. Zur Functionenlehre, *Monatsberichte der Akademie der Wissenschaften zu Berlin* 1880, 719–743. Here in: [Weierstrass 1894–1927 2, 201–223].

—— 1894–1927. *Mathematische Werke*, 7 volumes. Berlin: Mayer & Müller; Reprint Hildesheim: Georg Olms and New York: Johnson Reprint.

—— 1986. *Einführung in die Theorie der analytischen Funktionen, nach einer Vorlesungsmitschrift von Wilhelm Killing aus dem Jahr 1868.* Schriftenreihe des Mathematischen Instituts der Universität Münster, 2. ser., vol. 38.

—— 1988. *Einleitung in die Theorie der analytischen Funktionen, Vorlesung Berlin 1878, in einer Mitschrift von Adolf Hurwitz, bearbeitet von Peter Ullrich.* Dokumente zur Geschichte der Mathematik, vol. 4. Braunschweig/Wiesbaden: Vieweg.

Weil, A. 1979. Riemann, Betti and the Birth of Topology, *Archive for History of Exact Sciences* **20**, 91–96.

Eliakim Hastings Moore (1862–1932)
Courtesy of R. C. Archibald,
A Semicentennial History of the American Mathematical Society, 1888–1938

Maxime Bôcher (1867–1918)
Courtesy of R. C. Archibald,
A Semicentennial History of the American Mathematical Society, 1888–1938

A Profile of the American Mathematical Research Community: 1891–1906

Della Dumbaugh Fenster* and Karen Hunger Parshall*[†]

*Department of Mathematics, [†]Corcaron Department of History, University of Virginia,
Charlottesville

When the British mathematician James Joseph Sylvester accepted an appointment to the first professorship of mathematics at the newly founded Johns Hopkins University in 1876, he adopted a country in which mathematics at a research level had essentially gone uncultivated. Although American institutions of higher education had included mathematics in their curricula from the beginning, the level of sophistication of their programs only rarely exceeded that of the calculus throughout the nineteenth century. Although efforts to sustain specialized mathematical journals had repeatedly been mounted, they had inevitably failed due both to a scarcity of interested subscribers and to a dearth of qualified contributors. Finally, although mathematicians like Robert Adrain, Nathaniel Bowditch, and Benjamin Peirce had achieved significant results prior to 1850, they had done so in relative isolation and largely without the stimulus of any broader scientific or mathematical support group [1].

As the century progressed, however, an American scientific community became more and more clearly focused around publications like Benjamin Silliman's *American journal of Science and Arts* (begun in 1818) as well as around alliances like the American Association for the Advancement of Science (started in 1847) and the National Academy of Sciences (legislated into existence in 1863). Concurrent with this ever-sharpening sense of identity, the scientists themselves settled upon a set of credentials which served to determine their standing professionally. Research, publication, and the domestic or—better yet—foreign doctoral degree increasingly separated the true scientist from the amateur and from the charlatan [2].

The adoption of this latter credential, the doctorate, went hand-in-hand with the post–1850 developments in American higher education [3]. As educators cast about for ideas to improve their institutions, they looked to Europe—and particularly to Germany—and borrowed ideas like *Lehr-* and *Lernfreiheit* which they then molded to fit their peculiarly American needs. Furthermore, by the century's closing decades they had begun to stress not only graduate education in general but also research and the training of future researchers in particular. Thus, whereas research was viewed as a calling in the German context, in its American manifestation it was, to some extent, an activity for which one could be and was trained [4]. The Ph.D. signified the achievement of original research work, generally of a highly specialized nature, and implied the capability for further production. Also, since only those who had actively contributed could possibly train those who aspired to this end, the doctorate became a foremost requirement for entrance into the professoriate.

These ideals of specialized graduate training culminating in the doctoral degree, together with original research, certainly took neither American science nor American higher education by storm. Although they only gradually replaced the more traditional view of the liberally educated gentleman–scientist, a key turning point came in 1876 with the opening of The Johns Hopkins University according to the vision and under the guidance of its first president, Daniel Coit Gilman. Based on his years of observation of the European academic scene, Gilman explicitly committed his university to both research and graduate training. The Johns Hopkins thus became a model for others to follow, and follow they did. New schools like Clark (opened in 1889) and the University of Chicago (opened in 1892) took graduate education and original research as part of their mission from the outset. Older institutions such as Harvard and Yale successfully moved with the changing times. Finally, state-supported and federally funded land-grant universities adeptly reconciled their ostensibly utilitarian concerns with the more rarefied objective of basic research. Relative to science in general and mathematics in particular, this trickle-down effect within higher education, coupled with the overall trend toward professionalization, differentiation, and specialization within the broader scientific context, produced rather dramatic results during the closing quarter of the nineteenth century [5].

The fact that the graduate ideal did only filter gradually through the system initially forced many students to seek their advanced training abroad. During the 10 years after Sylvester's 1883 departure from Hopkins, for example, aspiring American mathematicians turned primarily to Germany for their studies. Even after those students returned to set up shop in universities at home, often with a German doctorate in hand, many of their compatriots continued to opt for the as yet more prestigious foreign degree [6]. Still, by the mid 1890's, they no longer *had* to go abroad to obtain a solid graduate education, and many decided to do their work on native shores.

With their studies completed either at home or abroad, the fresh Ph.D.'s, particularly of the 1890's, found themselves a part of young but growing specialized communities of researchers. Although the chemists had formed their American Chemical Society as early as 1876, the mathematicians and physicists would only follow with their national societies in 1894 and 1899, respectively. These organizations, with their associated journals, served as critical focal points by providing important lines of communication and interaction [7]. Once again, taking mathematics as the case in point, the New York mathematical Society first met in 1888, started up its *Bulletin* in 1891, assumed a national orientation in 1894, and supported its *Transactions* by 1899. The Society's publications thus supplemented America's extant research journals, namely, the *American Journal of Mathematics* (founded by Sylvester at Hopkins in 1878) and the *Annals of Mathematics* (begun by Ormond Stone in 1884 at the University of Virginia). Even though the number of participants in the society and of contributors to the journals initially were small, they increased steadily up to and after 1900.

In their forthcoming book, entitled *The Emergence of the American Mathematical Research Community 1876–1900*, Karen Hunger Parshall and David E. Rowe discuss and analyze in detail the complex and interconnecting forces which we have cursorily sketched above [8]. In the present study, we supplement their arguments and findings with hard quantitative data that document the real growth of an American mathematical research community around the turn of the twentieth century. We isolate who constituted the American mathematical research community; we measure their participation within that community; we analyze where they came from and compare their educational and employment records; we survey their research interests; and, through supplementary anecdotal information, we give some indication of the depth and extent of those interests. Clearly, the kind of statistical study we present here does not and *cannot* speak directly to such issues as the overall *quality* of the mathematical research produced by Americans or the *comparative* strength of the American community internationally. It *does*, however, provide a firm quantitative foundation for studies of a qualitative and/or comparative nature. Furthermore, like the book by Parshall and Rowe, it disproves the prevailing folklore, which holds that a mathematical research community came to existence in the United States only with the influx of talented European refugees in the 1930's.

In compiling our data, we focus on the 15-year period from 1891 to 1906, that is, the first 15 years of publication of the *Bulletin of the New York* (then, after 1894, *American*) *Mathematical Society*. As a source for isolating turn-of-the-century participants and for determining the extent of their involvement, this journal would seem unparalleled. Coming out in ten numbers in each academic year, the *Bulletin* not only carried original research papers, survey articles, and works of an historical nature, but also contained detailed accounts of the Society's monthly meetings (including lists of attendees along with the titles and abstracts of the talks presented), personal notes submitted by Society members and others, announcements of course offerings at domestic and foreign universities, general news of the Society and its business, and reports on the meetings of such foreign organizations as the London Mathematical Society (founded in 1865) and the Deutsche Mathematiker–Vereinigung (begun in 1890). A careful culling of these items from each of the first fifteen volumes of the journal uncovers the names of those participating—in one way or another and more or less—in the American mathematical research community. We further identify these participants using the available biographical entries in James McKeen Cattell's *American Men of Science* and use this personal information to obtain a geographic and educational profile of the American mathematical community [9]. Finally, we combine this with publication data taken from the *American Journal*, the *Annals*, the *Transactions* of the Society, and the *Mathematical Papers* of the Chicago Congress of 1893 [10] to gain an idea of both the special technical interests of the community and the actual level of activity of its members. This process also reveals an unexpectedly deep foreign participation in the mathematical life of the United States. The overall picture which emerges is one of an extensive, highly educated, geographically diverse body of men and women engaged in the mathematical endeavor at a research level and attracting a significant amount of foreign interest and support.

BULLETIN OF THE AMERICAN MATHEMATICAL SOCIETY

When Thomas Fiske, Edward Stabler, and Harold Jacoby of Columbia College solicited interested parties to meet on November 24, 1888 to discuss the feasibility of forming a New York Mathematical Society, only three other people answered their call [11]. Undaunted, by year's end these six enthusiasts had met again, adopted a constitution, and launched their fledgling organization. In an effort to increase their membership, they decided to target not only mathematicians living in and around New York City but also teachers of mathematics, actuaries, engineers, and others who might share their mathematical interests. Their recruitment strategy resulted in their growth to sixteen members by the end of 1889 and to twenty-three by the close of 1890. Falsely impressive in terms of percentage-wise increase, these numbers reflect very limited enthusiasm for the endeavor even within the restricted geographical area of greater metropolitan New York. Still, the early participants shared a strong sense of commitment and resolved to forge ahead and publish a bulletin, provided they could further strengthen their ranks. Using mailing lists gathered from then prominent science publishing houses like Macmillan and Co., John Wiley and Sons, and Ginn and Co., the Society succeeded in boosting its roster to a more respectable eighty-nine and in so doing extended its geographical reach at least as far south as South Carolina and as far west as Michigan. In response to this relatively strong show of support, plans proceeded to "...publish a periodical of a critical and historical nature, called the *Bulletin of the New York Mathematical Society*," and the first number appeared in October 1891 [12].

Coming out monthly throughout the academic year, that is, from October through June, the *Bulletin* sought not to compete with journals like the *Annals* and the *American Journal*, but rather to supplement their more properly research-oriented offerings with "...historical and critical articles, accounts of advances in different branches of mathematical science, reviews of important new publications, and general mathematical news and intelligence" [13]. To this end, it carried such articles as "Kronecker and his Arithmetical Theory of Algebraic Equation" by Princeton's Henry Burchard Fine, reports like that of Alexander Ziwet on "The Annual Meeting of the German Mathematicians," and reviews such as "Edwards' Differential Geometry" by Charlotte Angas Scott of Bryn Mawr [14]. It also served as a current directory of both the membership and colleagues abroad via personal notices. Subscribers during that first year thus learned that "Professor W. H. Echols, Jr., recently Director of the Missouri School of Mines, has been called to a chair at the University of Virginia" [15] and that "Professor H. Weber of Marburg has accepted a call to Göttingen to fill the post vacated by Professor H. A. Schwartz [sic] [while] Professor Frobenius of Zurich has accepted a call to Berlin" [16]. Furthermore, they received news of the Society's monthly New York-based meetings, like this report of the gathering on March 5, 1892:

> The following persons having been duly nominated, and being recommended
> by the council, were elected to membership: Professor Arthur Cayley,
> Cambridge University, England; Professor J. de Mendizábal Tamborrel,
> Military College, Mexico; Professor Truman Henry Safford, Williams Col-
> lege; Professor Edmund A. Engler, Washington University. The secretary
> read letters from Professor Cayley and Professor Sylvester expressing inter-
> est in the work of the Society.—The following original papers were read:
> "On exact analysis as the basis of language" by Professor Alexander Macfar-
> lane; "A geometrical construction for finding the foci of the sections of a
> cone of revolution" by Professor Edmund A. Engler. Mr. Maclay made some
> remarks upon the locus of the centers of curvature of parallel section of a
> ruled surface at points upon the same generatrix [17].

The *Bulletin*'s expansion from a relatively slender first volume of 246
pages to volumes of consistently double that length from 1898 on reflected
not only the concurrent growth of the Society but also the depth and
extent of the journal's coverage of the mathematical scene both at home
and abroad. Its reporting of such events as promotions, academic moves,
and individual trips abroad enabled even the most far-flung of the Society's
membership to communicate and to keep abreast of the community's
development. For the historian seeking to map and track that community,
then, the *Bulletin* serves as a sort of microscope through which even the
more obscure members of the population may be observed.

THE "MOST ACTIVE" PARTICIPANTS

Viewing the American mathematical community first through the lowest
power of this microscope reveals, quite naturally, the most visible, most
active members. Participants who faithfully published journal articles, gave
talks at meetings of the Society or elsewhere, and/or served to promote
the cause of mathematics through their involvement in such activities as
elected office and committee or editorial work, the community's most
active sustained it in its early formative years.

In the present study, we use a dual selection criterion in defining the
"most active" members of American mathematics. First, based on information gathered from the *Bulletin*, the "most active" contributed—through
their publications, talks, or service—at least once a year, on average, over
our 15-year period from 1891 to 1906. Second, these "most active" members have an entry in either the first, second, or third edition of *American*

Men of Science. Thus, while the *Bulletin* allows us to isolate the principals within the American mathematical community, the *American Men of Science* entries—with their information on place of birth, educational and employment history, and areas of research interest—permit us to compare and analyze this group in the most uniform and meaningful way possible. The names of the 60 men and two women, Ida May Schottenfels and Charlotte Angas Scott, who comprise this "most active" contingent appear in Table 1 along with numbers representing their respective talks, published papers, service-related activities, and total participation level. (See the annotations to Table 1 for the precise meanings we assign to the terms "talks," "papers published," and "service.") The companion Table 2 displays a collation of this same information and gives a clear indication of the relative breakdown over the three areas of activity isolated. Not surprisingly, these two tables reflect a broad range of participation levels even among those whom we label "most active."

At the highest end of the spectrum, the University of Chicago's first mathematics Ph.D., and one of America's most prolific mathematicians, Leonard Eugene Dickson averaged almost five talks and five published papers a year over the fifteen years. Taking into account the fact that he only earned his doctoral degree in 1896 and so essentially made all 143 of his contributions in the final ten years covered by our study, his output actually ran to seven talks and papers annually. And this accounts only for his contributions to the American mathematical scene. Dickson also published the results of his research, work primarily on linear groups and the theory of algebras during this era, in such foreign journals as the *Journal für die reine und angewandte Mathematik, Mathematische Annalen*, and *Proceedings of the London Mathematical Society* [18].

While Dickson's record is clearly exceptional, that of Alexander Chessin assumes more human proportions. Born and educated in Russia, Chessin received his doctoral degree in 1889 at the University of St. Petersburg. He took a lectureship in celestial mechanics at Johns Hopkins in 1894 and rose to an associate professorship there before tendering his resignation effective 1899. After spending the intervening years in and around New York City, he proceeded to the mathematical chair at Washington University in St. Louis in 1901. Chessin averaged one talk and one paper a year, mostly in his specialties of function theory and differential equations, and served the broader mathematical community as a member of the organizing committee for the American Mathematical Society's (AMS) Eleventh Summer Meeting held in his new hometown of St. Louis in 1904. With a

Table 1.

The 62 Most Active Members of the American Mathematical
Community 1891–1906

	T	P	S	Total		T	P	S	Total
Allardice, R. E.	8	4	4	16	Lovett, E. O.	31	35	6	72
Blake, E. M.	13	5	0	18	McClintock, E.	6	11	10	27
Blichfeldt, H. F.	14	13	0	27	Macfarlane, A.	11	3	4	18
Bliss, G. A.	14	8	5	27	McMahon, J.	6	6	10	22
Bôcher, M.	27	46	18	91	Maschke, H.	15	13	4	32
Bolza, O.	10	15	5	30	Mason, C. M.	9	7	1	17
Brown, E. W.	14	18	13	45	Miller, G. A.	48	62	10	120
Chessin, A. S.	13	17	1	31	Moore, E. H.	27	18	25	70
Cole, F. N.	2	7	32	41	Morley, F.	26	18	29	73
Curtiss, D. R.	9	6	0	15	Newcomb, S.	1	2	29	32
Davis, E. W.	11	9	0	20	Newson, H. B.	20	9	2	31
Dickson, L. E.	74	67	2	143	Osgood, W. F.	14	22	6	42
Echols, W. H.	6	22	7	35	Pierpont, J. P.	11	23	4	38
Eisenhart, L. P.	17	19	0	36	Porter, M. B.	9	11	3	23
Emch, A.	9	8	0	17	Schottenfels, I. M.	17	3	0	20
Espteen, S.	14	12	0	26	Scott, C. A.	12	19	10	41
Fiske, T. S.	2	6	33	41	Shaw, J. B.	25	8	0	33
Fite, W. B.	9	6	0	15	Smith, D. E.	0	4	14	18
Halsted, G. B.	8	4	4	16	Snyder, V.	24	28	3	55
Haskell, M. W.	13	9	1	23	Stecker, H. F.	13	5	0	18
Hathaway, A. S.	12	8	3	23	Stone, O.	1	2	21	24
Hedrick, E. R.	14	8	6	28	Stringham, W. I.	9	6	6	21
Hill, G. W.	5	7	7	19	Taber, H.	11	6	0	17
Holgate, T. F.	4	6	9	19	Van Vleck, E. B.	14	15	4	33
Huntington, E. V.	21	18	5	44	Veblen, O.	8	7	0	15
Hutchinson, J. I.	14	12	1	27	White, H. S.	22	23	12	57
Jacoby, H.	2	5	9	16	Wilczynski, E. J.	19	18	2	30
Johnson, W. W.	5	9	5	19	Wilson, E. B.	9	19	2	30
Kasner, E.	33	21	6	60	Woodward, R. S.	19	4	15	38
Keyser, C. J.	11	7	1	19	Young, J. W.	7	7	2	16
Lehmer, D. N.	11	9	1	21	Ziwet, A.	0	9	33	42

T = talks, P = Papers Published, and S = Service. "Talks" denotes papers read either personally or *in absentia* at AMS meetings, AAAS meetings, the Chicago Congress, etc. "Papers" tallies articles published in *Bull. AMS*, *Trans. AMS*, *Annals of Mathematics*, *American Journal of Mathematics*, or *Papers Read at the Chicago Congress*. "Service" refers to holding office in or serving on committees of the AMS or AAAS (1 = one year's service), serving on the editorial board of one of the above journals (1 = one year's service), serving as an officer or organizer of a teachers' group, etc. "Totals" sums the number of talks, papers published, and service for each individual. "Most active" members of the community have totals which are greater than or equal to 15. Of these people, only D. R. Curtiss, S. Epsteen, A. Macfarlane, and I. M. Schottenfels were not starred in the first edition of *American Men of Science*.

Table 2.

Participation Record of the American Mathematical Community's
62 Most Active Members 1891–1906

	Talks		Publications		Service	
0	2	3.2%	0		13	21.0%
1	2	3.2%	0		6	9.7%
2	3	4.8%	2	3.2%	5	8.1%
3	1	1.6%	1	1.6%	3	4.8%
4	1	1.6%	4	6.5%	6	9.7%
5	2	3.2%	3	4.8%	4	6.5%
6–10	15	24.2%	24	38.7%	13	21.0%
11–14	19	30.6%	7	11.3%	3	4.8%
15	1	1.6%	2	3.2%	1	1.6%
16–20	5	8.1%	9	14.5%	1	1.6%
21–25	4	6.5%	5	8.1%	2	3.2%
26–30	3	4.8%	1	1.6%	2	3.2%
31–35	2	3.2%	1	1.6%	3	4.8%
36–40	0		0		0	
41–45	0		0		0	
46–50	1	1.6%	1	1.6%	0	
51–55	0		0		0	
56–60	0		0		0	
61–65	0		1	1.6%	0	
66–70	0		1	1.6%	0	
71–75	1	1.6%	0		0	

Total			Total			Total		
15	3	4.8%	41–45	7	11.3%	91–100	1	1.6%
16–20	17	27.4%	46–50	0		101–110	0	
21–25	7	11.3%	51–60	3	4.8%	111–120	1	1.6%
26–30	8	13.0%	61–70	1	1.6%	121–130	0	
31–35	7	11.3%	71–80	2	3.2%	131–140	0	
36–40	4	6.5%	81–90	0		141–150	1	1.6%

"Talks," "Papers," "Service," "Totals," and "Most Active" are defined as in Table 1. To interpret this and similarly constructed tables, read across a line as follows: Line 2—Of the "most active," 2 people or 3.2% gave 1 talk; 0 people had 1 publication; and 6 or 9.7% people earned 1 service unit. Percentages are based on the 62 members of the "most active" sample space and have been rounded to the nearest 0.1%.

total of 31 activities, Chessin's participation level falls roughly at the median of the "most active" sample space [19].

Finally, at the lowest end of the spectrum, two sorts of career patterns emerge. Five of the twenty mathematicians with totals of twenty or less, and all three with totals of fifteen, received their doctorates during the

final five years of our study. Thus, since David Curtiss, William Fite, Max
Mason, Oswald Veblen, and John Young only really entered the mathematical community after 1901, their totals actually reflect the much higher
average of three activities per year through 1906 [20]. The other 15
participants at this lowest "most active" level range from Nebraska's Ellery
Davis, who averaged one talk and one paper every 18 months, to the
distinguished mathematical astronomer, George William Hill, who only
spoke or published within the American mathematical setting roughly once
every three years but who served the Society faithfully as a member of
Council, as a Vice President, and as the third President.

Whether prolific in talks and published research like Dickson, George
A. Miller, and Edgar O. Lovett, whether giving of service like Frank
Nelson Cole, Thomas S. Fiske, Simon Newcomb, Ormond Stone, and
Alexander Ziwet, or whether highly productive in all three categories like
Maxime Bôcher, William Fogg Osgood, Eliakim Hastings Moore, and
Henry Seely White, the "most active" faction delineated through our study
formed a solid and vibrant core around which mathematics at a research
level grew in the United States [21]. They built the American Mathematical Society; they edited the country's research journals; and they advanced
the frontiers of the mathematical sciences. Surrounding, supporting, and
complementing them, the extensive second echelon of "active" participants
made a community—rather than merely an élite clique—out of American
mathematics.

THE "ACTIVE" PARTICIPANTS

Readjusting our historical microscope to its next highest power, we uncover an additional 258 people who played an "active" role mathematically. These 242 men and 16 women each satisfy the *American Men of
Science* criterion in addition to having three "activities" or, on average, at
least one activity every five years [22]. Here, "activity" takes on a broader
meaning than "service" as construed relative to the "most active" sample
space. Encompassing "service," "activities" also include election to the
American Mathematical Society and attendance at meetings of the
AMS, the American Association for the Advancement of Science, or congresses [23]. By widening the study's purview in this way, the faithful
listeners—if not active creators—are rescued from oblivion. The combined
data for all "active" members, including the 62 "most active," appear in
Table 3. (Clearly, the data pertaining exclusively to the 258 newly defined

Table 3.

Participation Record of the American Mathematical Community's 320 Active Members 1891–1906

	Talks		Publications		Service	
0	103	32.2%	125	39.1%	215	67.2%
1	60	18.8%	56	17.5%	25	7.8%
2	39	12.2%	28	8.8%	17	5.3%
3	23	7.2%	22	6.9%	16	5.0%
4	20	6.3%	18	5.6%	8	2.5%
5	13	4.1%	9	2.8%	7	2.2%
6–10	26	8.1%	35	10.9%	21	6.6%
11–15	21	6.6%	9	2.8%	4	1.3%
16–20	5	1.6%	9	2.8%	1	0.3%
21–25	4	1.3%	5	1.6%	2	0.6%
26–30	3	0.9%	1	0.3%	2	0.6%
31–35	2	0.6%	1	0.3%	3	0.9%
36–40	0		0		0	
41–45	0		0		0	
46–50	1	0.3%	1	0.3%	0	
51–55	0		0		0	
56–60	0		0		0	
61–65	0		1	0.3%	0	
66–70	0		1	0.3%	0	
71–75	1	0.3%	0		0	

Total			Total			Total		
0	56	17.5%	16–20	17	5.3%	71–80	2	0.6%
1	32	10.0%	21–25	7	2.2%	81–90	0	
2	50	15.6%	26–30	8	2.5%	91–100	1	0.3%
3	19	5.9%	31–35	7	2.2%	101–110		
4	21	6.6%	36–40	4	1.3%	111–120	1	0.3%
5	16	5.0%	41–45	7	2.2%	121–130	0	
6–10	46	14.4%	46–50	0		131–140	0	
11–14	18	5.6%	51–60	3	0.9%	141–150	1	0.3%
15	3	0.9%	61–70	1	0.3%			

"Talks," "Papers," "Service," and "Totals" are defined as in Table 1. "Active" members of the community satisfy the following criteria: 1) entry in either the first, second, or third edition of *American Men of Science*; 2) at least three activities, or, on average, at least one activity every five years. Since five Canadians—Alfred T. DeLury, J. Charles Fields, William Findlay, John C. Glashan, and Norman R. Wilson—also satisfy these criteria, we have included their data consistently in deriving the figures for the "active" sample space. "Activities" are defined as "service" as well as election to AMS membership and attendance at AMS meetings (after 1894), meetings of the AAAS, or congresses. Percentages are based on the 320 members of the "active" sample space and have been rounded to the nearest 0.1%. The figure of 320 naturally includes the 62 "most active" participants. Data for the remaining 258 may be gotten by subtracting the figures from Table 2 from those presented here.

"active" members may be recovered by subtracting corresponding figures in Table 2 from those presented in Table 3.) As in the "most active" category, various levels of participation also emerge within the less refined "active" category.

At the bottom of this group, 56, or 17.5%,of the 320 active contributors to American mathematics offered neither talks nor publications nor service to their community. These people participated rather through their membership in the Society or through their presence at its meetings and at other conferences. Consider, for example, the case of Joseph Allen. After earning both a Bachelor's and a Master's degree at Harvard by 1894, Allen held an instructorship in mathematics at Cornell from 1894 to 1897 but never completed a Ph.D. On leaving Cornell, he assumed the post of Tutor at the College of the City of New York in 1897, eventually rising to the rank of Assistant Professor during the period ending in 1906. His New York City base enabled him to attend the Society's gatherings with relative convenience, and he want to at least five meetings after his induction in 1898 [24]. On the other hand, the University of Michigan's Joseph Markley had nontrivial geographical barriers to overcome in order to attend even functions of the Chicago Section of the Society after its founding in 1897. Nevertheless, Markley, an 1889 Harvard Ph.D., managed to get to the Chicago Section's second meeting in 1897 as well as to the Sixth, Eighth, and Eleventh AMS Summer Meetings in 1899, 1901, and 1904, respectively [25].

Within the "active" segment of the community, though, the records of Allen and Markley contrast decidedly with that of someone like Stanford's William A. Manning. On receiving his doctorate from Stanford in 1904, that is, virtually at the close of the period under consideration, Manning proceeded for a year of study abroad in Paris before returning to an assistant professorship at his *alma mater*. In attendance regularly at the meetings of the Society's San Francisco Section after its organization in 1902, Manning's participation record—seven talks and six published papers—in his specialty of substitution groups translates into an average of roughly two talks and publications a year over his three academic years in residence from 1902 to 1906 [26].

While Manning's career stands out as exceptional within this group, that of Henry T. Eddy falls in right at the median. An older member of the overall sample space, Eddy was born in 1844 and was active throughout the period from 1891 to 1906. As Professor of Engineering and Mechanics at the University of Minnesota from 1894, Eddy concentrated primarily on mathematics of an applied nature and leaned perhaps as much toward physics as toward mathematics. Thus, he belonged to the ranks of the mathematically interested, if not the actively mathematically involved. In spite of his split allegiance, within our time frame Eddy gave two talks and published one paper, this a work entitled "Modern graphical developments" in the volume from the 1893 Chicago Congress [27]. He also attended at least seven society-sponsored events ranging from its first Colloquium held in Boston in 1896 to various of its geographically dispersed summer Meetings.

As with most members of this early "active" cross-section (215 out of 320, or 67.2%), however, neither Eddy nor Manning performed services directed toward the society's broader goals. Obviously, this kind of participation was limited both by the number of available posts one could hold and by the geographical locations of the seats of power and influence. Nevertheless, as the case of William Dennett proves, even the merely "active" could serve. A surgeon at the New York Eye and Ear Infirmary in New York City from 1888 to 1901 and honorary surgeon there after 1907, Dennett received his M.D. from Harvard in 1874. Although he never spoke before the AMS and never published in the mathematical journals under consideration here, he joined the society in 1891, attended at least 28 of its meetings (or about twice yearly), and served on its Council and as its Treasurer for a combined total of seven years [28]. Thus, if anyone in our sample space merits the title "mathematical enthusiast," it must surely have been William Dennett!

With members like Allen, Markley, Manning, Eddy, and Dennett running the gamut from casual but interested observers to real research-level contributors, the cross-section of the 258 properly "active" members of American mathematics formed the audience for much of the work of the 62 "most active." They attended meetings; they read journals; and they occasionally contributed. In short, they represented a reasonably solid bedrock of support for mathematics in the United States. Less solid but nevertheless present, the "rank and file" completed the profile of American-based participants in the American mathematical community [29]. To focus on this segment of the population, we must turn the historical microscope to its highest possible power.

THE "RANK AND FILE"

For the most part, the "rank and file" comprise the most obscure, least visible segment of the community. The 563 men and 51 women in this group either: 1) have fewer than three "activities" (as defined above for the "active" subspace) over the period from 1891 to 1906, or 2) do not have entries in any of the first three editions of *American Men of Science*. Although 39 men and 11 women (or 8.1% of the 614 total) would have fallen in the "active" category except for the latter (technical) deficiency, the vast majority of the "rank and file" have fewer than two activities to their credit over the 15 years under scrutiny [30]. Perhaps they joined the Society but then did not contribute further; perhaps they spoke or published once or twice; or perhaps they submitted a personal note recording a career milestone such as a change in academic affiliation or the earning of a doctoral degree. On the community's fringes by-and-large, the "rank and file" comprised a sizeable, virtually silent, majority of Americans who had some training or interest in mathematics but who were not in a position to participate more than marginally in the mathematical endeavor. Before trying to come to terms with those truly in this latter category, however, the 50 otherwise "active" members of the "rank and file" merit closer examination.

First of all, several important figures in the history of nineteenth-century American mathematics fall into the present study's "rank and file" because they had died prior to the 1906 publication of the first edition of *American Men of Science*. For instance, Yale's internationally renowned theoretical physicist and mathematician, Josiah Willard Gibbs died in 1903 at the age of sixty-four, while his friend, teacher, and Yale colleague in mathematics, Hubert Anson Newton, died at sixty-six in 1896. Although the reclusive Gibbs joined the AMS just months before his death and did not participate in its activities, Newton served as the Society's Vice President for the almost two years from 1895 through to the time of his death [31].

As an extreme in this category, though, consider James Joseph Sylvester's first student and ultimate successor at Hopkins, Thomas Craig, who died prematurely in 1901 at the age of 44. A specialist in the theory of differential equations, Craig earned his doctorate in 1878 and, in the years following, published extensively not only in the *American Journal* but also in such foreign periodicals as the *Comptes rendus* of the French Academy of Sciences and the *Journal für die reine und angewandte Mathematik*. He joined the Society in 1891, took part in its meetings, served for three years on its Council, and spent five years at the helm of the *American Journal* as its editor-in-chief. In all, he read two papers, published five, and devoted the equivalent of 12 years of service to the body politic for a combined total of 19. Craig should therefore stand side by side with the 62 "most active" distinguished in our study, but his early death barred him from that number for reasons of statistical uniformity [32].

Even though Craig is the only member technically in the "rank and file" who should have fallen in with the "most active," people like Albert Munroe Sawin and Roxanna Vivian are more typical of this category's 50 exceptional members. Sawin, who, it would seem, never earned a doctoral degree, first joined the AMS in 1891 and then rejoined in 1899 after apparently letting his membership lapse. On the faculties of the University of Wyoming (in 1891), Clark University of Atlanta, Georgia (in 1899), and Syracuse University (in 1900), he published four papers but gave neither talks nor service within our time period. Thus, his career falls squarely in the middle of the "active" category [33]. Roxanna Vivian, on the other hand, belonged to the Society and attended at least three AMS meetings, but her combined total of talks, publications, and service was *zero* [34]. Her case thus exemplifies the problems involved in isolating the "rank and file" of the American mathematical research community at the turn of the century: while they may have attended meetings of the Society, they tended not to leave a written record through mathematical publication. Furthermore, she fits the general profile of women in American science around 1900: she had earned a higher degree; she found employment at a women's college; she devoted herself (apparently) to teaching; and she published relatively little (if at all) [35]. Because, like fully two-thirds of the 71 women uncovered by our study, Vivian does not have an entry in *American Men of Science*, her career could easily have slipped into obscurity had it not been for the sensitivity of the *Bulletin* as a microscope for viewing the community down to its deepest levels [36].

What conclusions now surface from this morass of data and anecdotal information on the levels of participation within the three tiers of American mathematics? To begin with, by the period under consideration mathematics in the United States had grown from an almost embarrassingly small to a respectably-sized endeavor. In his book on *American Science in the Age of Jackson*, George H. Daniels presents the findings of his numerical study of 16 national scientific journals published during the period from 1815 to 1844, discovering a mere 22 men engaged in mathematical research, or a slight 3.44% of the 638 scientists (three of whom were women) in his sample space. Looking for mathematical articles actually published in these periodicals, he finds only 58, or 2.60% of the recorded total of 2225 [37]. Following Daniels' data up with figures gathered by Robert V. Bruce for the years from 1846 to 1876 reveals that roughly 44 people entered the field of mathematics during these decades, or 9.3% of the 477 scientists he identifies in the pages of the *Dictionary of American Biography* [38]. Granting that Daniels' sample space is larger and less selective, hence deeper, than Bruce's, a comparison of their figures points to a significant percentage-wise increase in mathematical activity in the United States over the 60-odd years from 1815 to 1876 but nonetheless to a minuscule group of people in real terms.

Although unable to draw from comparative data relative to the full American scientific community from 1891 to 1906, our study does lay out for examination a relatively large and extended group of individuals participating in American mathematics. Concentrating on Table 3, we find 320 "active" members of the mathematical community during our 15-year period as opposed to Bruce's 44 representatives over his 30-year time span. Furthermore, based on our sample space of four specialized mathematical research journals—entities almost, but not quite, unheard of in Daniels' time frame—we find upwards of 1110 exclusively mathematical papers published in 15 years in contrast to the 58 appearing in 16 journals between 1815 and 1844. Mirroring the general growth of American science and the ever-improving conditions within higher education as much as the growth of the mathematical research community, these figures reflect only real number increases and not the percentage-wise growth of mathematics as a subdiscipline of American science. Still, in sheer numbers, there were enough mathematically inclined people in this country by the turn of the twentieth century to form a self-sustaining community of researchers and devotés, almost fully separate from those general scientific organizations like the American Association for the Advancement of Science which had supported it over the 30 years from 1846 to 1876 and before.

GEOGRAPHICAL DEMOGRAPHICS

As evidenced by many of the remarks above, the broad and deep base of support for research-level mathematics, which made an American mathematical research community a reality in the 15 years from 1891 to 1906, extended from coast to coast and from north to south. Although by no means uniformly distributed, no area of the country went without at least some mathematical representation, either as the birthplace of one of the community's members-to-be or as the locale for all or part of his or her career.

In his study of 477 scientists active during the 30-year period from 1846 to 1876, Bruce calculates 1831 as the median birth date of the 415 American-born or 36 years prior to the 1867 median birth date for our group of "active" mathematicians [39]. He finds that within his time frame the Northeast produced 18 times as many mathematicians as the Midwest and five and a half times more than the South, while the South exceeded the Midwest by a margin of 3.3 to 1 [40].

In our sample, the Northeast (with 34.6% of the total population of the United States and its territories in 1870) produced the most future mathematicians, or 42.8% of those in the active category [41]. (See Table 4.) As runner-up, the rapidly expanding Midwest (with 35.8% of the overall population in 1870) generated a full 30.0%, or 96 of the 320 total, and the South (with 27.1% of the U.S. population in 1870) came in at a poor third

Table 4.

Birthplaces of the Active Members of the American Mathematical Community 1891–1906

	Most Active		Active	
Northeast	24	38.7%	137	42.8%
South	2	3.2%	18	5.6%
Midwest	20	32.3%	96	30.0%
West	1	1.6%	6	1.9%
Territories	0		0	
Canada	2	3.2%	19	5.9%
Others	13	21.0%	44	13.8%

	Population 1870	% of Total Population	Population 1890	% of Total Population
Northeast	13,336,000	34.6%	22,699,000	29.8%
South	10,432,000	27.1%	19,034,000	25.0%
Midwest	13,786,000	35.8%	29,451,000	38.6%
West	903,000	2.3%	4,495,000	5.9%
Territories	102,000	0.3%	536,000	0.7%

Northeast: Connecticut, Delaware, Maine, Maryland, Massachusetts, New Hampshire, New Jersey, New York, Pennsylvania, Rhode Island, Vermont, Washington, D.C.

South: Alabama, Arkansas, Florida, Georgia, Kentucky, Louisiana, Mississippi, North Carolina, South Carolina, Tennessee, Virginia, West Virginia.

Midwest: Illinois, Indiana, Iowa, Kansas, Michigan, Minnesota, Missouri, Nebraska, Ohio, Oklahoma, Texas, Wisconsin.

West: California, Colorado, Idaho, Montana, Nevada, North Dakota, Oregon, South Dakota, Utah, Washington, Wyoming.

Territories: Alaska, Arizona, Hawaii, New Mexico.

Canada: Provinces represented—Nova Scotia, Ontario.

Percentages are based on the 62 members of the "most active" and the 320 members of the "active" sample spaces. They have been rounded to the nearest 0.1%. Population information compiled from U.S. Bureau of the Census, *Historical Statistics of the United States: Colonial Times to 1970*, 2 vols. (Washington, D.C.: U.S. Government Printing Office, 1975).

at 18 of 320 or 5.6%. Thus as a birthplace of future mathematicians, the Northeast led the Midwest by a margin of 1.4 to 1 while it outdistanced the South by 7.6 to 1. The Midwest also exceeded the South, but by 5.3 to 1.

A comparison of these margins with those of Bruce's time period undoubtedly reflects not only the changing population demographics of a country moving westward but also the relative educational opportunities at the elementary and secondary levels in the various regions of the continental United States. Most notable here is the strong surge of the Midwest as a birthplace of mathematicians, the concomitant slide in the Northeast, and the dramatic drop in the South. Given the median birth date of our sample, however, the South, ravaged by the Civil War, would not have been expected to have fared too well.

Also of interest, especially in comparison with the figures representing the geographical distribution of the mathematical community (in Tables 5 and 6), is the weak showing of the West as a birthplace in 1867 but its spectacular climb as a place to work over our 15-year period. Producing only six of the 320 "active" participants or, 1.9%, the West later went from supporting 30 academic careers over the five years from 1891 to 1896 to supporting 71 from 1901 to 1906 [42]. The Midwest also had impressive gains in this regard, sustaining a total of 154 mathematicians during the first five-year period and increasing to a total of 298 by 1906. Once again, the South made the smallest strides (of these three regions), starting roughly even with the West in 1891–1896 but improving only slightly by 1906. (In overall population, the West grew from 2.3% of the total U.S. population in 1870 to 5.9% in 1900; the Midwest picked up another 2.6% of the total to account for 38.6%; and the South slipped 2.1% to 25.0%. The Northeast shrank the most percentage-wise, 4.8%, declining to 29.8% of the U.S. population in 1900 (see Table 4).

Going hand in hand with these figures on geographical distribution, the data in Table 7 on the number of academic or government affiliations of the "active" members of American mathematics indicate a fairly high degree of mobility within their ranks. Although 47.2%, or almost half of the "active" sample space, held an academic post at only one institution, fully half moved at least once or as many as four times during the years from 1891 to 1906. Thus, while reasonably stable, the American mathematical research community was by no means static. Jobs opened up, particularly in the West and Midwest, and people moved around to fill them.

Table 7 also shows that the federal government did not offer very strong support for "active" mathematicians over the 15-year period. Whereas from 1846 to 1876, the mathematically inclined might well have found jobs in the United States Coast Survey under the guidance first (in this period)

Table 5.

Geographical Distribution of the American Mathematical
Community 1891–1906

1891–1896	Most Active	Active	Rank + File	Total
Northeast	19	107	155	262
South	3	11	22	33
Midwest	18	70	84	154
West	3	12	18	30
Territories	0	0	1	1
Canada	0	3	4	7
Foreign	0	3	0	3

1896–1901	Most Active	Active	Rank + File	Total
Northeast	28	144	194	338
South	2	13	27	40
Midwest	24	94	118	212
West	8	23	31	54
Territories	0	0	2	2
Canada	0	5	1	6
Foreign	1	2	0	2

1901–1906	Most Active	Active	Rank + File	Total
Northeast	37	174	266	440
South	2	13	25	38
Midwest	24	108	190	298
West	9	25	46	71
Territories	0	0	2	2
Canada	0	9	2	11
Foreign	1	2	0	2

The geographical areas are defined as in Table 4. The "Rank and File" comprises the 563 men and 53 women who fell short of the criteria for the "active" classification. (See Table 3.) The geographical data presented here were based on a sample space of the 524 men and 45 women for whom at least some geographical data is available. Due to the facts that: 1) not all 320 "active" members were active throughout this 15-year period, and 2) some participants were in more than one region in a given five-year period, the figures in these columns do not sum to any fixed numbers. The figures in the "Total" columns represent the sum of the corresponding figures in the "Active" and "Rank and File" columns.

of Alexander Dallas Bache and then of Benjamin Peirce, in the Nautical Almanac directed by Peirce's brother-in-law, Charles H. Davis, or at Washington's Naval Observatory, the figures given here clearly suggest that these no longer represented very attractive or very abundant job options for an increasingly research-oriented community of scholars. Over

Table 6.

Growth by Geographical Area of the American Mathematical Community 1891–1906

	1891–1896/1896–1901		1896–1901/1901–1906		1891–1896/1901–1906	
Northeast	+76	29.0%	+102	30.2%	+178	67.9%
South	+7	21.2%	−2	−5.0%	+5	15.2%
Midwest	+58	37.7%	+86	40.6%	+144	93.5%
West	+14	46.7%	+17	31.5%	+41	136.7%
Territories	+1	100.0%	0		+1	100.0%
Canada	−1	−14.3%	+5	83.3%	+4	57.1%
Foreign	−1	−33.3%	0		−1	−33.3%

The geographical areas are defined as in Table 5. These figures represent the changes in the "Total" columns of Table 5, first by five-year periods and finally over the entire 15-year period. The percentages have been rounded to the nearest 0.1%.

Table 7.

Number of Academic and Government Affiliations of the Active Members of the American Mathematical Community 1891–1906

Academic*

	Most Active		Active	
0**	2	3.2%	22	6.9%
1	29	46.8%	151	47.2%
2	16	25.8%	92	28.8%
3	10	16.1%	45	14.1%
4	4	6.5%	9	2.8%
5	1	1.6%	1	0.3%
	62		320	

	Most Active		Active	
Coast Survey	1	1.6%	7	2.2%
Nautical Almanac	3	4.8%	3	0.9%
Naval Observatory	0		4	1.3%
Weather Bureau	0		1	0.3%

*Here, only the academic (including high school) affiliations have been counted. In cases where a person has left and then returned to the same institution, this has been counted as one affiliation.

**Among those with no academic affiliations during the period 1891–1906, employment ranged from unemployment to posts as actuaries, doctors, engineers, private tutors, private investigators, lawyers, computers, and government employees. The percentages are based on the respective totals of 62 and 320.

the 15 years beginning in 1891, only 4.7% of the 320 mathematically "active" had earned their living in the employ of these government agencies [43]. Rather, American mathematicians overwhelmingly pursued academic careers during a time of relative expansion within higher education [44].

The numbers presented in Tables 5, 6, and 7, reflective as they are of patterns of movement within the mathematical community, further point to why formal Sections of the AMS sprang up first in Chicago (the Midwest) in 1897 and later in San Francisco (the West) in 1902. As more and more mathematicians concentrated in these regions, the Society could not provide adequate levels of real support from its headquarters on the East Coast. Since it proved geographically impossible for Society participants in the West and Midwest to attend the monthly gatherings in New York City, these men and women organized themselves into regional Sections which met regularly and adopted the same format as their parent organization. Because they campaigned for and won official sanction for their activities, the reports of their meetings, and especially the abstracts of their talks, routinely appeared on the pages of the *Bulletin*. Thus, the Sections served to bring into the fray mathematicians living and working in areas of the country remote from New York City. The Sections also marked the beginning of a gradual dispersal of power and influence away from the Northeast and toward the rest of the United States [45].

EDUCATION OF PARTICIPANTS

Just as the Northeast lost its total dominance over the mathematical scene, so it saw its educational monopoly crumble with the rise of land-grant institutions after 1862 and with the era of great fortunes at the end of the nineteenth century. As various midwestern states, in particular, built up facilities with federal land-grant funds and as entrepreneurs like Jonas Clark, Ezra Cornell, Johns Hopkins, John D. Rockefeller, Leland Stanford, Cornelius Vanderbilt, and others endowed their own schools, students desirous of higher education no longer felt compelled to attend the traditional colonial and early nineteenth-century colleges. In fact, those mathematicians active during the period from 1891 to 1906 almost all earned a doctoral degree, and those who earned it at home were rather more likely to have attended (recently founded) Chicago, Hopkins, or Clark than (long-established) Yale, Columbia, or Harvard (see Tables 8 and 9).

Table 8.

Highest Degree Earned by Members of the American Mathematical Community 1891–1906

	Most Active		Active		Women	
No Degree	0		6	1.9%	1	1.4%
B.A. or B.S.	3	4.8%	14	4.4%	5	7.0%
C.E. or E.M.*	3	4.8%	7	2.2%	0	
M.A. or M.S.	6	9.7%	54	16.9%	6	8.5%
M.D.	0		2	0.6%	0	
Ph.D. or D.Sc. (domestic)	31	50.0%	180	56.3%	23	32.4%
Ph.D. (foreign)	20	32.3%	63	19.7%	3	4.2%
Unknown	0		0		33	46.5%

*C.E. denotes a first degree in civil engineering, and E.M. denotes a first degree in mining engineering.

In the "Most Active" column, Edgar O. Lovett has been counted twice—once for his domestic Ph.D. from the University of Virginia and once for his Leipzig Ph.D. under Sophus Lie. Additionally, in the "Active" column, T. Proctor Hall has been counted three times—once for an M.D. and twice for domestic Ph.D.'s from Illinois Wesleyan and Clark; Harris Hancock earned two foreign Ph.D.'s (Paris and Berlin); Robert E. Moritz is credited for one domestic (Nebraska) and one foreign (Strasbourg) Ph.D.; and Henry Dallas Thompson received two marks—one for a domestic (Princeton) and one for a foreign (Göttingen) doctorate. Taking these duplications into account, the columns add correctly 62, 320, and 71, respectively. Since the percentages have been calculated relative to these totals, they add up to slightly more than 100% in the first two columns due to the above noted duplications. Rounding has been done to the nearest 0.1%.

Totally in step with the advances in American higher education and the general trend toward professionalization within American science, the members of the growing mathematical community shared high levels of educational achievement. In our period, 76.0% of those "active" had attained the doctorate with 16.9% stopping at the Master's and a slight 6.6% going only as far as the bachelor's level. Among the "most active," the data are even more impressive, with 82.3% holding the doctorate. These figures stand in striking contrast to comparable statistics for that earlier mathematical generation active between 1846 and 1876 surveyed by Bruce. He finds that 75.0% of his group reached the Bachelor's level but went no further, an additional 12.5% proceeded only to the Master's degree, and a scant 10.0% actually earned the Ph.D. [46]. By the close of the nineteenth century, the broader mathematical community had clearly accepted the doctoral degree as its legitimizing credential and had taken advantage of educational opportunities both at home and abroad.

Table 9.

American Colleges and Universities Awarding Doctoral Degrees to Members of
the American Mathematical Community 1891–1906

	Most Active		Active		Women	
Yale	2	6.5%	26	14.4%	5	21.7%
Chicago	5	16.1%	24	13.3%	2	8.7%
Johns Hopkins	5	16.1%	24	13.3%		
Clark	1	3.2%	16	8.9%		
Columbia	5	16.1%	15	8.3%	4	17.4%
Harvard	3	9.7%	15	8.3%		
Cornell	2	6.5%	12	6.7%	2	8.7%
Pennsylvania	0		11	6.1%	4	17.4%
Princeton	0		5	2.8%		
Syracuse	1	3.2%	4	2.2%		
Bryn Mawr	0		2	1.1%	5	21.7%
Kansas	1	3.2%	2	1.1%		
Michigan	1	3.2%	2	1.1%		
Nebraska	0		2	1.1%		
Others*	5	16.1%	20	11.1%	1	4.3%

*Each school in this category produced only one doctorate in the sample. The five granting
degrees to "most active" members were: Cumberland (Kentucky), Ohio Wesleyan, Purdue, Virginia,
and Wisconsin. The other 15 included: Colorado, Dartmouth, Haverford, Illinois, Illinois Wesleyan,
Lafayette, New York University, North Carolina, Rutgers, Stanford, Tulane, Union, Vanderbilt,
Washington (St. Louis), and College of Wooster (Ohio). Wisconsin granted the one degree noted in
the "other" column to a woman.

The percentages have been calculated relative to the total number of domestic degrees awarded
or 31, 180, and 23, respectively, and have been rounded to the nearest 0.1%. (See Table 8.)

As an extreme case of this new zeal for degrees, consider the "active,"
Canadian-born, Thomas Proctor Hall. Beginning his career with a Bachelor's degree from the University of Toronto, Hall moved to the United
States for first a Master's and then a doctorate (in either mathematics or
physics in 1888) from Illinois Wesleyan [47]. Although nominally a Ph.D.,
this degree did not necessarily result from the highest standards of
scholarship or originality. Especially in the 1880's, a fair number of
American schools began to award doctoral degrees in spite of the fact that
they often lacked staff qualified to supervise students at that level. Be this
as it may, Hall followed up his first Ph.D. with a second from Clark
University in 1893. A newly founded, then exclusively graduate-level institution, Clark boasted a strong and talented faculty which initially included
the physicist, A. A. Michelson; the mathematicans William E. Story and
Oskar Bolza; the anthropologist Franz Boaz; and the psychologist (and

President) G. Stanley Hall. Thus, the student Hall undoubtedly had to meet higher standards in order for his dissertation on "New Methods of Measuring the Surface Tension of Liquids" to earn him the first Clark Ph.D. in physics [48]. After leaving Worcester, he held various professorships in the Midwest and eventually took yet another degree, this time an M.D., from the National Medical University. Although atypical, Hall's educational career nevertheless reflects the increased emphasis placed upon the highest terminal degree during our time period.

Perhaps more telling of the changing educational and professional environments, however, our turn-of-the-century sample of mathematically "active" also contains 18 women, with 71 in the total space, "rank and file" inclusive. Once more, these figures sharply contrast with the complete absence of women in Bruce's roughly mid-century group of 40 mathematicians. Although we lack educational information on 33, or 46.5% of the women isolated, our findings on the remaining 38 do not run counter to those for the full "active" contingent. Of the 71 women, at least 36.6%—or more than a third—held the Ph.D. Clearly not as high as the comparable figure for all "active" participants, this percentage does represent a significant gain, especially in light of the fact that opportunities for women to pursue higher studies in the United States were few even in the 1890's. Many universities of the day either refused to admit women altogether (like Harvard and Princeton) or they denied them degrees in spite of permitting them to attend courses (like Johns Hopkins) [49]. Still, American institutions like Bryn Mawr and Yale, and, somewhat later, schools abroad such as Göttingen, opened their doors to American women and gave them the chance to earn that increasingly necessary and desirable credential, the doctorate [50].

While American women only became eligible for the German doctorate in the mid to late 1890's, the mystique of the degree had attracted American men in fields like chemistry since the 1850's [51]. Beginning in the 1880's, though, the "German bug" began to bite more and more would-be American mathematicians [52]. As evidenced in Table 10, Göttingen ran a strong first in attracting young Americans, owing primarily to the presence of Felix Klein there after 1885. Leipzig, which came in second, boasted Klein on its faculty prior to 1885 and then Sophus Lie as his successor from 1886 to 1899. Finally, the likes of Georg Frobenius, Lazarus Fuchs, Hermann Amandus Schwarz, and Kurt Hensel also attracted a fair number of Americans to Berlin in the mid-nineties. In all, of the 20 "most active" participants in American mathematics who earned a foreign doctorate, 14 earned it at a German university. Germany produced

Table 10.

Foreign Universities Awarding Doctoral Degrees to Active Members of the
American Mathematical Community 1891–1906

	Most Active		Active	
Göttingen	9	45.0%	22	34.9%
Leipzig	2	10.0%	10	15.9%
Berlin	1	5.0%	8	12.7%
Munich	0		3	4.8%
Erlangen	1	5.0%	3	4.8%
Strasbourg	1	5.0%	3	4.8%
Cambridge	2	10.0%	2	3.2%
Zürich	1	5.0%	2	3.2%
Edinburgh	1	5.0%	2	3.2%
London	1	5.0%	2	3.2%
Bonn	0		1	1.6%
Bordeaux	0		1	1.6%
Kiel	0		1	1.6%
Königsberg	0		1	1.6%
Paris	0		1	1.6%
Vienna	1	5.0%	1	1.6%

Three women received foreign doctorates, two at Göttingen and one at the University of London.

The percentages have been calculated relative to the total number of foreign doctorates awarded, or 20 and 63, respectively, and rounding has been done to the nearest 0.1%. (See Table 8.)

52 of the 63 foreign doctorates earned by members of the full "active" contingent. One of those in this latter category, James Morris Page, studied under Lie at Leipzig and brought research mathematics back to his native South after earning his Ph.D. in 1888.

Page, who was born and educated in Virginia, received a Master's degree at Ashland's Randolph–Macon College in 1885. Moving on to Leipzig for the Winter Semester of 1886–1887, he enrolled in Lie's courses through the Winter Semester of 1887–1888. Thus embued with Lie's theory of transformation groups, Page wrote a dissertation entitled "On the Primitive Groups of Transformations in Space of Four Dimensions," which won him the 1888 doctorate [53]. This thesis clearly impressed its director, for in his official report on Page's work, Lie wrote that it "...deals with and resolves an important and rather difficult problem in the theory of continuous transformation groups. It is true that the author was largely shown a way that offered good prospects for success. Nevertheless, the task of carrying the entire investigation through required not only considerable energy but also a significant amount of finesse in dealing with the computational techniques.... I, recommend the grade of II^a (admondum laudabilis)" [54]. Of Lie's five American students—Page, Edgar O. Lovett, Charles Leonard Bouton, David Andrew Rothrock, and John van Elten Westfall—Page earned the highest mark. This made him, in some sense, Lie's "best" American student.

Apparently resolved to return home to Virginia instead of seeking a university-level post elsewhere, Page held a headmastership at a private school from 1888 to 1895. Page's teaching job, although undoubtedly far from a stimulus to his mathematical research, nonetheless put him in close proximity to the University of Virginia. There, he reorganized the moribund Mathematical Club in 1893 and served as its President. In addition to prevailing upon the Professor of Mathematics, W. H. Echols, to speak on his research on the theory of functions of a real variable, Page introduced his audience to his own investigations of Lie's theory of infinitesimal transformations [55]. The Club's student secretary, Edgar O. Lovett, must have taken Page's lectures to heart, for after earning his Virginia Ph.D. in 1895, he followed Page's example and proceeded to Lie in Leipzig. As mentioned earlier, Lovett took his doctorate under Lie in 1898 for a thesis on "The Theory of Perturbations and Lie's Theory of Contact Transformations" [56]. Perhaps owing to his success with Lovett, Page was offered and assumed an adjunct professorship at Virginia in 1896 and had risen to a full professorship by 1901. Although not terribly active during our period, Page did publish six papers and spent two years as an associate editor of *Annals of Mathematics* before the journal moved from Virginia to Harvard in 1899. In particular, he continued to follow the German scene, focusing at the turn of the century on Georg Scheffer's geometrical researches.

Like Page, 62 other members of our "active" sample (or 19.7% of the total) actually came back to the United states with a foreign doctorate in hand, but an even greater number (112 or 35.0% of the total) reported spending at least some time studying there (see Table 11). As with the Ph.D.'s, the vast majority of America's mathematical tourists, or 98 out of 112, went to Germany. France came in a weak second, drawing only 14 of the 112 (and those due primarily to the presence of Camille Jordan in Paris), and England showed at third with 11 of 112.

These figures clearly represent only a sort of greatest lower bound on the actual number of Americans who pursued their studies abroad, however. Because the data on the "rank and file" necessarily remains incomplete, we have no way of securing an accurate count of the number in that segment of the mathematical population who may have indulged their

Table 11.

Study Abroad by Members of the American Mathematical
Community 1891–1906

	Most Active		Active	
Germany	24	38.7%	98	30.6%
France	3	4.8%	14	4.4%
Great Britain	4	6.5%	11	3.4%
Switzerland	1	1.6%	5	1.6%
Austria	1	1.6%	1	0.3%
Russia	0		1	0.3%
Norway	0		1	0.3%
Sweden	0		1	0.3%
Total Abroad*	29	46.8%	112	35.0%

*Of the "Most Active," four people reported having studied in two countries, hence the total of 29. Of the "Active," 18 people reportedly studied in two countries and one studied in three, making the total 112. These numbers have been drawn exclusively from *American Men of Science* entries and so form a sort of greatest lower bound on the actual number of people who studied abroad. The percentages have been calculated relative to the total number of "most active" and "active" participants of 62 and 320, respectively, and rounding has been carried out to the nearest 0.1%.

Wanderlust. For example, one T. M. Blakeslee, Professor of Mathematics at Des Moines College in Iowa in 1893, shows up on Felix Klein's class rosters in Göttingen during the Winter Semester of 1890–1891 but enters our "rank and file" only because of his attendance at the Chicago Congress [57]. Similarly, a Paul Arnold took a professorship at the University of Southern California in 1901, thereby forming part of the "rank and file," but only after he had spent the Winter and Summer Semesters of 1897–1898 in Sophus Lie's courses at Leipzig [58]. This circumstantial evidence, coupled with the figures in Table 11, reflects the seemingly strong emphasis placed on the foreign study tour as an educational supplement by the American mathematical community during our period. Like the doctorate, first-hand exposure to European mathematics, especially for those who had done their degrees at home, represented an important professional credential [59]. The reasons for this are apparent.

First, thanks to their training, the Americans appreciated the quality of the mathematical work issuing from Europe—and particularly from Germany—and they realized how highly men like Klein, Lie, Frobenius, and David Hilbert merited their international reputations. They understood that although a growing number of people in the United States had mastered aspects of these men's research, the principals themselves would provide the surest way to the deepest possible insights. Second, the Americans also recognized that, in spite of the tremendous strides made at schools like Chicago and Harvard, the level and extent of the higher education available in mathematics on their own shores was generally lower than that in Germany or France. A comparison of the course offerings both at home and abroad points out this difference quite distinctly.

Consider what was perhaps America's best program at the time. At the University of Chicago during the Autumn, Winter, and Spring Quarters of 1895–1896, E. H. Moore offered a seminar on group theory and courses on elliptic functions and the theory of functions of a complex variable; Oskar Bolza lectured on algebraic functions and their integrals, linear differential equations, and advanced integral calculus; Heinrich Maschke taught invariant theory, the theory of the icosahedron, differential geometry, and analytic mechanics; Jacob Young covered the theory of equations; James Boyd handled the course on partial differential equations; Harris Hancock offered lectures on minimal surfaces and solid analytics; and, finally, Herbert Slaught taught differential equations [60]. Although mostly at the properly graduate level, at least four of these 16 courses aimed at the advanced undergraduate audience, and only one covered an applied topic. Still, this compared favorably with the situation at Germany's best, Göttingen, during the Winter Semester of 1895–1896. There, the faculty gave the following:

> Professor Schering: Potential function; Magnetic observations; Mathematico-physical seminarium; —— Professor Klein: Theory of numbers; Mathematical seminarium; —— Professor Schur: Spherical astronomy, I; Practical work with instruments in the observatory; Method of least squares; Astronomical problems in seminarium; —— Professor Hilbert: Integral calculus; Theory of partial differential equations; Mathematical seminarium; —— Professor Schoenflies: Theory of functions; Descriptive geometry; Mathematical proseminarium; —— Dr. Pockels: Electromagnetic theory of light; Fundamental principles of modern meteorology; —— Dr. Bohlmann: Homogeneous linear differential equations; Life insurance; —— Dr. Sommerfeld: Projective geometry; Exercises in descriptive geometry [61].

Obviously, Göttingen offered a broader range of courses in applied mathematics, but its coverage of the purer realms only outshone that of Chicago by the relatively greater distinction of its faculty [62].

Choosing programs more representative of both the American and German scenes, however, clearly underscores the real differences in depth between the two systems. Also during the 1895–1896 academic year, New York City's Columbia College, which awarded 8.3% of the American doctorates to those in the "active" sample, ran ostensibly graduate courses in the general theory of functions, the theory of substitutions, the analytical theory of curves of double curvature and surfaces, ordinary and partial differential equations, and advanced differential and integral calculus [63]. In this list of five courses, though, only the first three really represented research-level mathematics. By way of comparison, the University of Leipzig offered a mathematical seminar as well as courses on hypergeometric series and Eulerian integrals, mathematical physics, metrical geometry, differential equations which admit infinitesimal transformations, differential and integral calculus, analytical mechanics, and kinematics [64]. Richer at both the elementary and graduate levels, the curriculum also included Sophus Lie's courses on his ground-breaking research, something which Columbia's Thomas Fiske and Frank Cole obviously could not match.

Because of the recognized differences in quality and quantity highlighted in this last comparison, the AMS published on the pages of its *Bulletin* the course offerings of most of the German universities and of many other foreign schools alongside comparable notices for the American institutions. By providing such information with issue-to-issue regularity, the Society not only facilitated but also tacitly encouraged its membership to undertake study abroad. As the figures, particularly in Tables 8, 10, and 11 show, the American mathematical public read and heeded its advice. Yet Tables 8 and 9 also indicate that it viewed the United States as a viable mathematical training ground. Judging by the combined educational profile presented here, American mathematics, while coming into its own, still depended on Europe. Despite their gains, the Americans continued to regard the Europeans—but principally the Germans—as their mathematical superiors.

INTERESTS OF THE PARTICIPANTS

In its efforts to emulate the trendsetters, whether at home or abroad, the turn-of-the-century American mathematical community followed closely the work done on both sides of the Atlantic and chose its areas of specialization accordingly. To begin with, its members by and large considered themselves "mathematicians," with 86.3% of the "active" participants

designating pure or applied mathematics as their primary area of specialization. While such a percentage certainly fails to shock, its interest lies
not so much in its numerical magnitude as in the attitudes it mirrors. By
the end of our period, when James McKeen Cattell was busy collecting the
data for his *American Men of Science*, American mathematicians had
achieved the sense of identity and professional common cause crucial for
the establishment and maintenance of a community. They may have
earned their livings as professors or as actuaries or even as government
functionaries, but they styled themselves mathematicians first and foremost.

Of perhaps more interest than the large number of mathematicians in
our sample, however, almost 15% of the mathematical community did not
consider mathematics as their main specialty. Among the non-mathemati-
cians, astronomers participated the most, comprising 6.3% of those "active," and physicists—both pure and applied—made up 4.1% of the 320.
Since neither the American Astronomical Society nor the American Physical Society was founded until 1899, the presence of at least some of the
more mathematically-inclined astronomers and physicists within the properly mathematical community during our period seems natural [65]. In fact,
although relatively few in number, the astro-physical contingent made
important contributions to the AMS, supplying three of the Society's first
five presidents [66]. As active as the astronomers and physicists may have
been, however, mathematics proper naturally dominated the Society, and
its many subspecialties each found representation within the group.

As Table 12 shows, the fields of specific interest to the Americans
spanned the mathematical landscape from Lie's fashionable theory of
continuous groups to the more classical theory of elliptic functions and
integrals, from algebraic and differential geometry to the history of mathematics, and from number theory to the latest in foundational studies.
Although, as folklore has it, Americans did slightly favor algebraic topics
with some 26.9% of the "active" sample working in those areas, they
almost equally pursued researches in analysis (22.5%) and geometry
(21.7%) [67]. Furthermore, while group theory did engage a substantial
14.1% of those "active," it did not outdistance algebraic and differential
geometry as a subspecialty (14.7%). In order to get a better idea of the
actual research done by our constituency over the years from 1891 to 1906,
consider the work of three members of our "most active" sample space,
George A. Miller, Maxime Bôcher, and Oswald Veblen. Their cases
represent the best of American mathematics during this time period as
they illustrate the breadth and extent of the mathematical endeavor in the
United States.

Table 12.

Mathematical Areas of Interest of the Active Members of the American
Mathematical Community 1891–1906

	Most Active		Active	
Algebra				
–Finite, Substitution, and Continuous Groups	16	25.8%	45	14.1%
–Invariant Theory	7	11.3%	18	5.6%
–Linear Associative Algebra	6	10.0%	14	4.4%
–Quaternions and Vector Algebra	4	6.5%	9	2.8%
Analysis				
–Automorphic Functions	4	6.5%	5	1.6%
–Calculus of Variations	4	6.5%	8	2.5%
–Differential Equations	0		19	5.9%
–General Function Theory	10	16.1%	29	9.1%
–Hyperelliptic and Elliptic Functions and Integrals	3	4.8%	11	3.4%
Analysis Situs	1	1.6%	4	1.3%
Elementary Mathematics and Pedagogy	3	4.8%	41	12.8%
Engineering and Applied Mathematics	0		10	3.1%
Foundations and Logic	9	14.5%	17	5.3%
Geometry				
–Projective and Non-Euclidean	9	14.5%	22	6.9%
–Algebraic and Differential	13	21.0%	47	14.7%
History and Philosophy of Mathematics	2	3.2%	13	4.1%
Mathematical Physics and Astronomy	13	21.0%	56	17.5%
Number Theory	5	8.1%	29	9.1%
Probability	0		1	0.3%
Other Interests*	0		4	1.3%
No Specified Interests	2	3.2%	26	8.1%

*The "other interests" included: mathematical economics, biology, ancient history, and oph-thamology!

"Interests" were determined according to listings in the first, second, or third edition of *American Men of Science* and/or titles of published papers or delivered talks.

Note: Since many individuals listed more than one interest, the numbers in these columns do not add up to 62 and 320, respectively, and the percentages add up to more than 100%. They have been rounded to the nearest 0.1%.

Beginning with algebra, the work of that prolific mathematician, G. A. Miller, did much to establish America's notoriety in the field of finite group theory [68]. Born in Pennsylvania in 1863, Miller followed up his 1887 Bachelor's degree from his home state's Muhlenburg College with a doctorate from Cumberland (Kentucky) College in 1893. Like Thomas Proctor Hall's Ph.D. from Illinois Wesleyan, Miller's Cumberland credential by no means reflected the best possible graduate training then available in the United States. To supplement his education, and probably at the instigation of his colleague at the University of Michigan, Frank Nelson Cole, Miller left his Michigan instructorship after two years for a study tour in Leipzig and Paris from 1895 to 1897. With Sophus Lie at Leipzig and Camille Jordan in Paris, these two cities represented crucial stops in the tour of any traveling algebraist. Oddly enough, Miller does not show up on the extant rosters for Lie's courses in 1895 and 1896, even though several of his published papers from these years give Leipzig as his home base [69].

Whether or not Miller actually studied under Lie, he returned from his trip abroad to a four-year stint as an Instructor at Cornell, moved to an assistant and then associate professorship at Stanford from 1901 to 1906, and finally took the position at the University of Illinois he would hold until his retirement in 1931. Undaunted by the relative transience of his early mathematical career, Miller managed to present 48 papers before the AMS, publish no fewer than 62 others, and earn ten service marks during the period of our study, for a total of 120 [70]. His voluminous, although not always profound, researches focused on the theory of groups in general and on the theory of finite groups in particular.

Approaching group theory more as a series of curious examples than as a unified general theory, Miller worked on the classification of such group-theoretic creatures as "Non-Abelian Groups In Which Every Subgroup Is Abelian," "The Groups of Order p^m Which Contain Exactly p Cyclic Subgroups of Order p^a," and "The Non-Regular Transitive Substitution Groups Whose Order Is the Cube of Any Prime Number" [71]. Clearly intrigued by theorems like those of Sylow, which guarantee certain subgroups within groups on the basis of combinatorial considerations, Miller literally made a career out of dissecting finite groups in virtually every way conceivable.

Not as prolific as Miller but perhaps more influential ultimately, Maxime Bôcher made his mark both nationally and internationally in the branch of classical analysis known as potential theory. After earning his Bachelor's degree from Harvard in 1888 at the age of twenty, Bôcher won a Parker Fellowship for study abroad. He took this to Göttingen, where he earned his doctorate under Felix Klein in 1891. Bôcher returned to the United States to assume an instructorship at his *alma mater* in 1891 and steadily moved up the ranks there, becoming a full professor in 1904. At Harvard, he, together with his friend and colleague, William Fogg Osgood, defined one of the key spheres of Felix Klein's influence in the United States. Furthermore, by 1910, Bôcher and Osgood had established one of the three strongest, research-oriented mathematics programs in the United

States. (The University of Chicago and Princeton claimed the other two.) [72].

As a researcher, Bôcher concentrated initially on the geometrical approach to potential theory, which he had first learned about in his adviser's course at Göttingen during the 1889–1890 academic year. Requiring a firm command of the theories of elementary divisors, boundary-value problems in partial differential equations, and Lamé polynomials, this area demanded much mathematically. Bôcher proved equal to the task, however, owing partly to the fact that he had already been schooled in the basics of potential theory by his undergraduate professors, William E. Byerly and Benjamin O. Peirce. By 1894, he had expanded his award-winning dissertation into a classic book, *Ueber die Reihenentwicklungen der Potentialtheorie*, in which he gave an exhaustive analysis of the special kind of coordinate systems Klein advocated for use in the potential-theoretic setting [73]. Following the publication of this book, Bôcher frequently represented the fruits of his continued researches, particularly in the theory of linear differential equations, to the American mathematical community. During the 15 years from 1891 to 1906, in fact, he gave 27 talks and published 46 papers in the American context. His broader influence within the community was further reflected in the 18 service marks he earned during that same time period.

While Bôcher worked in New England to extend the frontiers of classical analysis, another young entrant onto the American scene, Oswald Veblen, made significant strides in geometry from his academic base in Chicago. Veblen, who earned an A.B. at the University of Iowa in 1898 and repeated the degree at Harvard in 1900, studied at the University of Chicago from 1900 to 1903. During this time period, Veblen's teacher and adviser, E. H. Moore, had recognized the importance of Hilbert's 1899 *Grundlagen der Geometrie* and had thus shifted his primary research interests from group theory to these new foundational ideas. He guided Veblen in this same direction, and by 1903 the latter had written his thesis on "A System of Axioms for Geometry" [74]. There, Veblen presented a system of twelve independent axioms for geometry—distinct from that given by Hilbert—and dependent only on the undefined notions of point and order [75]. After completing his degree, Veblen stayed on at Chicago at the rank of Associate from 1903 to 1905 and did important collaborative work in 1905 with the visiting Scottish algebraist, Joseph H. M. Wedderburn. In their one joint paper, Wedderburn and Veblen pooled their respective mathematical talents and gave to the *Transactions* the first constructions of finite geometries which failed to satisfy the theorems of both Desargues and Pascal [76]. Veblen continued his work in projective geometry following his departure from Chicago for a preceptorship at Princeton University. From the vantage point, first of Princeton and then of the Institute for Advanced Study, Veblen pushed his interests in projective geometry into the newer area of topology—or analysis situs as it was then called—and established himself as an international leader in the field.

As exemplified by the cases of Miller, Bôcher, and Veblen, the American mathematical community covered newer as well as more classical areas in the years from 1891 to 1906. Its members did not shy away from those topics with a long history and much background to master in favor of then shallower but deepening domains. Through their various pursuits, they began to make their presence known within international mathematical circles. Although the Sylvester school at Hopkins in the early 1880's awakened Europe to the mathematical potentiality of the United States, by 1910 Europe could no longer ignore that that potentiality had become actuality, and while the Americans surely did not challenge the German mathematical superiority within this time frame, they clearly and respectably held their own.

FOREIGN PARTICIPATION

While Europeans like André Michaux and Thomas Nuttall had come to America in search of new and exotic plants in the early years of the Republic, and while the Swiss natural historian, Louis Agassiz, had played a key role in the development of American science beginning at mid-century, the mathematicians of Europe had generally regarded the New World as a mathematical wasteland and had barely given it a second thought. This situation began to change after Sylvester's successful experiment at The Johns Hopkins University from 1876 to 1883 and as a result of his promotion of the *American Journal* abroad [77]. In fact, Sylvester had so energetically solicited research papers for his journal from foreign sources that, in its early years, the supposedly *American* journal drew as many as a third of its submissions from non-Americans. Quite naturally, most of these foreign contributions came from Sylvester's friends back in England, men like Arthur Cayley and Percy McMahon. Through them and others, word of a mathematical awakening in the United States gradually spread over the British Isles. Similarly, early contributors from France and Italy, men such as Charles Hermite and Francesco Faà di Bruno, brought

American mathematics to the attention of their countrymen. Conspicuously absent from the tables of contents of those first issues of the *American Journal*, however, were the names of German researchers. Not until the 1890's would the Germans begin to view the American mathematical endeavor in a serious light, a turnabout resulting not only from the influx of American students to the German universities but also from the active interest Felix Klein took in the American scene. Furthermore, the Mathematical Congress held in Chicago in 1893 in conjunction with the World's Columbian Exposition as well as the increasingly visible American Mathematical Society, forced the rest of the mathematical world to take notice of the Americans [78]. Indicative of this growing awareness, European mathematicians began to participate in ever greater numbers in the American mathematical community during the years covered by our study. As recorded in the *Bulletin*, they joined the AMS, and they attended mathematical meetings, gave talks, and published papers within the American context (see Table 13).

Given the commonalty of language and heritage, it comes as no surprise that the British Isles contributed the most foreign participants, 42 out of 132 or 31.8%, as well as the largest number of actual foreign members of the AMS, 24 out of 46 or 52.2%. Following in second-place relative to participation, 25 Germans or 18.9% of the 132 either spoke, published, or attended in the American context, but only two of them formally joined the Society. By contrast, 15 mathematicians from neighboring Canada, or just 11.4% of the total, took part, yet 11 of them subscribed to membership to make up 23.9% of the 46 total. Finally, France, Italy, Austria, and Russia also showed measures of support with 17, 10, 7, and 7, respectively.

Since, except for the Canadians residing in the East, the members of the foreign sample found themselves geographically far removed from the activities of the AMS, their participation tended to take the form of publications rather than talks or attendance. As Table 14 shows, while 66.7% of the foreigners gave the American audience at least one published paper, 81.8% of them never attended a meeting, and 59.8% never spoke, even *in absentia*, before the Society [79]. As Table 14 also makes clear, however, several glaring cases broke this general rule.

Most notable among these exceptions, Göttingen's Felix Klein gave 19 talks, published three papers, and attended three separate mathematical events for a total of 25, the highest of any foreign participant. Twice in America during our period, Klein attended both the Chicago Congress and the Evanston Colloquium in 1893 and then returned for Princeton's sesquicentennial celebrations in 1896. On each of these visits, but espe-

Table 13.

Foreign Participation in the American Mathematical
Community 1891–1906

	No. of Participants		No. of AMS Members	
Australia	1	0.8%	0	
Austria	7	5.3%	0	
British Isles	42	31.8%	24	52.2%
Canada	15	11.4%	11	23.9%
France	17	12.9%	2	4.3%
Germany	25	18.9%	2	4.3%
Greece	1	0.8%	0	
India	2	1.5%	2	4.3%
Italy	10	7.6%	4	8.7%
Mexico	1	0.8%	1	2.2%
Russia	7	5.3%	0	
Sweden	2	1.5%	0	
Switzerland	1	0.8%	0	
Country Unknown*	1	0.8%	0	
Totals	132		46	

*A Professor A. W. Scott attended the Fourth Summer Meeting of the AMS at the University of Toronto. He may have been either Canadian or American. "Foreign participants" are defined here as those who were born *and* who pursued their careers outside of the United States during all or most of the period 1891–1906. Thus, the German-born mathematician and chess champion, Emmanuel Lasker, has been counted as an "American" since he emigrated to the United States in 1890 and lived in this country during our period. Although born in Nova Scotia and a professor at Dalhousie from 1901–1907, D. A. Murray has been counted as an "American" since he spent the years from 1890–1894, 1894–1901 on the faculties of New York University and Cornell, respectively. Finally, Ontario-born Thomas Proctor Hall spent the years from 1893 to 1905 in various posts in the United States before returning permanently to Canada, so we have numbered him among the Americans. Those Canadians—DeLury, Fields, Findlay, Glashan, and Wilson—included in the active sample have also been counted here, since they also satisfy our definition of "foreign participants." Percentages have been calculated relative to the respective totals of 132 and 46 and have been rounded to the nearest 0.1%.

cially on the first, he lectured extensively on the latest developments in mathematical research. At the Chicago Congress, Klein, as the featured speaker, gave survey talks on "The Present State of Mathematics" and "Ueber die Entwicklung der Gruppentheorie während der letzten Jahre" [80]. Moving on to the colloquium organized in his honor in nearby Evanston, he continued with a series of 12 lectures carefully crafted to give his American listeners the broadest possible exposure to the then current

Table 14.

Participation Record of the 132 Foreign Participants in the American
Mathematical Community 1891–1906

	Talks		Papers Published		Attendance		Totals	
0	79	59.8%	44	33.3%	108	81.8%	26	19.7%
1	36	27.3%	49	37.1%	16	12.1%	36	27.3%
2	8	6.1%	19	14.4%	3	2.3%	34	25.8%
3	4	3.0%	10	7.6%	3	2.3%	7	5.3%
4	2	1.5%	2	1.5%	1	0.8%	11	8.3%
5	0		4	3.0%	0		3	2.3%
6	1	0.8%	3	2.3%	0		6	4.5%
7	0		0		0		1	0.8%
8	0		1	0.8%	0		1	0.8%
9	1	0.8%	0		0		0	
10	0		0		0		1	0.8%
11	0		0		0		3	2.3%
12	0		0		0		0	
13	0		0		0		0	
14	0		0		0		1	0.8%
15	0		0		0		0	
16	0		0		0		0	
17	0		0		1	0.8%	0	
18	0		0		0		0	
19	1	0.8%	0		0		1	0.8%
20	0		0		0		0	
25	0		0		0		1	0.8%

Professor A. W. Scott, with 0 talks, 0 papers published, and 1 attendance has been included in these figures.

The percentages have been calculated relative to the total of 132 foreign participants and rounding has been done to the nearest 0.1%.

research areas in mathematics [81]. Finally, the New York Mathematical Society convened a special meeting in Klein's honor prior to his return home. On that occasion, he returned to the geometrical themes he had expounded upon in Evanston and lectured on recent work in spherical trigonometry. As reported in the *Bulletin*, "[i]n concluding he deprecated the opinion obtaining among many persons that the so-called elementary subjects offer to the mathematician no opportunities for investigation" [82]. When Klein returned three years later for the festivities at Princeton in October of 1896, the by then American Mathematical Society honored him once more, this time by moving its regular meeting from New York

City to Princeton where Klein gave four lectures on the "Theory of the Top." This seems to have been the first time a regular meeting of the Society was ever held outside of New York [83].

The AMS did not make quite as much fuss when the French mathematician Jacques Hadamard came to the United States as the University of Paris's official representative to Yale's bicentennial in 1901, but it did invite him to speak at its regularly scheduled October meeting in New York. Explaining his recent findings "On the Theory of Elastic Plates," Hadamard later published them in the pages of the *Transactions* [84]. In addition to this official presentation before the Society, he also went on a short lecture tour to Columbia, Princeton, and Yale, speaking "On the Mathematics of Physics," his area of specialization [85]. In all, Hadamard gave four talks, published two papers, and attended four American mathematical functions for a total of ten in our study.

As Continental mathematicians, Klein and Hadamard owed their high levels of participation in American mathematics primarily to the fact that they actually journeyed to the United States. Once here, their relatively short stays were packed with mathematical activities as different isolated groups of Americans sought to benefit and learn from them firsthand. However, another group of foreign participants, the Canadians living in the East, had the opportunity to contribute to the American scene in a more regular and leisurely fashion.

Of the 15 Canadians who appear in our sample, only one was active outside of the easternmost provinces during our period and most of the rest lived or worked in Ontario [86]. Thus, by and large, the Canadians were close enough that they could take part in mathematical activities across the border with some regularity. University of Toronto professor J. Charles Fields, for example, gave six talks—either in person or *in absentia* —at AMS functions, published five papers, and attended three meetings for a total of 14 in our count. A specialist in the theory of Abelian integrals and of algebraic functions, Fields had earned his Ph.D. at Johns Hopkins in 1887 and had stayed on there as a postgraduate fellow until 1889. After three years at Pennsylvania's Allegheny College, he spent seven years abroad, studying at Paris, Göttingen, and Berlin [87]. On his return to this continent, he used his family home in Hamilton, Ontario as the base from which he traveled to attend meetings of the Society held at Chicago, Columbia, and Cornell. He continued to remain visible within the AMS after he secured the post at Toronto in 1902. (The Fields Medal bears his name today.) Another highly visible Canadian, William Findlay, taught at McMaster University in Hamilton. With the second highest total

(19) in our study, Findlay only contributed one talk and one published paper to the American mathematical community during our time frame, but he attended at least 17 separate meetings of the Society. After taking his Ph.D. in 1901 under E. H. Moore at Chicago, Findlay moved to a tutorship at Barnard College in New York City until he accepted a position at McMaster in 1905 [88]. Since he lived in New York City for four of the 15 years from 1891 to 1906, his was another exceptional case for a foreign participant.

While as Table 14 indicates, most foreign mathematicians fell far short of the participation levels of Klein, Hadamard, Fields, and Findlay, foreigners nonetheless made up 132 or 12.4% of the 1061 people associated in some way with American research-level mathematics between 1891 and 1906. Through their interaction, whether great or slight, these foreigners helped not only to inform the American audience of then current researches abroad but also to pass on to their respective communities the news and advances of American mathematics. Furthermore, such notables as Klein, Hadamard, Elie Cartan, Maurice Fréchet, Paul Gordan, Emile Goursat, David Hilbert, Emile Picard, and Henri Poincaré, among others, took the United States seriously enough to present their work to the broader public through its journals. This, too, reflects the growing credibility of the American mathematical endeavor. In the 30 years from 1876 to 1906, and particularly in the last half of this period, the foreign perception of mathematical America had clearly shifted from "wasteland" to "fertile field" worthy of cultivation.

CONCLUSION

As should now be evident, dramatic changes took place in American mathematics over the 15 years from 1891 to 1906. The New York Mathematical Society counted barely 100 on its rosters when it launched its *Bulletin* in 1891. Starting as a volume thin in mathematics as well as news, the *Bulletin* rapidly grew into the major clearinghouse of information for a geographically dispersed American mathematical community numbering over a thousand. A kind of historical microscope, this source records the involvement of some 1061 mathematically interested individuals who contributed in some way to the viability and vitality of the community. Of these, 60 men and two women sustained the endeavor through their high visibility, appearing before the society, publishing their research, and rendering service to the common mathematical cause. Furthermore, these individuals, whom we term the "most active," molded the graduate programs at the nation's colleges and universities and set the country's standards for mathematical research.

Following their work and their examples, 242 men and 16 women maintained less visible but still "active" careers. Owing to their high levels of educational achievement, these "active" participants appreciated the researches that the 62 "most active" presented to the Society and sometimes contributed researches of their own. Thus, they were more than just mathematically literate, they were mathematically productive, and they pursued their studies and their work in virtually every corner of the United States.

Although the Northeast and Midwest supported them in the largest concentrations, the West expanded rapidly over the 15-year period, further tilting the demographics of mathematics westward and dispersing the power base more evenly across the country. The society may have continued to maintain its headquarters in New York City, but the Chicago and San Francisco Sections increasingly challenged the Northeasterners both in influence and in mathematical output. Also, these multiple focal points, coupled with communication outlets such as the *Bulletin* and the *Transactions*, rendered American mathematics accessible to an interested, but not necessarily productive, "rank and file." Since these participants—563 men and 51 women—tended not to engage in research themselves, they by and large left no published legacy of their involvement in American mathematics. Thus, their presence might well have passed unnoticed. Set on its highest power, however, our microscope reveals the extent and depth of this "rank and file" and thereby uncovers the real, and surprising, extent and depth of the turn-of-the-century American mathematical community.

The prevailing folklore is simply wrong. American mathematics in 1900 was much more than a few German-trained men at the University of Chicago, at Harvard, and perhaps at Columbia, Yale, and Princeton, who made some modest contributions to mathematics. It was a broadly-based endeavor with active contributors all over the country, further legitimized by 132 participants from abroad. It encompassed all branches of the mathematical discipline: algebra, analysis, geometry, number theory, mathematical physics and astronomy, engineering. Thus, when those talented, world-class refugees arrived in America in the 1930's, they did not create an American mathematical community from nothing. Rather, they stepped into a strong, active, deeply-rooted community which, as our study shows, came into being over the period from 1891 to 1906.

NOTES

[*] The authors wish to thank the referees of an earlier version of this paper for their helpful and insightful comments and suggestions.

[1] On nineteenth-century developments in American mathematics, see Karen Hunger Parshall and David E. Rowe, *The Emergence of the American Mathematical Research Community 1876–1900: James Joseph Sylvester, Felix Klein, and Eliakim Hastings Moore* (Providence: American Mathematical Society), forthcoming. Chapter 1 deals with the period from 1800 to 1876. For a contemporaneous view of this period, see Florian Cajori, *The Teaching and History of Mathematics in the United States* (Washington, D.C.: Government Printing Office, 1890). The dismal situation relative to journals has been outlined in Karen Hunger Parshall, A Century-Old Snapshot of American Mathematics, *The Mathematical Intelligencer* **12** (1990): 7–11.

[2] On the general developments in science in nineteenth-century America, see John C. Greene, *American Science in the Age of Jefferson* (Ames: The Iowa State University Press, 1984); George H. Daniels, *American Science in the Age of Jackson* (New York: Columbia University Press, 1968); and Robert V. Bruce, *The Launching of Modern American Science, 1846–1876* (New York: Alfred A. Knopf, 1987).

[3] See, for example, Laurence R. Veysey, *The Emergence of the American University* (Chicago: University of Chicago Press, 1965).

[4] Joseph Ben-David, The Universities and the Growth of Science in Germany and the United States, *Minerva* **7** (1968–1969):1–35, esp. 6–9.

[5] Karen Hunger Parshall, America's First School of Mathematical Research: James Joseph Sylvester at the Johns Hopkins University 1876–1883, *Archive for History of Exact Sciences* **38** (1988):153–196; Eliakim Hastings Moore and the Founding of a Mathematical Community in America, 1892–1902, *Annals of Science* **41** (1984):313–333; Karen Hunger Parshall and David E. Rowe, American Mathematics Comes of Age: 1875–1900, 3–28, in Peter Duren, et al., eds., *A Century of Mathematics in America—Part III* (Providence: American Mathematical Society, 1989); and Parshall and Rowe, *The Emergence of the American Mathematical Research Community*.

[6] David E. Rowe, Die Wirkung Deutscher Mathematiker auf die Amerikanische Mathematik, 1875–1900, *Mitteilungen der Mathematischer Gesellschaft der DDR* 3–4 (1988):72–96.

[7] On the histories of these various societies, see Charles Albert Browne and Mary Elvira Weeks, *A History of the American Chemical Society: Seventy-Five Eventful Years* (Washington, D.C.: American Chemical Society, 1952); Raymond Clare Archibald, ed., *A Semicentennial History of the American Mathematical Society: 1888–1938*, 2 vols. (New York: American Mathematical Society, 1938); and Daniel J. Kevles, *The Physicists: The History of a Scientific*

Community in Modern America (New York: Alfred A. Knopf, 1978). Two recent books also deal with the emergence of research-level work in other sciences in the United States. See Jane Maienschein, *Transforming Traditions in American Biology, 1880–1915* (Baltimore: The Johns Hopkins University Press, 1991); and John W. Servos, *Physical Chemistry from Ostwald to Pauling: The Making of a Science in America* (Princeton: Princeton University Press, 1990).

[8] See note 1 above.

[9] We have drawn from James McKeen Cattell, *American Men of Science: A Biographical Directory*, 1st ed. (New York: The Science Press, 1906); 2nd ed. (New York: The Science Press, 1910); and 3rd ed. (New York: The Science Press, 1921). In statistically analyzing the American scientific community in the first half of the twentieth century, R. H. Knapp and H. B. Goodrich utilize the third (1921) and the seventh (1944) editions of *American Men of Science*. They discuss the usefulness, validity, and shortcomings of this reference as a source. See R. H. Knapp and H. B. Goodrich, *Origins of American Scientists: A Study Made under the Direction of a Committee of the Faculty of Wesleyan University* (Chicago: University of Chicago Press, 1952).

In 1936, then Secretary of the AMS, R. G. D. Richardson, prepared a statistical analysis of American mathematical research production (1862–1934) based on the *Annual List of Papers Read before the Society and Subsequently Published* and other sources. He published his findings in The Ph.D. Degree and Mathematical Research, *American Mathematical Monthly* **43** (1936):199–215; reprinted in Peter Duren, et al., eds., *A Century of Mathematics in America—Part II* (Providence: American Mathematical Society, 1989), 361–378. More recently, Daniel Kevles contributed a comparative study, entitled The Physics, Mathematics, and Chemistry Communities: A Comparative Analysis, to Alexandra Oleson and John Voss, eds., *The Organization of Knowledge in America, 1860–1920* (Baltimore: The Johns Hopkins University Press, 1979), 139–172. There, he surveys the "productive" members of the respective communities over the period 1870–1914 and defines productivity solely in terms of publication.

[10] Eliakim Hastings Moore, Oskar Bolza, Heinrich Maschke, and Henry S. White, eds., *Mathematical Papers Read at the International Mathematical Congress Held in Connection with the World's Columbian Exposition: Chicago 1893* (New York: Macmillan and Co., 1896).

[11] This account of the early history of the Society follows Archibald, ed., 1:4–5.

[12] Archibald, ed., 1:48. For instance, Sylvester's student, Ellery William Davis, then at Columbia College in South Carolina, joined the Society as a result of this drive as did Alexander Ziwet of the University of Michigan. (Davis moved to the University of Nebraska in 1893.) The Mathematical Society had an essentially open membership policy, unlike some scientific societies of the time.

[13] *Ibid.*

[14] Henry B. Fine, Kronecker and his Arithmetical Theory of Algebraic Equation, *Bulletin of the New York Mathematical Society* **1** (1892):173–184 (hereinafter cited as *Bull. NYMS*, or *Bull. AMS* after 1894; Alexander Ziwet, The Annual Meeting of the German Mathematicians, *Bull. NYMS* **1** (1892):96–101; and Charlotte A. Scott, Edwards' Differential Calculus, *Bull. NYMS* **1** (1892):217–223.

[15] *Bull. NYMS* **1** (1891):32.

[16] *Bull. NYMS* **1** (1892):215.

[17] *Ibid.* 169.

[18] For a complete bibliography and biographical sketch, see A. Adrian Albert, "Leonard Eugene Dickson: 1874–1954," *Bull. AMS* **61** (1955):331–345; or A. Adrian Albert, ed., *The Collected Mathematical Papers of Leonard Eugene Dickson*, 5 vols. (New York: Chelsea Publishing Co., 1975).

[19] *Bull. AMS* **10** (1904):518. Also on the organizing committee were Missouri's Earle R. Hedrick and Northwestern's Henry S. White.

[20] After earning his Harvard Ph.D. in 1903, David Curtiss proceeded to a year abroad in Paris before taking up an instructorship in mathematics at Yale's Sheffield Scientific School for the academic year 1904–1905. He spent more than the rest of our period on the faculty at Northwestern where he pursued his researches in function theory and differential equations. William Fite received his doctorate in finite group theory in 1901 from Cornell and spent the years from 1892 through 1906, first at the Michigan Military Academy and later Cornell. Max Mason graduated from Göttingen in 1903 and returned to MIT for a year before taking a position at the Sheffield School from 1905 to 1908. He specialized both in the theory of differential equations and in the calculus of variations. Getting his Ph.D. at Chicago in 1903 for a thesis on the foundations of geometry, Oswald Veblen stayed on at his *alma mater* for two years before accepting the call to Princeton in 1905. He remained on the faculty there until the opening of the Institute for Advanced Study in 1931. Finally, John Wesley Young, like Fite, got his doctorate at Cornell, but somewhat later in 1904. The brother-in-law of E. H. Moore, Young pursued his interests in group theory and automorphic functions during our period, first at Northwestern (1903–1905) and then at Princeton (1905–1908).

[21] After Dickson, the most prolific mathematician in our sample, George A. Miller, worked on the theory of finite groups and had a career which took him from the University of Michigan (1893–1895) to Cornell (1897–1901) to Stanford (1901–1906) and ultimately to Illinois. (See the section, "Interests of the Participants," below.) Edgar O. Lovett enjoyed the distinction of being one of Sophus Lie's American students at Leipzig and earned his second Ph.D. under Lie's direction there in 1896. (The first came from the University of Virginia in 1895.) Shortly after his return to the United States, he went to Princeton where he remained until 1908. University of Virginia astronomer, Ormond Stone, founded and then edited the *Annals of Mathematics* from 1884

to 1899 while directing the Leander McCormick Observatory. Finally, Alexander Ziwet, a mathematical physicist at the University of Michigan, participated continuously as a member of the Society's committee of publication in addition to pulling yeoman's duty for the Chicago Section. Since the careers of the others presented in this list have been well documented in many sources, we shall not sketch them here but refer the interested reader to, for instance, Archibald, ed., 1:99–103, 124–166.

[22] Here, we focus on the American mathematical research community as a whole, and so mention the participation of women only within that broader context. We deal with the women separately and in greater detail in our companion article, Della Dumbaugh Fenster and Karen Hunger Parshall. Women in the American Mathematical Research Community: 1891–1906 [this volume, 229–261].

[23] Attendance at AMS meetings was not carefully documented until 1894 and after. Thus, we have been unable to include data on attendance at the early meetings.

[24] *Bull. AMS* **5** (1899):325; *Bull. AMS* **7** (1901):373; *Bull. AMS* **10** (1904):221 and 485; and *Bull. AMS* **11** (1905):201.

[25] *Bull. AMS* **4** (1898):182; *Bull. AMS* **6** (1899):1; *Bull. AMS* **8** (1901):1; and *Bull. AMS* **11** (1905):55. Somewhat surprisingly, Clark University's William E. Story also had a combined total of zero and attended only five (but perhaps more) AMS meetings. Story, who specialized in invariant theory and geometry after earning his Leipzig Ph.D. in 1875, was Sylvester's right-hand man from 1876 on and accepted the call to Clark for its opening in 1889.

[26] See various numbers from *Bull. AMS* **8** (1902) through *Bull. AMS* **12** (1906).

[27] Henry T. Eddy, Modern graphical developments, 58–71 in E. H. Moore, et al., eds., *Mathematical Papers: Chicago 1893*. For a discussion of this and other papers in this volume, see Parshall and Rowe, *The Emergence of the American Mathematical Research Community*, Chapter 7.

[28] Dennett served on the auditing committee as early as 1893 (*Bull. AMS* **2** [1893]:111) and was first elected treasurer in 1901 (*Bull. AMS* **7** [1901]:201).

[29] We discuss the foreign participants in American mathematics below.

[30] We decided to exclude these 50 people from the "active" (or, in one case, the "most active") category for reasons of statistical uniformity. The *American Men of Science* entries lacking, we would have been unable to include data on them in the various tables which follow. Were they added in, however, the "active" sample space would increase to 370, perhaps a more representative figure.

[31] On Gibb's election to membership, see *Bull. AMS* **9** (1903):393. Announcements of Gibb's many awards, both foreign and domestic, did appear in the *Bulletin*. In addition to teaching and inspiring Gibbs, Newton also figured prominently as E. H. Moore's mentor and doctoral adviser at Yale. For more

on Newton's life, see Andrew W. Phillips, Hubert Anson Newton, *Bull. AMS* **3** (1897):169–173.

[32] On Craig's life, see David Eugene Smith's article in the *Dictionary of American Biography*.

[33] Albert M. Swain, Lagrange's Sextic. *Annals of Mathematics* **6** (1891):14; The Algebraic Solution of Equations, *Annals of Mathematics* **6** (1892):169–177; The Rational Functions of the Cubic, *Annals of Mathematics* **9** (1894–1895):158–162; and The Algebraic Solution of Equations, p. 330 in E. H. Moore, et al., eds., *Mathematical Papers: Chicago 1893*. This latter paper was presented before the Congress but only appeared by title in the congress volume since it had already been published under the same title in the *Annals*.

[34] Prior to joining the AMS in 1900, Vivian won the University of Pennsylvania's Alumnae Fellowship for Women in 1898. Holding it for the three years of her immediately pre-doctoral research, she earned her degree in 1901 for a dissertation on "The poles of a right line with respect to a curve of order n" and proceeded immediately to an instructorship in mathematics at Wellesley. She remained in this post through 1906, but in the spring of that year announced that she would be on leave from Wellesley for the 1906–1907 academic year in order to teach at the American College for Girls in Constantinople. See *Bull. AMS* **5** (1898):110; *Bull. AMS* **7** (1901):201 and 326; *Bull. AMS* **8** (1901):35; and *Bull. AMS* **12** (1906):510.

[35] For an in-depth discussion of women in American science, see Margaret Rossiter, *Women Scientists in America: Struggles and Strategies to 1940* (Baltimore: Johns Hopkins University Press, 1982), especially Chapters 1–2.

[36] See note 22 above.

[37] George H. Daniels, *American Science in the Age of Jackson* (New York: Columbia University Press, 1968), 6–33, esp. 22–23.

[38] Robert V. Bruce, A Statistical Profile of American Scientists, 1846–1876, 63–94, esp. 71 in George H. Daniels, ed., *Nineteenth-Century American Science: A Reappraisal* (*Evanston: Northwestern University Press*, 1972).

[39] We calculated the median birth date of the "active" sample space based on information from the first three editions of *American Men of science*. The median for the "most active" was 1865. Thus, the median ages of the two samples in 1900 were 33 and 35, respectively.

[40] Bruce, "A Statistical Profile," 74. We have deduced these margins from Bruce's Table 4. Since his data reflects only the native-born, it is not strictly comparable to our figures. He does not give a full breakdown of his sample space of *mathematicians* by birth place, that is, he does not provide figures on foreign-born mathematicians separately. Nevertheless, at least some indication of the productivity over time of the various regions emerges by comparison.

[41] See the annotations to Table 4 for the definitions we have adopted for the various geographical regions. Note that these and the percentages which

follow are calculated with respect to the total number of "active" mathematicians, not relative to the total number of native-born (257). Population data for the various states come from U.S. Bureau of the Census, *Historical Statistics of the United States: Colonial Times to 1970*, 2 vols. (Washington, D.C.: U.S. Government Printing Office, 1975).

[42] As remarked in the annotations accompanying Table 5, these figures should not be taken as absolute. Especially relative to the "rank and file," our information is incomplete at best. These numbers merely reflect identifiable trends.

[43] On the government as an employer of mathematicians, see A. Hunter Dupree, *Science in the Federal Government: A History of Policies and Activities* (Baltimore: Johns Hopkins University Press, 1986); and Bruce, *The Launching of Modern American Science: 1846–1876*.

[44] We have not isolated separate figures for mathematicians in industry in Table 7 due to the small numbers involved. Of the 320 in the "active" category, five earned a living, at least at some point during our 15-year period, in the insurance industry and four in engineering. (Here, we have included neither professors of engineering or actuarial science nor consultants who otherwise held teaching jobs.) Most notable among the engineering contingent, the German-born electrical engineer Charles Proteus Steinmetz gave four talks and published two papers for a total of six relative to our count. These nine (except Steinmetz, who did hold one academic post at Union College at the end of our period) are included in the figure of 22 representing zero academic affiliations.

[45] On the history of these Sections, see Archibald, ed., 1:8–9, 74–81. Arnold Dresden organized statistics on the first 25 years of the Chicago Section in A Report on the Scientific work of the Chicago Section, 1897–1922, *Bull. AMS* **8** (1922):303–307. On the San Francisco Section, see E. J. Wilczynski, The First Meeting of the San Francisco Section of the American Mathematical Society, *Bull. AMS* **8** (1902):429–437. In all, 16 papers were presented on this occasion.

[46] Bruce, A Statistical Profile, 89.

[47] As one of our "active" members, Hall gave five talks and published one paper in the *American Journal* on The Projection of Fourfold Figures Upon a Three-Flat [**15** (1893):179–189].

[48] Louis N. Wilson, List of Degrees Granted at Clark University and Clark College: 1889–1914, Publications of the Clark University Library **4** (1914):1–52, esp. p. 7.

[49] Although Hopkins refused to grant Christine Ladd-Franklin her Ph.D. in mathematics, which she had earned under Sylvester's direction in the 1880's, it made an exception for Florence Bascom and awarded her a Ph.D. in geology in 1893. Subsequently, women were admitted first to the Hopkins Medical School and only later to the University proper. Hopkins did confer the

doctorate on Ladd-Franklin belatedly in 1926. For more on American universities and the doctorate for women, see Rossiter, Chapters 1 and 2.

[50] Göttingen awarded its first doctorate to a woman in 1895, when Felix Klein's British student, Grace Chisholm (later Young), satisfied the requirements for her degree in mathematics. Mary Frances "May" Winston (later Newson) was the first American woman to receive a Göttingen doctorate in mathematics (also under Klein) in 1897. See our companion paper (referred to in [22] above) for more details.

[51] On the role of Justus Liebig in training American agricultural chemists in the 1850's, see Charles E. Rosenberg, *No Other Gods: On Science and American Social Thought* (Baltimore: The Johns Hopkins University Press, 1976), Chapter 8; and Margaret Rossiter, *The Emergence of Agricultural Science: Justus Liebig and the Americans, 1840–1880* (New Haven: Yale University Press, 1975). See also Servos's book referred to in note 7 above.

[52] Mary Frances "May" Winston, the first American woman to earn a Göttingen doctorate in mathematics, used this phrase in describing her feelings toward a German education. See the biographical article typewritten by her daughter, Mrs. Caroline Beshers, and held in the Smith College Archives.

[53] The logbooks recording the enrolled students in the various courses at Leipzig are held in official archives in Dresden. We thank David Rowe for sharing the information provided to him by Professor Walter Purkert of Leipzig.

[54] *Ibid.*

[55] *Bull. NYMS* **3** (1893):66.

[56] See [53].

[57] The logbooks recording the course enrollments for Felix Klein's courses are held at the University of Göttingen Archives. We thank David Rowe for sharing this information with us. See, also, E. H. Moore, et al., eds., *Mathematical Papers: Chicago 1893*, x.

[58] *Ibid.*, and *Bull. AMS* **8** (1901):38.

[59] The *American Men of Science* entries clearly reflect the importance placed on foreign travel, since the entrants fairly carefully documented their various trips abroad chronologically in their write-ups.

[60] *Bull. AMS* **1** (1893):260.

[61] *Bull. AMS* **2** (1895):83.

[62] Chicago did, in fact, offer these more applied courses, but through its departments of physics and through its astronomy courses.

[63] *Bull. AMS* **1** (1895):210.

[64] *Bull. AMS* **2** (1895):24–25.

[65] Furthermore, at this time, those schools which counted an astronomer among their faculties tended to have only one and counted him or her administratively as a member of the Mathematics Department. Thus, the theoretical as opposed to observational astronomers of the day belonged to the mathematical community by association. Also worth noting, even though the official

national societies for astronomy and physics came along only in 1899, these scientists fit quite comfortably into the AAAS.

[66] Of the first five presidents, John Van Amringe, Emory McClintock, George Hill, Simon Newcomb, and Robert Woodward, Hill and Newcomb were internationally renowned astronomers and Woodward was a well-known physicist. For more on the careers and publications of these men, see Archibald, ed., 1:117–144.

[67] These percentages represent upper bounds of interest within the "active" sample, since some people have been counted more than once within, say, the analysis category, having interests in various of the subspecialties isolated in Table 12. As pointed out by one of the referees of an earlier version of this paper, we would have been equally justified in classifying Lie groups under "analysis." This would have completely negated the slight "algebraic" propensity which seems to emerge. The "folklore" thus seems to be fairly conclusively refuted. We thank the referee for pointing this out to us.

[68] Here, we could have chosen to feature Chicago's Leonard E. Dickson. While Dickson was most definitely the better mathematician of the two, we focus on the less well-known Miller and refer the reader to Parshall and Rowe, *The Emergence of the American Mathematical Research Community* for a discussion of Dickson's early work in an historical context. See Chapter 9.

[69] See [53].

[70] Miller's service primarily revolved about the San Francisco Section of the Society, an organization in which he was quite active after his move to Stanford.

[71] These papers appeared in *Trans. AMS* **4** (1903):398–404; *Trans. AMS* **7** (1906):228–232; and *Annals of Mathematics* **10** (1895–1896):156–158, respectively.

[72] See Parshall and Rowe, *The Emergence of the American Mathematical Research Community*, Chapters 6 and 10 for a discussion of Harvard's mathematical development.

[73] Maxime Bôcher, *Ueber die Reihenentwicklungen der Potentialtheorie* (Leipzig: Verlag B. G. Teubner, 1894). See Chapter 5 in Parshall and Rowe, *The Emergence of the American Mathematical Research Community*, for a discussion of this work in context.

[74] David Hilbert, *Grundlagen der Geometrie* (Leipzig: B. G. Teubner, 1899). Veblen's thesis appeared under this title in *Trans. AMS* **5** (1904):343–384.

[75] Hilbert's system required eight undefined terms and had twenty axioms, not all of which were independent.

[76] Oswald Veblen and Joseph H. M. Wedderburn, Non-Desarguesian and Non-Pascalian Geometries, *Trans. AMS* **8** (1907):379–388. For a complete discussion of this work in its broader historical context, see Karen Hunger Parshall, In Pursuit of the Finite Division Algebra and Beyond: Joseph H. M. Wedder-

burn, Leonard E. Dickson, and Oswald Veblen, *Archives internationales d'Histoire des Sciences* **33** (1983):274–299.

[77] For an indication of Charles Hermite's changing attitude on this issue, see Parshall, America's First School of Mathematical Research, 189–190.

[78] Parshall and Rowe discuss this impact of the Chicago Congress on American mathematics in detail in *The Emergence of the American Mathematical Research Community*, Chapter 7.

[79] For some of the more notable foreigners who published in the United States, see the discussion which follows.

[80] E. H. Moore, et al., eds., *Mathematical Papers: Chicago 1893*, 133–135, 136, respectively. Both articles were merely presented in abstract form.

[81] His lectures were published as Felix Klein, *The Evanston Colloquium: Lectures on Mathematics* (New York: Macmillan and Co., 1894).

[82] *Bull. NYMS* **3** (1893):22.

[83] *Bull. AMS* **3**(1896):30.

[84] Jacques Hadamard, La Théorie des Plaques élastiques planes. *Trans. AMS* **3** (1902):401–422. See also *Bull. AMS* **8** (1901):96.

[85] *Bull. AMS* **8** (1901):131–132.

[86] The one Canadian active outside of the eastern provinces was Norman Richard Wilson. From 1900 to 1914, he served on the faculty of the University of Manitoba's Wesley College. He earned his Ph.D. under Oskar Bolza's direction at the University of Chicago in 1908. Two of the 132 foreigners were women and both of them, Harriet Brooks and May E. G. Waddell, were Canadian.

[87] During the Winter Semester of 1894–1895, in fact, Fields took Felix Klein's course at Göttingen. See [57].

[88] Findlay's thesis appeared in print as William Findlay, The Sylow Subgroups of the Symmetric Group, *Trans. AMS* **5** (1904):263–278.

Mary F. Winston Newson
Courtesy of Handschriftenabteilung der Niedersächsische
Staats-und Universitätsbibliothek Göttingen

Women in the American Mathematical Research Community: 1891–1906

Della Dumbaugh Fenster* and Karen Hunger Parshall*[†]

*Department of Mathematics, [†]Corcoron Department of History, University of Virginia, Charlottesville

An American mathematical *research* community in 1891–1906? In the mathematical folklore, such a community only came into existence during the 1930's largely thanks to the influx of talented refugees fleeing the worsening political situation in Europe. *Women* in an American mathematical research community in 1891–1906? Prior to the 1970's, when research in the history of science began to focus on women, only a handful of *European* women in *all* scientific disciplines taken together received recognition in the literature. *American* women went essentially unnoticed; and American women in *mathematics* seemed to define an empty set. But folklore only imperfectly reflects the past, and trends in historiography change to uncover new aspects of the historical record. Here we show that far from absent in a nonexistent research community, women formed a visible part of a viable organization committed to research, the American Mathematical Society.

In our companion paper, "A Profile of the American Mathematical Research Community: 1891–1906," [this volume] we uncovered an American mathematical community consisting of 1061 American and foreign men and women [1]. Using the issues of the *Bulletin of the New York* (later *American*) *Mathematical Society* as our primary source of information, we traced not only the Society's membership but also those on its periphery. A unique blend of mathematical research and news, the *Bulletin* functioned in our study as a sort of historical microscope which enabled us to view American mathematics at various depths. When focused at its lowest setting, the microscope revealed 60 men and 2 women in the most visible group. These individuals, whom we have termed the "most active," comprised the core of the mathematical community from 1891–1906. Following the examples of the "most active," 242 men and 16 women maintained less prominent but still "active" careers within the mathematical constituency. These "active" participants were not only mathematically literate but also mathematically productive, occasionally contributing original work of their own. Moreover, these "active" members pursued their studies and their work in virtually every corner of the United States. The highest power of the microscope ultimately unveiled a group of 563 men and 53 women as interested, but not necessarily productive, in American mathematics. These "rank and file" participants tended not to engage in research themselves, but rather seized other opportunities for involvement as the focal points of mathematics spread westward across the country. At each of these levels of participation, women came to the fore. Here, we examine this generally ignored subspace of the American mathematical community and analyze the depth and extent of the participation of women during the period from 1891–1906.

In his *A Semicentennial History of the American Mathematical Society 1888–1938*, Raymond C. Archibald briefly traced the events which led to the organization of the New York (later American) Mathematical Society in the late 1880's. "The time was ripe," he wrote, "for an organization to draw together many people scattered throughout the country who were especially interested in mathematical pursuits" [2]. This same sense of timeliness proved even more crucial in the emergence of women onto the mathematical scene less than a decade later [3].

It was no mere coincidence that women contributed at all levels of the American mathematical community in the 15 years of our study. The time was indeed ripe, as three efforts converged in this time period to create a climate conducive to women in mathematics. First, more than 150 years of advocacy and advancement in women's education culminated in a general opening of graduate schools and the subsequent awarding of doctorates to women. Second, women desirous of active participation in the mathematical community not only possessed the intellect but also the stamina to withstand the hardships of being the "first." Thirdly, and perhaps most important, women operated in an arena with many influential mathematicians—and others—who not only sympathized with but also supported their active participation. To understand the favorable environment created by these opportunities for education as well as by the internal and external efforts on behalf of women, we must begin our investigation some years earlier.

By the early eighteenth century, women participated in a society which largely adhered to the opinion expressed by Rousseau that "[t]he whole education of women ought to be relative to men. To please them, to be useful to him, to make themselves loved and honoured by them, to educate them when young, to care for them when grown, to counsel them, to console them, and to make life sweet and agreeable to them—these are the duties of women at all time and what should be taught them from their infancy" [4]. The early nineteenth-century New York educator, Emma Hart Willard, exploited precisely this sentiment in her arguments and positive action for the education of women.

Willard's success, in part, hinged on her presentation of higher education for women as a means of creating better mothers. In her highly acclaimed *Women Scientists in America*, Margaret Rossiter noted that, to Willard's advantage, she had also learned that "...if one wishes to introduce seemingly radical reforms it helps to be personally discrete and socially conservative" [5]. A sense of discretion, combined with a unique teaching style and overall verve, allowed Willard to advance the cause of women's education locally to an unprecedented level with the 1821 opening of the Troy Female Seminary [6].

Willard's work at the Seminary set an important precedent which soon translated into opportunities for women at state-run normal schools, coeducational institutes, and, of course, women's colleges [7]. At all of these levels, however, an enormous gap separated the ideal of women's education and the actual implementation of that ideal. As a result, the establishment of these institutions did not take place overnight. Consider, for instance, the case of Oberlin College. In 1833, its founders explicitly articulated as part of the College's mission "the elevation of the female character." They sought to bring "within the reach of the misjudged and neglected sex all the instructive privileges which hitherto have unreasonably distinguished the leading sex from theirs" [8]. In actuality, Oberlin never attained these noble objectives. Operating under the assumption that a woman's mind did not function like that of a man, the college slightly altered its curriculum to compensate for this perceived deficiency [9]. This attitude created and promoted an environment which neither elevated the female character nor provided the unbiased education so ambitiously described by the founders. Moreover, Oberlin apparently advocated women's education, not for its own sake, but for the tacit purpose of training women "for intelligent motherhood and a properly subservient wifehood" [10].

The single most significant step in advancing women's education beyond the realm of motherhood and housewifery came with Mary Lyon and the 1837 establishment of Mount Holyoke Seminary (later College). Through the strength of her will and in a New England environment where expanding opportunities for women were a part of both daily thought and life, Lyon carried women's education to a new height [11]. Under her guidance, Mount Holyoke implemented a tougher admission policy, more rigorous academic requirements, and a longer course of study (three years as opposed to two). Admission to Mount Holyoke depended not only on age (16 or above) and on the ability to complete entrance examinations, but also on a separate demonstration of "maturity and promise of intellectual growth" [12]. Furthermore, the elevated course requirements set the tone for the new educational opportunities available to women: the three years of study included courses in English Grammar, Geography, History, Botany, Euclid, Algebra, Natural Philosophy, Philosophy of Natural History, Intellectual Philosophy, Chemistry, Astronomy, Geology, Religion, Logic, and Rhetoric [13]. By highlighting the ability and capacity of women's minds, Mount Holyoke set the stage for the next major development in women's education, namely, the true women's college. In the two decades following the 1865 opening of Vassar College, the creation of other women's colleges, such as Smith, Wellesley, Bryn Mawr, and the Baltimore College for Women (later Goucher), would finally entrench his ideal, particularly relative to the sciences [14].

Ostensibly dedicated to the equal education of young women, the women's colleges fulfilled yet another important role by providing educated women the opportunity for college-level teaching and research. Consonant with the educational progress of the day, especially with regard to the doctorate as the increasingly important professional credential, the standards for faculty positions rose. Rossiter well characterized the difficulties inherent in this development when she wrote, "The coming of the doctoral degree to higher education in the 1880's created a certain dilemma for the presidents of women's colleges: which was the best way to give their students a top-quality education—by hiring the best women available ... or by giving these precious jobs to persons with doctorates, when they would, because of discriminatory practices at the major American and German (but not Swiss) graduate schools, almost always be men?" [15]. The only way out of this dilemma was for more women to obtain doctorates, and this could happen only if universities opened their doors to women at the doctoral level. In the late nineteenth century, however, most Ph.D. granting institutions in America and abroad did not support coeducation and so effectively blocked this educational avenue to women. But "most" was not "all," and the few schools which did admit women at the graduate level—Bryn Mawr, Cornell, and others—set an example for other institutions to follow. That more and more graduate schools did change their policies and admit women students marked what has been termed "one of women scientists' greatest triumphs" [16].

The opening of the graduate schools, described by Rossiter as a tripartite process, began in the two decades preceding the 1890s with a phase of so-called "unrealized potential" during which women primarily gained admission thanks to a "special student" status [17]. This category of attendance allowed the student to sit in on the lectures of a particular professor, usually without official admittance to the university, and hence, without eligibility to obtain a degree [18]. An explosive period from 1890–1892 followed when at least six prominent graduate schools—Yale, Brown, Columbia, Stanford, and the Universities of Pennsylvania and Chicago (the latter welcoming women from its inception in 1892)—opened their doors to women [19]. Finally, from 1893 to 1907 (and on), the process entered an "embattled stage" when, largely forced by peer pressure, lagging universities not only admitted women to their graduate programs but also awarded them doctorates. Given the intimate connections at the end of the nineteenth century between the trends in higher education and the development of science in America, it comes as no surprise that the involvement of women in mathematics, in particular, closely paralleled this educational profile.

Equally as important as the educational progress of the day were both the women who surmounted gender barriers and the men and women who encouraged them in their courageous pursuits. Like most pioneers, the women mathematicians needed an undaunted, ever-faithful leader to sustain and encourage them as the seemingly impossible became reality. The women found their champion in Christine Ladd-Franklin. Capable, progressive, and well acquainted with the task at hand based on her own personal experience, Ladd-Franklin flourished in this role. She initially promoted women in mathematics from the "inside" as a student trying to bypass seemingly inflexible rules by becoming an "exception."

Following her 1869 graduation from Vassar and nine-year-long tenure as a secondary-school teacher, the then Christine Ladd decided to pursue her mathematical studies under the best teacher America had to offer at the time, James Joseph Sylvester at Johns Hopkins University. Writing directly to Sylvester on March 27, 1878, she confided that "[i]t is my desire to listen next year to such of your mathematical lectures as I may be able to comprehend" and asked "[w]ill you kindly tell me whether the Johns Hopkins University will refuse to permit it on account of my sex?" [20]. Seeing nothing wrong with her plan and ever desirous of good, new students, Sylvester wrote to Hopkins President, Daniel Coit Gilman, several days later on April 2, 1878 to inform him that

> I have written to Miss Ladd saying that I did not personally anticipate that her sex would be an objection when attending lectures at our University and that I should rejoice to have her as a fellow worker among us—but that on my return to Baltimore I would bring the matter officially before the Authorities of the University and acquaint her with the result. I happened to mention the matter to Dr. Thomas and Mr. King whom I met in the railway train: and they seemed to favor the notion and to be inclined to give her every facility for carrying out her wishes on the subject. My own impression is that her presence among us would be a source of additional strength to the University. I regard her as more than another Miss Somerville in prospect and I cannot but think that with your fertility of resource you would hit upon some plan of utilizing her for the purposes of the University [21].

Even though the two trustees, James Carey Thomas and Francis T. King, may have looked favorably upon the idea when confronted by the ebullient Sylvester on the train from Baltimore to New York, other members of the University's Board of Trustees like George William Brown and Reverdy Johnson opposed setting a precedent for co-education [22]. After a month's worth of deliberation between the President, the Trustees, and the Professor, Gilman wrote back to Ladd on April 26, 1878 with the University's verdict:

> In reply to your inquiry, I am instructed by the Executive Committee to say that the Board of Trustees is not favorable to the admission of women as students of the University in the ordinary acceptation of that same* ... While they are thus governed by the general principles adopted, they recognise in you such exceptional attainments as a mathematician that they consent simply to your attendance upon Professor Sylvester's lectures in accordance with your original request addressed to him. As this is an exceptional recognition of your mathematical scholarship, no charge will be made for tuition & your name will not be enrolled on the annual Register.

> *and therefore cannot grant your application in the Extended form in which it comes before them [23].

Thus, Ladd was permitted to come and study at Hopkins but as an officially "invisible" student.

After one year, Gilman and the Board relented somewhat, permitting her to sit in on the lectures of the logician Charles S. Peirce and of the Leipzig-trained William E. Story as well as allowing her to take the stipend, though not the title, of a fellow. Here, Gilman's progressiveness ended. When Ladd finished all of the requirements for the doctorate in 1882, he refused to confer her degree [24]. In actuality, Ladd was only

Table 1.

Women in the American Mathematical Community
Who Received Doctorates Prior to 1900

Name	Year of Doctorate	Institution	Adviser
Charlotte Angas Scott	1885	London	
Winifred Edgerton =	1886	Columbia	
(Mrs. Frederick J. H. Merrill)			
Ida Martha Metcalf	1893	Cornell	Oliver or Wait*
Annie Louise MacKinnon =	1894	Cornell	Oliver or Wait*
(Mrs. Edward Fitch)			
Charlotte Cynthia Barnum	1895	Yale	
Agnes Sime Baxter =	1895	Cornell	Oliver
(Mrs. Albert Ross Hill)			
Elizabeth Street Dickerman[†]	1896	Yale	
Ruth Gentry	1896	Bryn Mawr	Scott
Isabel Maddison	1896	Bryn Mawr	Scott
Mary Frances "May" Winston =	1897	Göttingen	Klein
(Mrs. Henry B. Newson)			
Leona May Peirce[†]	1899	Yale	

*James E. Oliver and Lucien A. Wait served on the committees of both Metcalf and MacKinnon. Based on the titles of their theses and the course offerings at the time, it appears that Oliver served as the formal adviser with Wait as the second reader.

[†]The names of these two women did not emerge from our compilation from the *Bulletin of the American Mathematical Society* but come rather from Eells's article (see below). We include them here for completeness. We do not include Anne L. Bosworth (Mrs. Theodore M. Focke), who *is* listed as an 1899 Göttingen Ph.D. in *American Men of Science*. While she successfully defended her thesis in 1899, she officially received her degree in 1900.

Compiled from James McKeen Cattell, *American Men of Science: A Biographical Directory;* Walter Crosby Eells, Earned Doctorates for Women in the Nineteenth Century, *Bulletin of the American Association of University Professors* **42** (1956): 649–651; Amy C. King and Rosemary McCroskey, Women Ph.D.'s in Mathematics in USA and Canada: 1886–1973, *Philosophia Mathematica* **13/14** (1976–1977): 89–129; Kathleen Jacklin, Cornell University Archives to Della Dumbaugh Fenster, 14 June, 1990; Diane E. Kaplan, Yale University Archives to Della Dumbaugh Fenster, 21 May, 1990; and Helmut Rohlfing, Handschriftenabteilung, Niedersächsische Staats- und Universitätsbibliothek Göttingen to Karen Hunger Parshall, 31 October, 1991.

In general, Eells's article is reliable. He does, however, incorrectly list Margaretta Palmer as an 1894 doctorate in mathematics from Yale. Palmer's degree was actually in astronomy. Eells's notation also leaves the degrees from Cornell of Louise Hannum and Agnes Simes Baxter ambiguous. Hannum's degree was in metaphysics and Baxter's in mathematics. The King and McCloskey article is incomplete.

slightly ahead of her time. In 1886 Columbia awarded the first doctorate in mathematics to an American woman, Winifred Edgerton [25]. As Table 1 shows, 11 women had shared this distinction by 1900.

Thwarted in her own personal aspirations, Ladd was halted but not defeated, and she began to direct her unique talents into the more general effort for women in science and, in particular, for women in mathematics. Her focus thus turned to encouraging the effort from the sidelines rather than engaging as a player on the field. Though the details surrounding these earliest doctorates remain obscure, it is clear that Christine Ladd-Franklin and the Association of College Alumnae (ACA) contributed, directly and indirectly, to many of the doctorates in the 1890's. Designed to unite the college-educated women of the day, the ACA was formed in 1882 by Mrs. I. Tisdale Talbot. As she conceived it, the ACA should serve not only as a support group for women in their quests to use the gifts of a college education but also as an inspiration for later women in pursuit of college degrees [26]. Relative to this dual purpose, the associated members opened discussion as early as their second meeting in May of 1882 on the opportunities, or lack thereof, for post-graduate study for American women [27]. By 1888, with the introduction of Christine Ladd-Franklin's brain-child, the ACA European Fellowship, the women had launched one of their most ambitious projects directly toward correcting this clearly perceived deficiency. Their simple plan had boundless implications: by gaining ground at foreign universities, and particularly at the prominent and influential German ones, American women would set a precedent for the universities at home to follow.

Ruth Gentry, the first mathematician but second recipient of the ACA European Fellowship, accomplished precisely what Christine Ladd-Franklin and her cohorts had intended. After graduating from the University of Michigan in 1890, Gentry proceeded to Bryn Mawr for her graduate work before receiving the Fellowship for 1891–92. Apparently, she first tried to take her stipend to the University of Heidelberg but was denied admission there on October 20, 1891. She then applied to the University of Berlin, having secured temporary approval from two professors and from the University Rector but not from the University Senate. After submitting her application to the Minister of Education on January 26, 1892, she finally received permission to attend lectures but not to enroll for a degree [28].

Gentry's persistence led directly to advances in higher education for women in Germany [29]. Influenced by the cases of both Gentry and the German, Marie Gernet [30], five members of the Faculty of Science and Mathematics at the University of Heidelberg recommended that the University allow women to sit in on courses at the discretion of the instructor. Though the University Senate denied this proposal, on November 23, 1891, Baden's Ministry of Education overruled its decision and granted women the right to audit courses in both mathematics and the sciences. Meanwhile, only a month after Gentry's application to Berlin, the Prussian Minister of Education formally queried his faculties on the issue of the admission of women [31]. Although the responses were largely negative, four medical faculties (Bonn, Breslau, Königsberg, and Marburg) did open their doors to women with appropriate secondary degrees [32]. While hardly representing overwhelming successes, these events brought the women of both Germany and America one step closer to a conducive environment for their higher education.

The turning point in Prussia—and, in some sense, America—ultimately occurred at the University of Göttingen. Once again, Christine Ladd-Franklin served as a catalyst. In 1891, she accompanied her husband, Fabian Franklin, to Germany for a year of sabbatical leave from his post at the Johns Hopkins University. She had hoped to use this opportunity to obtain the doctorate which Hopkins had denied her almost ten years earlier. As in the Baltimore of 1882, however, the climate in Germany in 1891 did not lend itself to the successful realization of her goal. During her year abroad, she succeeded only in conducting experimental research on color vision in the laboratory of G. E. Muller and in auditing the mathematical lectures of Felix Klein. Still, both Felix Klein's sympathy and his assurances that "it was only the beginning for women at Göttingen" did provide cause for future optimism [33]. In fact, Klein resolved to push the Prussian Ministry of Education by "seek[ing] out women who might be interested in pursuing a doctorate at Göttingen" during his trip to the United States for the Chicago World's Fair in 1893 [34]. Mary Frances "May" Winston, a graduate student at the University of Chicago smitten with the idea of studying in Germany, met Klein during this visit.

Winston had graduated from the University of Wisconsin with a classical degree in 1889. She had then applied for a graduate fellowship at Wisconsin but was denied, presumably on the basis of her gender [35]. After teaching for two years at Downers College in Fox Lake, Wisconsin, she received a fellowship at Bryn Mawr for the 1891–1892 academic year. Although her year in Pennsylvania proved rewarding, she could not afford to complete her Ph.D. there and so returned home to Illinois. Fortunately, the nearby University of Chicago opened in the fall of 1892, and Winston was able to enter as a member of its first class. According to her daughter, May Winston

   ...had no intention of staying at Chicago above one year, for as she put it "she was bitten by the German bug." A number of the math[ematics] faculty [at the University of Chicago] were Germans [Heinrich Maschke and Oskar Bolza]...They tried to encourage May...to stay with them and work for a Ph.D. there but they sympathized with her dream of a German degree. So when Professor Felix Klein of Göttingen University was sent by his government to the World's Fair in Chicago to represent German leadership in mathematics, they saw that May was asked to his famous symposium in Evanston. There she was introduced to Klein who told her that if she could come he would do everything he could to help her. He felt Germany...must mend its ways [36].

By the fall of 1893, the Prussian Ministry of Education had indeed begun to "mend its ways" by granting Winston, along with ACA Fellow Margaret Maltby and the Englishwoman Grace Chisholm permission to audit courses at Göttingen [37]. Although Winston herself had competed for the ACA Fellowship that year, she did not receive it, owing to her lack of original research [38]. Given her tight financial circumstances, this could have proved disastrous for her plans to study abroad. Fortunately, though, the ever-vigilant Christine Ladd-Franklin heard of her need and advanced $500.00 toward her expenses. This contribution, along with the ACA fellowship she did win for 1895–1896 and whatever monies her family could send, financially sustained Winston through her three years at Göttingen [39]. According to her daughter, Caroline Beshers, Winston's parents, Thomas and Caroline Winston, greatly encouraged their daughter in her endeavor "...and would have starved rather than bring her home without a degree" [40]. Furthermore, her family, together with her best friend, Grace Chisholm, gave Winston the moral support she needed to deal with the uncertainties faced by a woman at a major German university. These first three women, Chisholm, Maltby, and Winston, proved equal to the challenge, and their numbers grew to at least 14 by the winter semester of 1894–1895 [41]. Seven of this total studied mathematics, astronomy, and physics. Among these seven, at least Winston, Annie MacKinnon and Isabel Maddison had also studied in America.

Having finished her Ph.D. at Cornell in the spring of 1894, MacKinnon attended Göttingen as the 1894–1895 ACA European Fellow. In a letter of introduction to Klein in July of 1894 she summarized her understanding of the opportunities for women at Göttingen. She wrote:

Last winter [1893] I heard through Prof. Oliver that you had obtained permission for certain women to attend your lectures. Later I saw a notice in the newspapers to the effect that admission for the coming year had been given to three women whose names were mentioned. And this spring, there appeared in the *Nation*—one of our most reliable weekly papers—a statement that it had been decided that women may be admitted to the University of Göttingen and that degrees may be granted to them....

With many other Americans, I greatly appreciate your attitude toward the admission of women to your University [42].

Isabel Maddison, on the other hand, hailed originally from England and scored as the 27th Wrangler on the Mathematical Tripos in 1892. Moving on for graduate studies at Bryn Mawr, she won the first Mary E. Garrett European Fellowship in 1894–1895 and continued her work at Göttingen [43]. In a detailed account of the ambiguous position of women at Göttingen—from the individual admissions procedure to the perpetual question of whether an official degree would be conferred—Maddison assessed the overall situation from a student's perspective. She concluded that "if the restrictions appear at first irksome, we must remember that the question at issue is a very large one. Anything that has so far been done is in the nature of an experiment, and it is most desirable that the trial should be made with the best material, that only a few women should present themselves, and that those should be thoroughly qualified to make the best of their opportunities" [44].

These pioneers—Winston, MacKinnon, Maddison, and others—not only had their gender in common but also a mutual interest in mathematics. While the ACA encouraged their endeavors as women, the American Mathematical Society (AMS) provided a common ground for their mathematical pursuits. In fact, our companion study of the American mathematical research community over the period from 1891 to 1906, provided a particularly interesting and unusual glimpse at both the diversity and the activities of the women participating in American mathematics. Of the 1061 mathematically interested participants we uncovered, 71 women emerged, or 6.7% of the total (see Table 2) [45]. Of these 71 women, the AMS counted 50 among its membership including 29 who participated in at least three activities. (Here, as in our companion paper, we defined activities as attendance and/or talks presented at various mathematical gatherings, papers published in certain American journals, and service to the broader mathematical public [46].) Eighteen of these 29 women also had entries in *American Men of Science*, and thereby qualified for inclusion in our category of "active" participants. Finally, two of them, Charlotte Angas Scott and Ida May Schottenfels, filled out the ranks of the "most active" with their total participation levels of 56 and 43, respectively [47]. Table 3 below provides details on the 18 "active" women members of the 1891–1906 community. By focusing particularly on the attendance figures in both Tables 2 and 3, an especially telling observation emerges.

Table 2.

The 71 Women in the American Mathematical Community 1891–1906

Name	T	P	S	A	Name	T	P	S	A
Acer, Sara A.	0	0	0	0	Kelsey, H. M.	0	0	0	0
Alden, G. C.	0	0	0	0	Knepper, Myrtle	0	0	0	0
Anderson, Mary E.	0	0	0	0	McKelden, Alice M.	0	0	0	0
Andrews, Grace	0	0	0	12	MacKinnon, Annie L.	0	2	0	0
Barnum, Charlotte C.	0	0	0	3	Maddison, Isabel	0	2	0	3
Becker, A. F.	0	0	0	0	Martin, Emilie N.	1	1	1	4
Bosworth, Anne L.	0	0	0	1	Merrill, Helen A.	1	1	0	2
Brewster, Helen	1	0	0	0	Morrison, Bessie G.	0	1	0	0
Brooks, Harriet	0	0	0	0	Palmié, Anna H.	1	0	0	2
Buckingham, H. D.	0	0	0	0	Pendelton, E. F.	0	0	0	0
Burrell, Ellen L.	0	0	0	2	Pengra, Charlotte E.	0	0	0	0
Busbee, Christine	0	0	0	0	Pesta, R. A.	0	0	0	0
Byrd, Mary E.	0	0	0	0	Proctor, Mary	1	0	0	0
Carstens, R. L.	1	1	0	0	Ragsdale, Virginia	1	1	0	4
Coddington, Emily M.	0	0	0	1	Rayson, Amy	0	0	0	5
Cowley, Elizabeth B.	0	0	0	9	Richardson, S. F.	0	0	0	0
Cronin, S. E.	0	0	0	0	Schottenfels, Ida M.	17	3	0	23
Cummings, Louise D.	0	0	0	6	Scott, Charlotte A.	12	19	18	23
Cunningham, Susan J.	0	0	0	0	Sinclair, Mary E.	0	0	0	1
Davies, C. A.	0	0	0	0	Smith, Adelaide	0	0	0	0
Decherd, M. E.	0	0	0	0	Smith, Charlotte	0	0	1	0
Denis, B.	0	0	0	0	Smith, Clara E.	2	0	0	3
Doak, Eleanor	0	0	0	0	Sykes, Mabel	0	0	2	0
Ely, Achsah	0	0	0	0	Trueblood, Mary E.	0	0	0	6
Gates, Fanny C.	0	1	0	2	Underhill, Mary	0	0	0	7
Gentry, Ruth	0	0	0	0	Van Benshoten, Anna L.	0	0	0	2
Glazier, Harriet E.	0	0	0	0	Vivian, Roxana H.	0	0	0	3
Gould, Alice B.	0	0	0	4	Waddell, May E. G.	0	0	0	0
Gould, M. F.	0	0	0	0	Walker, M. S.	0	0	0	0
Griffiths, Ida	0	0	0	12	Wentz, Estella K.	0	0	0	1
Hammerslough, Carrie	0	0	0	8	Wilkinson, Anne L.	0	0	0	0
Hardcastle, Frances	1	1	0	1	Williams, Ella C.	1	0	0	23
Hayes, Ellen	2	0	0	4	Winston, Mary F. "May"	1	1	0	7
Hitchcock, Fanny R. M.	0	0	0	0	Wood, Ruth G.	2	1	0	11
Ingalls, N. S.	0	0	0	0	Young, M. M.	0	0	0	0
Johnson, Anna	0	0	0	0					

"T" denotes the number of talks presented; "P" denotes the number of papers published; "S" stands for service; and "A" refers to the number of meetings and/or colloquia attended. For the precise definitions of these terms, see [46] below. The names as provided here come from the listings in the *Bulletin of the American Mathematical Society* and *American Men of Science*. We have used maiden names in the cases of women who subsequently married.

Table 3.

The 18 Active Women in the American Mathematical Research Community 1891–1906

Name	Birth Place	Highest Degree	Abroad	T	P	S	A
Andrews, Grace	NY	Ph.D. Columbia 1901		0	0	0	12
Barnum, Charlotte C.	MA	Ph.D. Yale 1895		0	0	0	3
Cowley, Elizabeth	PA	Ph.D. Columbia 1908		0	0	0	9
Cummings, Louise D.	Ontario	Ph.D. Bryn Mawr 1914		0	0	0	6
Gates, Fanny C.	IA	Ph.D.* Pennsylvania 1909	Cav., G, Z	0	1	0	2
Hammerslough, Carrie	MD	M.A. Columbia 1897		0	0	0	8
Hayes, Ellen	OH	B.A. Oberlin 1878		2	0	0	4
Maddison, Isabel	England	Ph.D. Bryn Mawr 1896	G, D	0	2	0	3
Martin, Emilie N.	NJ	Ph.D. Bryn Mawr 1901	G	1	1	1	4
Merrill, Helen A.	NJ	Ph.D. Yale 1903	G	1	1	0	2
Palmié, Anna H.	NY	Ph.B. Cornell 1890	G	2	1	0	0
Schottenfels, Ida M.	IA	M.A. Chicago 1896		17	3	0	23
Scott, Charlotte, A.	England	Sc.D. London 1885		12	19	18	23
Smith, Clara E.	CT	Ph.D. Yale 1904		2	0	0	3
Trueblood, Mary E.	IN	Ph.M. Michigan 1896	G	0	0	0	6
Van Benschoten, A. L.	NY	Ph.D. Cornell 1908	G	0	0	0	2
Winston, Mary F.	IL	Ph.D. Göttingen 1897	G	1	1	0	7
Wood, Ruth G.	RI	Ph.D. Yale 1901	G	2	1	0	11

*Gates's Ph.D. was in physics, not mathematics.

"T," "P," "S," and "A" are defined as in Table 2. For the precise definitions of these terms, see [46] below. In the "Abroad" column, Cav. denotes England's Cavendish Laboratory, G stands for Göttingen, D refers to Dublin, and Z denotes Zürich. (By "Abroad" here we mean going from North America to a European country. Hence, Scott, for example, is not listed as having gone abroad). Ph.B. and Ph.M. denote Bachelor's and Master's degrees, respectively. All of the women designated mathematics as a first specialty in their *American Men of Science* entries, except Gates (who listed physics first and mathematics second) and Hayes (whose only specialty was astronomy). Cowley also listed astronomy as a second specialty.

These numbers indicate the consistent participation of women, at least in terms of presence. Isolating the attendance record, an investigation of the 18 "active" and 11 "almost active" women contributors reveals that nearly one-half (14) attended at least five meetings. Not surprisingly, the two women in the "most active" category, Scott and Schottenfels, account for two of the three women attending more than 20 meetings from 1891–1906. However, also at this high level of attendance, Ella Cornelia Williams failed to fall even within our "active" sample, owing to the absence of an *American Men of Science* entry. A resident of New York City, Williams attended at least 23 mathematical meetings, including the first Buffalo Colloquium in 1896. Moreover, she not only attended but also actively participated in Society gatherings as evidenced by her remarks at a November 5, 1892 meeting of the then New York Mathematical Society, and by her presentation of "An orthomorphic transformation of the ellipsoid" at the NYMS meeting on October 8, 1892 [48]. Beyond her participation in these meetings, however, Williams faded into obscurity.

Even more impressive, the attendance record of the Chicago-trained Ida May Schottenfels resulted from her participation in a wide range of geographically dispersed events. In 1902, for example, she was present at the tenth regular meeting of the Chicago Section at Northwestern on January 2–3, 1902. On February 22 and April 26 of that same year, she appeared at regular AMS meetings in New York City, presenting "On the definitional functional properties for the analytical functions $(\sin \pi z)/\pi$, $(\cos \pi z)/\pi$, $(\tan \pi z)/\pi$" at the latter of the two gatherings. By September, she was back in the Midwest for the ninth summer meeting of the AMS at Northwestern. Furthermore, the AMS pressed her to be in two places at once, or so it seemed, for less than a year later, on April 25, 1903, while she presented "On the simple groups of order $8!/2$" at a regular AMS meeting in New York City, the San Francisco Section read her "Generational definition of an abstract group simply isomorphic with the simple substitution group G_{20160}^{21}" at a regular meeting at Stanford [49]. In our period alone, she read 17 papers, published 33 articles and attended at least 23 mathematical events.

Schottenfels' record was only outclassed by that of the most visible and active woman in the American mathematical community, Charlotte Scott. Although educated at Girton, the women's college at Cambridge, she attended lectures by Arthur Cayley at Trinity and effectively did her graduate work under him. Since Cambridge refused to confer degrees upon women at this time, Scott received both her Bachelor of Science (1882) and her Doctor of Science (1885) degrees from the University of London [50]. Immediately upon earning this latter credential, M. Carey Thomas lured Scott to America with a position as head and sole member of the Mathematics Department at Bryn Mawr College [51]. While there, Scott advised seven doctoral students [52], received the college's first endowed chair on the basis of her fine teaching record, and, in general, according to the directors of Bryn Mawr College, made contributions "second only to that of President Thomas" during her 40-year tenure at the school [53]. The length and significance of her tenure at Bryn Mawr spurred one biographer to compare her life poetically to "a pebble dropped in a pond, with every-widening circles rippling out from the center as her influence expanded" [54]. Her early doctorate in mathematics served as one of the initial ripples with Scott's high and uncompromising standards perpetuating the effect. Furthermore, her influence seemed largely independent of her womanhood. Scott steadfastly refused to set herself apart on the basis of gender. In fact, according to the son of her friend, the mathematician, Frank Morley, Scott believed that "intelligent men would give her sex the credit that is due" [55]. The AMS justified this belief for, on the basis of her talent and energy, she became a relatively important force within the Society.

As Table 3 indicates, Scott read 12 papers before the Society, published 19 articles, and attended at least 23 mathematical meetings from 1891–1906. Her 18 service marks, however, reveal a great deal more about her influence and participation. She earned seven of these for the seven years she served on the (elected) Council of the AMS—the charter year of 1894, 1895–1897, and 1900–1902—and one for her year in the Society's Vice-Presidency in 1906 [56]. Her tenure as co-editor of the *American Journal of Mathematics* accounts for eight more, and the remaining two reflect her appointments both to the College Entrance Examination Board in 1903 with William H. Metzler and John S. French and as a collaborator on the French *Revue semestrielle des publications mathématiques* in 1898 [57].

By holding such important and visible positions, Scott undoubtedly encouraged women to participate in the mathematical community [58]. In fact, the lives of Scott and Christine Ladd-Franklin reflect the two most potent tactics employed by the promoters of women in mathematics [59]. Using a direct, forceful, and more externalized approach, Ladd-Franklin fought hard to gain ground for women. Although herself in the academic arena in the earliest stages of the effort, her greatest influence came later and from the outside of academe in the broader political sphere of the growing women's movement. She viewed the entry of women into mathematics as part of a much larger social issue and attacked it as such. By providing both financial and emotional support for those actively pursuing degrees, Ladd-Franklin substantially improved the climate for women in mathematics. Charlotte Scott, on the other hand, propelled women forward using more discreet methods. In striving for the highest standards in mathematics, both in terms of teaching and research, she advanced within the mathematical ranks, thereby gaining an otherwise unheard of position for a woman [60]. Her uncompromising standards and her steadfast refusal to expect less of a woman than of a man in the academic realm both challenged and advanced women in her field.

During the course of their careers, many of Scott's male counterparts in the AMS supported women in mathematics as well. The most prolific of the AMS members during the 15 years from 1891 to 1906, Leonard E. Dickson, supervised the dissertations of 18 women during his 36 years on the faculty at the University of Chicago [61]. Other prominent members of the AMS also promoted women in mathematics by directing their doctoral work. In particular, Gilbert Bliss of the University of Chicago had 12 female students; Virgil Snyder of Cornell supervised 13; and Frank Morley of Johns Hopkins University guided eight, to name but a few [62]. Although he directed no women students owing to his departure from Johns Hopkins in 1895, or 12 years before the official enrollment of women there, Fabian Franklin nevertheless joined his wife in promoting the cause of women in mathematics in particular and in higher education in general. Frank Nelson Cole, Secretary of the AMS for the quarter-century from 1895–1920, demonstrated his support for women in the form of collaboration. In 1917, he published "The complete enumeration of triad systems in fifteen elements" with Louise Duffield Cummings and Henry Seeley White. The three collaborated again in 1919 with the publication of "Complete classification of the triad systems on fifteen elements" [63]. Finally, Henry B. Newson supported the career of May Winston through the glowing letter of recommendation he wrote on her behalf in 1897 to the Kansas State Agricultural College. He subsequently married her three years later [64].

In addition to these men and their support, the women enjoyed the encouragement of Eliakim Hastings Moore, at once a backbone of the Mathematics Department at the University of Chicago and a mathematical leader during our time period [65]. In some sense, Moore's acceptance in 1892 of the chair of Chicago's Mathematics Department represented a hospitable view (at least publicly) toward women in mathematics. Given his university's open policy relative to women students at all levels, Moore and his department had no qualms about enrolling women in its graduate program in the fall of 1892. At least two women, Anna Johnson Pell [later Wheeler] and Mary Wells studied with Moore at Chicago as well [66].

With so many influential AMS members and other prominent mathematicians like Sylvester and Klein supporting the notion of women in mathematics, with Charlotte Scott setting the precedent for the possibilities available to women, and with the groundwork laid and supported by Christine Ladd-Franklin and others, it would appear that the turn-of-the-century mathematical community was, indeed, hospitable to women. The women in the American mathematical community enjoyed a most unusual environment compared to the other national professional organizations of the day. As Margaret Rossiter noted, the typical female participation in the organizations was

> ... at best restricted to a fraction of those relatively few women who already had the best credentials (American or European doctorates), the best positions (usually professorships at women's colleges), most publications, or, in lieu of these, as in engineering, strong personal contacts with prominent men in the field [67].

Although the AMS resembled other groups in terms of the miniscule number of élite women occupying elected offices, it differed markedly in that its hospitality was *not* reserved merely for a few première female members. The 71 women in our study—50 AMS members and 21 others with an interest in mathematics—formed a diverse group of contributors including high school teachers, graduate students, professors, women with doctorates, participants in other societies, foreigners, as well as the élite.

Although 13 of the 18 "active" female members of the Society earned doctorates, a doctorate did not guarantee above average, or even average, participation in the community. Charlotte Elvira Pengra, for example, received a Ph.D. in mathematics from Wisconsin in 1901 but only surfaced in our study in the yearly list of doctorates published by *Science* and included in the *Bulletin* [68]. After her splashing entry into the world of mathematics, Ruth Gentry appeared only twice in the *Bulletin*'s first 15 years, once for her 1894 induction into the AMS and once for her promotion to an associate professorship of mathematics at Vassar College in 1900 [69]. Anne Bosworth, a 1900 Göttingen doctorate, participated only by attending one AMS meeting and by speaking at one meeting of the Association of Ohio Teachers of Mathematics and Science [70]. Similarly, considering activity in the form of attendance, Ella Williams and Ida Griffiths attended at least 23 and 12 AMS meetings, respectively, including the first Buffalo Colloquium, although neither held a doctorate in mathematics.

Just as activity did not necessarily go hand-in-hand with a doctorate, it did not depend upon employment at an institution of higher education either. As seen in Table 3, Ida May Schottenfels presented 17 papers and published three, making her second only to Charlotte Scott in this category of participation. Schottenfels' highest earned degree, however, was a Master's and she spent her teaching career at grammar and high schools in Chicago and at the New York State Normal School.

Active high school teachers also emerged among the mathematically interested. Thus, the AMS counted A. F. Becker of Yeatman High School in St. Louis, Missouri and R. A. Pesta of Wendell Phillips High School in Chicago, Illinois among its numbers. Their memberships indicate that the Society did not limit itself to women whose mathematical interests fell only within the confines of collegiate teaching and research. Furthermore, a developing interest in mathematical pedagogy at the turn of the century introduced some participants to the American mathematical community who were otherwise beyond the typical bounds of the AMS in terms of membership. Their pedagogical pursuits led to the formation of regional teachers' associations. Prominent research mathematicians of the day, such as Chicago's E. H. Moore and Harvard's William Fogg Osgood, to name only two, participated in the organization and leadership of these associations, and the *Bulletin* carried brief records of their proceedings [71]. Because of this coverage, the activities of women like Mabel Sykes of South Chicago High School and Charlotte Smith of Girls' High School of Brooklyn came to the attention of the broader mathematical constituency [72].

This wider community also included a number of participants who made few contributions within the American mathematical context. Two foreign women—Adelaide Smith of Cape Colony and May E. G. Waddell of Orono, Canada—entered our sample space by virtue of their induction into the AMS [73]. Harriet Brooks surfaced thanks to the notice of her appointment to a tutorship at McGill University [74]. On the other hand, Bessie Growe Morrison published an article on binary quantics in the *American Journal of Mathematics* in 1901 but apparently had no further involvement in mathematics thereafter [75]. Finally, several American women also came to light purely on the basis of their induction into the Society [76].

The diversity reflected in this analysis of the full constituency of women in our sample diminishes somewhat when investigating only the 18 "active" women (see Table 3). Ten of these women (55.6%) were born in the Northeast, five in the Midwest (27.8%), one in Canada (5.5%), and two in England (11.1%) [77]. Furthermore, whereas only 35.0% of the full "active" contingent within American mathematics studied abroad, nearly half of the women took a foreign tour and all of these spent some time at Göttingen [78]. Most telling of all, perhaps, 15 (83.3%) of these women pursued careers at women's colleges [79]. Given the overwhelming preponderance of women mathematicians at women's colleges, Charlotte Cynthia

Barnum's curriculum vitae stood out as quite unique for the time. Her employment following her 1895 doctorate from Yale included wide-ranging teaching experiences at both the secondary and college levels, actuarial work at two insurance companies, and positions at both the U.S. Naval Observatory and the U.S. Coast and Geodetic Survey, in addition to various editorial endeavors [80]. Ruth Goulding Wood, however, pursued a much more representative career. After earning a Bachelor's Degree from Smith in 1898 and a doctorate from Yale in 1901, she taught for a year at Mount Holyoke and then began her 33-year tenure at Smith College. (Her time at Smith was only interrupted by a year of study [1908–1909] at Göttingen.) This sort of total dedication to the students and administration of primarily one institution was shared by all 15 of the women employed by women's colleges.

Needless to say, the consequences of such commitment extended far beyond the institutions the women called their own. Their students benefited not only from the mathematical instruction they provided but also from their exposure to these contemporary role models. The teaching abilities of at least three of our 15 women achieved positive notoriety. As previously mentioned, Charlotte Scott received Bryn Mawr's first endowed chair on the basis of her teaching record. Isabel Maddison, one of Scott's first Ph.D.'s, wrote that students "appealed to [her] for help in all kinds of difficulties. Her keen logical mind was brought to bear on any subject, no matter how far from her real interests . . . and the way in which she dragged the relevant facts to light, analyzed them and deduced the solution, was an object lesson in judicial reasoning" [81].

While Scott dedicated her efforts to Bryn Mawr, Helen Abbott Merrill spent 39 years (1893–1932) in the service of mathematics at Wellesley. The remarks of her Wellesley colleague (and one of the 71 women in our study), Mabel Young, attested to Merrill's gifts as a teacher. According to Young,

> Miss Merrill was deeply interested in mathematics, but rather as a teacher than as a research worker. She was preeminently a teacher. To share with her students subjects which she enjoyed in graduate work, she planned introductory courses in the Theory of Functions and Descriptive Geometry, work rarely open to undergraduates at that time. The girls responded and not a few went on in later years to explore these fields for themselves. She had the enviable power of suggesting to students that the subject in hand led up and on. She could detect promise in unskilled performance and led many a girl to discover unexpected abilities in herself. The high standard set became not a barrier but a challenge [82].

The women at Smith also benefited from excellent teaching, but there it came in the person of Ruth Wood. A former student best described Wood's uncanny teaching ability in these words:

> It is rarely that one encounters Miss Wood's peculiar ability to understand students, to know them often better than they know themselves, so see how their minds work, to draw them out and draw out of them qualities they never knew they possessed. While she held steadily to the subject in hand, adorning the blackboard with ellipses and parabolas, she used her own formulae for understanding people. Her clear-cut mind did not stop at lucid explanations of mathematical problems. It penetrated behind the fact of error in a student's work and found the cause of the error. Then, impersonally, objectively, but with quick sympathy she built confidence and self-respect into the befogged mental make-up of the least mathematical mind. And under her skillful guidance that mind would begin to grow up [83].

Although admittedly hagiographic, such testimonials nevertheless reflected the deep commitment with which women like Scott, Merrill, and Wood pursued their college-level teaching careers in mathematics.

Yet teaching did not define the only avenue these early women mathematicians followed en route to establishing an enduring legacy. From the beginning, those involved in the effort for women in mathematics shared high-minded ideals and a commitment not only to their own personal careers but also to the realization of broader goals for women. Annie MacKinnon, for example, not only had the necessary fortitude to pursue an early doctorate in mathematics, but she also participated in many organizations which encouraged women to take a public interest in local as well as national affairs [84]. M. F. Winston Newson's commitment to such broader principles manifested itself in a stand she took in the 1920s. Fully recognizing and accepting the responsibility of providing for a family of five at a time when job opportunities for women in mathematics were scant, Newson signed a petition in support of a controversial colleague in the Political Science Department at Washburn College. Dismissed for expressing his political views too freely in front of his students, this colleague's plight polarized the Washburn faculty. Within two years, Newson had joined the petition's other signatories by leaving the college in protest. She accepted a position as department head at Eureka College and remained there until her retirement in 1942 [85].

For still others in our sample, the commitment to causes did not end even at the times of their deaths. The will of Ruth Wood established a trust fund for augmenting the salary of at least one woman professor of mathematics to a level equal to that of the highest paid member of the Smith faculty [86]. Similarly, Isabel Maddison, following a 30-year-long career at Bryn Mawr, left $10,000.00 for the organization of a pension fund for non-faculty members (which tended to be women) at the college [87]. More than women in mathematics, these were women in higher education who recognized and fought to overcome the prejudices and double standards within the academic community.

As we showed in our companion study, the years from 1891 to 1906 witnessed a series of profound and important developments which resulted in the formation of a self-sustaining mathematical research community in America. In all, some 1061 people participated at various levels to make this community a reality. Of this number, 71—or 6.7% of the total—were women. Arriving on the scene as they did at century's close, these women mathematicians emerged from the rich heritage of multifaceted people and events characterizing the effort for women's education throughout the nineteenth century. Beginning effectively with Emma Hart Willard's creative and strategic approaches to education, this movement continued with the opening of women's colleges and with the increase of opportunities for doctoral pursuits (largely through the efforts of Christine Ladd-Franklin and others). In mathematics proper, these developments manifested themselves in what we have termed the "external" tactics of Christine Ladd-Franklin, the "internal" influences of Charlotte Scott, and the complete openness of the American Mathematical Society to women [88]. As predicted, all of these changes combined led, in mathematics as in other realms, to what Rossiter aptly described as "new roles and opportunities ... unfolding at the same time that new persons were becoming available to fill them" [89].

The women of our study frequently took advantage of the unique circumstances in which they found themselves. They earned their Ph.D.'s at home, but more importantly, abroad; they broke into the academic work force at both women's and co-educational institutions; and they participated in the newly forming American mathematical research community. While certainly neither as numerous nor on the whole as productive as their male counterparts, women nevertheless formed a vital and visible contingent within turn-of-the-century American mathematics. Once again, the prevailing folklore is simply wrong. American mathematics around 1900 was not made up merely of a handful of men like E. H. Moore. It was a broadly based community of men and women, who worked together at all levels to establish a tradition of research mathematics in the United States.

ACKNOWLEDGMENTS

The support of National Science Foundation grant #DIR-9011625 is gratefully acknowledged. We also thank the referees of a previous version of this paper for their remarks.

NOTES

[1] See our companion paper, Della Dumbaugh Fenster and Karen Hunger Parshall, A Profile of the American Mathematical Research Community: 1891–1906, in the present volume. The 1061 members of the mathematical community included 936 American and 125 foreign participants. Our study focuses on the American contributors.

[2] Raymond Clare Archibald, ed., *A Semicentennial History of the American Mathematical Society: 1888–1938,* 2 vols. (New York: American Mathematical Society, 1938), 1: 1–3, esp. 3.

[3] Dangerously ahead of her time, Prudence Crandall attempted the education of black women in the 1830's. For this woman's story—and a glimpse at the perils of being too progressive—see Eleanor Flexner, *Century of Struggle: The Woman's Rights Movement in the United States* (Cambridge: Harvard University Press, 1959), 38–40.

[4] Jean-Jacques Rousseau, *L'Émile or a Treatise on Education*, ed. W. H. Payne (New York and London, 1906), 263, as quoted in Flexner, 23–24. William Jay Youmans expressed similar sentiments in his editorial remarks entitled, A Profession for Women, *Popular Science Monthly* 38 (1890–1891): 701–702. Youmans concluded his piece with "...the modern woman should say of her home, 'This is my diploma'; and of her children, 'These are my degrees'." The social changes underway in the United States at the dawn of the nineteenth century began to undermine this philosophy. The Louisiana Purchase in 1803 and the subsequent westward expansion meant a greater need for teachers of the ever-increasing population. Moreover, the employment of women in, for example, the newly developed textile mills altered their position in the home.

[5] Margaret W. Rossiter, *Women Scientists in America: Struggles and Strategies to 1940* (Baltimore: The Johns Hopkins University Press, 1982), 4–6, esp. 6.

[6] For details, see Flexner, 25–26. When Willard did not have geometry textbooks, she carved cones and pyramids out of potatoes and turnips in order to illustrate the subject at hand.

[7] Rossiter, 5–6. Thus, as Rossiter notes, a letter of recommendation from Emma Hart Willard served as "the first teacher accreditation in America" [p. 6]. Note that Almira Hart Lincoln, Willard's sister, promoted science in the curriculum.

[8] Robert S. Fletcher, *History of Oberlin College to the Civil War*, 2 vols. (Oberlin: Oberlin College, 1943), 1:373, as quoted in Flexner, 29–30. Compare this stated purpose to that of the Johns Hopkins University as expressed by its President, Daniel Coit Gilman, in his inaugural address: "The object of the university is to develop character—to make men." See Francesco Cordasco, *The Shaping of American Graduate Education: Daniel Coit Gilman and the Protean Ph.D.* (Totowa, N.J.: Rowman and Littlefield, 1973), vi.

[9] Flexner, 30. Prior to 1841, the requirements for women students at Oberlin included an abbreviated "literary" course. Despite evidence which clearly indicated otherwise, this thought persisted even into the twentieth century as shown in W. D. Hyde, *The College Man and the College Woman* (New York: Houghton Mifflin and Co., 1906). Hyde wrote, "Supreme in acquisition, unequaled in transmission and distribution, when it comes to this distinctively creative act, this organizing of facts in the light of the universal principles which bind them into systematic unity, women, as a rule, have far less of this essential productive scholarship than men" [p. 207].

[10] As expressed by early women students at Oberlin and quoted in Flexner, 30.

[11] Flexner, 32.

[12] Flexner, 35.

[13] Flexner, 36. Mary Lyon also included "domestic duties" among the requirements for students at Mount Holyoke. Though Lyon intended this as a means of keeping costs low, this, perhaps, placated parents and others who were somewhat skeptical of such a bold leap for women.

[14] Flexner, 31–35, and Rossiter, 9–10. Flexner captured the freshness, zest, and unrestrained tolerance characteristic of the "first." These same characteristics, especially the latter, coupled with superb intellect sustained many American women earning the first doctorates in Germany. For example, physiology student Ida Hyde often had the strength to persevere solely because she believed she was the first American woman en route to a German doctorate. See Rossiter, 41–42 for details on Ida Hyde.

[15] Rossiter, 28.

[16] Ibid., 29.

[17] Ibid.

[18] Many major universities employed this strategy in an attempt to skirt the issue. The quintessential example, of course, was the Johns Hopkins University. A maverick institution in so many ways, Hopkins naturally attracted women seeking graduate education. Additionally, it provided opportunities unavailable elsewhere, and it was this distinction which forced the University to offer its sole pre-1911 doctorate to a woman. Florence Bascom received her Hopkins doctorate in geology because no other programs in geology existed elsewhere in the country. M. Carey Thomas and Christine Ladd-Franklin were among those who attempted, but failed, to break into the ranks at Johns

Hopkins University. See Hugh Hawkins, *Pioneer: A History of the Johns Hopkins University, 1874–1889* (Ithaca: Cornell University Press, 1960), 260, 266. As Hawkins aptly expressed it, "On the coeducational frontier, Johns Hopkins played the conservative" [p. 259].

[19] The opening of Yale University to women marked the milestone here. President Timothy Dwight downplayed the radical nature of this move, stressing instead its appropriateness in light of Yale's outstanding facilities and environment. Moreover, he saw this as a move to help the women's colleges since many of the women obtaining Yale doctorates would teach at such institutions. See Rossiter, 34. Also, note that Cornell opened in 1868 and welcomed women students at a separate college in 1874.

[20] Christine Ladd to J. J. Sylvester, March 27, 1878, Daniel C. Gilman Papers Coll. #1, Special Collections Division, Milton S. Eisenhower Library, The Johns Hopkins University (hereinafter denoted Gilman Papers). We thank the Department of Special Collections, as always, for its hospitality and The Johns Hopkins University for permission to quote from its archives. For more on graduate education at The Johns Hopkins, see Karen Hunger Parshall and David E. Rowe, *The Emergence of the American Mathematical Research Community 1876–1900: James Joseph Sylvester, Felix Klein, and Eliakim Hastings Moore*, AMS/LMS Series in the History of Mathematics, (Providence: American Mathematical Society, forthcoming), Chapters 2 and 3.

[21] J. J. Sylvester to Daniel C. Gilman, April 2, 1878, Gilman Papers. Here, Sylvester is referring to the British scientific writer and popularizer Mary Somerville.

[22] See George W. Brown to Daniel C. Gilman, April 6, 1878; Reverdy Johnson to Daniel C. Gilman, undated, in Gilman Papers.

[23] Daniel C. Gilman to Christine Ladd, April 26, 1878; Gilman Papers. For a description of the academic atmosphere Ladd found herself in at Hopkins during Sylvester's tenure there, see Karen Hunger Parshall, America's First School of Mathematical Research: James Joseph Sylvester at the Johns Hopkins University 1876–1883, *Archive for History of Exact Sciences* 38 (1988): 153–196; and Karen Hunger Parshall and David E. Rowe, *The Emergence of an American Mathematical Research Community* (Providence: American Mathematical Society, forthcoming).

[24] See, for example, Hawkins, 263–264; Walter Crosby Eells, "Earned Doctorates for Women in the Nineteenth Century," *Bulletin of the American Association of University Professors* 42 (1956): 644–651 on p. 650; and Edward T. James, ed., *Notable American Women 1607–1950: A Biographical Dictionary*, s.v. "Ladd-Franklin, Christine," by Dorothea Jameson Hurvich. In 1882, Christine Ladd married Fabian Franklin, an associate in mathematics at the Johns Hopkins University. Given this circumstance and the protocol for married women of the day (see note 74), the fact that she managed to remain active in academics merits note. Also, at its semicentennial celebration in

1926, Hopkins awarded Christine Ladd-Franklin her degree, forty-four years late. Then aged 79, she accepted it.

[25] Eells, 649–651. Edgerton married Frederick James Hamilton Merrill in 1887. For additional information on Edgerton, see Judy Green and Jeanne La Duke, Contributors to American Mathematics: An Overview and Selection, 117–146 (esp. 123–124) in *Women of Science: Righting the Record,* ed. Gabriele Kass–Simon and Patricia Farnes (Bloomington: Indiana University Press, 1990).

[26] See 3–14 in Marion Talbot and Lois Kimball Matthews Rosenberry, *The History of the American Association of University Women, 1881–1931* (New York: Houghton Mifflin Company, 1931) for a delightful account of the origins of the organization. The ACA (after 1921, the AAUW) was not solely responsible for forging the way for women scientists in America. It benefited from the organizational and leadership abilities, most notably of Christine Ladd-Franklin and Ellen Richards. Like Ladd-Franklin, Richards pursued graduate work slightly ahead of her time. Richards applied to the then five-year-old MIT for a graduate degree in chemistry in 1870. Admitted as a special student and without charge—to prevent her name from appearing in the catalog—MIT awarded her a second bachelor's degree in 1873. Though she continued her studies another two years, MIT did not grant her the doctorate for which she had hoped. Both Richards and Ladd-Franklin, in their mid-forties in the last decade of the nineteenth century, knew that the task at hand was to find and encourage women who were both scholarly and form-idable enough to handle the adversities of pioneering. See Rossiter, 42.

[27] See Talbot and Rosenberry, 143–144. Helen Magill, the first American woman to obtain the doctorate, raised the issue with the ACA.

[28] Thus, Gentry was the first woman to attend lectures at Berlin. James C. Albisetti, *Schooling German Girls and Women: Secondary and Higher Education in the Nineteenth Century* (Princeton: Princeton University Press, 1988), 225–226.

[29] In fact, Albisetti claims Gentry was "[t]he American woman who may have had the most impact in Germany" [p. 225].

[30] In 1895, Marie Gernet became the first woman to receive a doctorate in mathematics from the University of Heidelberg. Albisetti, 226.

[31] Ibid.

[32] Ibid., 225–226.

[33] Rossiter, 40. See also Judy Green, "Christine Ladd-Franklin (1847–1930)," 121–128 (esp. 122–123), in *Women of Mathematics: A Biobibliographic Source-book*, eds. Louise S. Grinstein and Paul J. Campbell (New York: Greenwood Press, 1987); and Albisetti, 225. It is worth noting that she also gained access to Hermann von Helmholtz's laboratories at Berlin. For more on Klein's role in educating women at the graduate level, see Parshall and Rowe, Chapter 5.

[34] Albisetti, 227. Albisetti conjectures that Friedrich Althoff, the university affairs official within the Prussian Ministry of Education, may have wanted to test the waters with foreign women students before opening universities to qualified Prussian women. This was, in some ways, a safe way to introduce the doctorate to Prussian universities since these test cases would return to America for employment. This was not at all unlike the reasoning behind the decision to open the graduate programs at Yale to women. See note 19 above.

[35] Caroline N. Beshers to Della Dumbaugh Fenster, 21 January 1990; 17 February 1990.

[36] Caroline N. Beshers to Della Dumbaugh Fenster, 21 January 1990. The Chicago World's Fair also marked an important turning point in May Winston's personal life since it was there that she met Henry B. Newson.

[37] Albisetti, 227. An auditor referred to a non-matriculating student. Maltby studied physics, and not mathematics, as her primary area of concern.

[38] Caroline N. Beshers to Della Dumbaugh Fenster, 21 January 1990.

[39] Ibid; Rossiter, 41; and Betsey S. Whitman, Mary Frances Winston Newson: The First American Woman to Receive a Ph.D. in Mathematics from a European University, *Mathematics Teacher* **76** (1983): 576–577, esp. 576. For an expanded version of the Whitman article, see Mary Frances Winston Newson (1869–1959), 161–164, in *Women of Mathematics*. Although Christine Ladd-Franklin was surely disappointed when Göttingen would not confer the doctorate upon her in 1892, she drew from this experience in her fight for women to gain access to foreign universities.

[40] Caroline N. Beshers to Della Dumbaugh Fenster, 17 February 1990.

[41] Albisetti claimed there were 14 (see p. 227); Isabel Maddison (see below in text) asserted that 15 women studied at Göttingen that fall—three English, one German, and 11 American.

[42] Annie L. MacKinnon to Felix Klein, July 9, 1894, Klein Nachlass X, Niedersächsische Staats- und Universitätsbibliothek, Göttingen. We thank the library for permission to quote from its archives. The second author particularly thanks Dr. Helmut Rohlfing, the director of the library's Handschriftenabteilung, for his help and hospitality during her research trip to Göttingen in 1990.

[43] Mary E. Garrett, the Baltimore heiress to the B & O Railroad fortune, founded this $500.00 scholarship for Bryn Mawr graduate students to pursue a year of study abroad (see also [51]).

[44] Clipping (source unknown) from the Archives, Mariam Coffin Canaday Library, Bryn Mawr College. We thank Susan Shifrin for her outstanding help in locating information on the women associated with Bryn Mawr and Bryn Mawr College for permission to quote from its archives.

[45] For more details on the 1061 members of the sample space, see Fenster and Parshall, A Profile of the American Mathematical Research Community:

1891–1906. The *Bulletin of the NYMS* (after 1894, *the AMS*) provided minutes of AMS meetings, including attendance records, papers read, names and locales of new members, articles of mathematical interest, and, perhaps most important to this study, personal notices exchanged by the members. As such, the *Bulletin* held stores of information about the early mathematical community. The mathematically interested group of 1061 was culled from the issues spanning the years from 1891 to 1906. The present study's time constraints 1891–1906 prevented the inclusion of contributions of some women. Anna Johnson (later Anna Johnson Pell Wheeler) belonged to the sample space only by virtue of her induction into the AMS in February, 1906. Mary Emily Sinclair, the first woman to receive a doctorate in mathematics from Chicago (1908), does not appear prominently here for the same reason.

[46] Levels of participation divided the sample space into smaller constituencies for more focused study. Those with an average of at least one activity per year over the 15-year period from 1891–1906 and included in either the first, second, or third edition of *American Men of Science* comprised the "most active" group of mathematicians. (See James McKeen Cattell, *American Men of Science: A Biographical Directory*, 1st ed. [New York: The Science Press, 1906]; 2nd ed. [New York: The Science Press, 1910]; 3rd ed. [New York: The Science Press, 1921].) The "active" members of the community included those with at least one activity every five years and an entry in *American Men of Science*. All others, i.e., those participants with two or fewer activities, or contributors not found in *American Men of Science*, composed the "rank and file." "Activities" were categorized as "talks," "papers published," "service," "attendance," or membership in the AMS. "Talks" referred to papers read either in person or in absentia at meetings of the AMS or AAAS, the Chicago Congress, etc. "Papers" represented articles published in the *Bulletin of the AMS*, *Transactions of the AMS*, *Annals of Mathematics*, *American Journal of Mathematics*, or Eliakim Hastings Moore, Oskar Bolza, Heinrich Maschke, and Henry S. White, eds., *Mathematical Papers Read at the International Mathematical Congress Held in Connection with the World's Columbian Exposition: Chicago 1893* (New York: Macmillan and Co., 1896). "Service" measured activity in terms of holding office in, or serving on committees of, the AMS or AAAS (1 = one year's service), serving on the editorial board of one of the journals listed above (1 = one year's service), serving as an officer or organizer of a teachers' association, etc. Attendance represents the minimum number of meetings or colloquia attended between 1891–1906. Given our use of the *Bulletin of the AMS* and the *American Men of Science*, these counts represent greatest lower bounds in the given areas. In particular, inclusion in *American Men of Science* occurred only if a solicited scientist submitted the required information about him or herself. Hence, it is possible that women in our study either were not asked for information or did not complete the designated form. Moreover, given that the AMS began in New York City and

later formed sections in Chicago and San Francisco, women residing in or near these geographical areas had a greater opportunity to attend meetings regularly. These are the most reliable attendance figures, nonetheless, and although perhaps not perfect, they accurately depict the trends of the day. See note to Table 1 in A Profile of the American Mathematical Research Community: 1891–1906.

[47] The "active" participants thus included 18 women, two of whom, Schottenfels and Scott, were also among the "most active." As noted, 11 mathematically interested "almost active" women were excluded from the "active" sample space since they did not appear in *American Men of Science*. Among these 11 were: Achsah Ely, one of four women in attendance at the Chicago Congress and a member of the AMS since 1891; Annie Louise MacKinnon, an early doctorate (see Table 1); and Roxana Vivian, a 1901 mathematics Ph.D. from the University of Pennsylvania.

[48] A brief but interesting summary of Williams' life can be found in Betsey S. Whitman, Women in the American Mathematical Society before 1900: Part One, *Association for Women in Mathematics Newsletter* **13** (1983): 10–14, esp. 14.

[49] For these references see *Bull. AMS* **9** (1902): 198, 271, 367, 368; *Bull. AMS* **9** (1902): 73; and *Bull. AMS* **9** (1903): 525, 537. Schottenfels' work on the groups of order 8!/2 was actually published as Ida May Schottenfels, "Two Non-Iso-morphic Simple Groups of the Same Order 20160." *Annals of Mathematics* **1** (1900): 147–152. She showed by brute force that the alternating group on eight letters A_8 and what we now call the projective special linear group $PSL_3(4)$ were not isomorphic. This is the only instance of two non-isomorphic finite simple groups of the same order. See, for example, Leonard Eugene Dickson, *Linear Groups With an Exposition of the Galois Field Theory* (Leipzig: B. G. Teubner, 1901; reprint ed., New York: Dover Publications, Inc., 1955), 308–310.

[50] See the following two articles by Kenschaft: Patricia C. Kenschaft, Charlotte Angas Scott, 1858–1931. *College Mathematics Journal* **18** (1987): 98–110, esp. 102; and Charlotte Angas Scott (1858–1931), 193–203, esp. 193–195 in *Women of Mathematics*. Both of these articles provide a more detailed look at the life of Charlotte Scott. Sonya Kovalevsky received her doctorate (in absentia) from the University of Göttingen in 1874, the only woman to precede Scott with a doctorate in mathematics. See, for example, Lynn M. Osen, *Women of Mathematics* (Cambridge, MA: MIT Press, 1974), 126; and Ann Hibner Koblitz, Sofia Vasilevna Kovalevskaia (1850–1891), 103–113, esp. 104–105 in *Women of Mathematics* for more information on Sonya Kovalevsky's doctorate.

[51] Thomas actually recruited Scott in 1884 but Bryn Mawr did not open until 1885. A significant figure in the advancement of higher education for women, Thomas acted as dean of Bryn Mawr from its inception until her inauguration as president of the college in 1894. She occupied this position for 28 years.

William Welch, the first dean of the School of Medicine at the Johns Hopkins University, paid tribute to Thomas upon her retirement at the 1922 commencement exercises at Bryn Mawr. Given that Thomas' persuasive arguments and Mary E. Garrett's funds were primarily responsible for opening the Johns Hopkins Medical School to *both* men and women, Welch's address merits note. See William H. Welch, Contribution of Bryn Mawr College to the Higher Education of Women, *Science* **56** (July 1922): 1–8.

[52] Scott advised four of these students, Louise Duffield Cummings, Ruth Gentry, Isabel Maddison, and Virginia Ragsdale, in the period 1891–1906. Our sample space includes all four of these women.

[53] As expressed by the directors of Bryn Mawr College and quoted in Kenschaft, Charlotte Angas Scott, 198 in *Women of Mathematics*.

[54] Patricia C. Kenschaft, The Students of Charlotte Angas Scott, *Mathematics in College* (Fall 1982): 16–20, p. 16.

[55] Kenschaft, Charlotte Angas Scott, 105.

[56] Archibald, ed., 1:96, 106. She was the only woman to occupy an elected position within the AMS in our time period. In fact, almost 20 years passed before the AMS elected another women to the Council; more than 70 until the next woman became vice president.

[57] Scott coedited the *American Journal of Mathematics* from 1899 to 1926. Only eight of these years, 1899–1906, fell within our period. For more on Scott's influence on the College Entrance Examination Board, see Kenschaft, Charlotte Angas Scott, 104 and idem., Charlotte Angas Scott, 200 in *Women of Mathematics*.

[58] Kenschaft credits Scott's "dynamic" leadership for the progress of women in American mathematics at the turn of the century in Women in Mathematics around 1900, *Signs* **7** (1982): 906–909, esp. 908. The results of Green and LaDuke, which encompass nearly a century's worth of comparative information, seem to substantiate this idea in terms of a much larger picture. See Judy Green and Jean LaDuke, Women in the American Mathematical Community: The Pre-1940 Ph.D.'s, *The Mathematical Intelligencer* **9** (1987): 11–23, esp. 12.

[59] These are somewhat similar to the two strategies early women scientists utilized in the broader sense. See Rossiter, xvii–xviii.

[60] In fact, she was the only woman mathematician starred in the first edition of *American Men of Science*. A star indicated one whose work was deemed "most important" by the ten leading scholars in a particular field. One thousand stars appeared in the first edition of *American Men of Science*, with eighty going to mathematicians.

[61] Archibald, ed., 1:185 and Green and LaDuke, Pre-1940 Ph.D.'s, 20.

[62] A complete list of Bliss' students can be found in L. M. Graves, Gilbert Ames Bliss: 1876–1951, *Bull. AMS* **58** (1952): 261–264. The counts for Snyder and Morley were taken from Green and LaDuke, Pre-1940 Ph.D.'s, 19–20. Green

and LaDuke provide intriguing statistics on the women who earned doctorates in mathematics before 1940. Though the time of their study extends 34 years beyond ours, it is interesting to note that eight mathematicians, including Bliss, Snyder, and Morley, supervised one-third of the women receiving Ph.D.'s before 1940. This group also included Arthur Byron Coble, Leonard E. Dickson, A. Landry, Anna Johnson Pell Wheeler, and Charlotte Scott.

[63] Archibald, ed., 1:101–103; as well as Frank N. Cole, Louise D. Cummings and Henry S. White, The complete enumeration of triad systems in fifteen elements, *Proceedings of the National Academy of Science* **3** (1917): 197–199; and Frank N. Cole, Louise D. Cummings, and Henry S. White, Complete Classification of the triad systems on fifteen elements, *Memoirs of the National Academy of Science* **14** (1919): 89.

[64] Newson and Winston met thanks to mathematics (see [36]) and courted while in its service, as evidenced by a few attendance records of Chicago sectional meetings. May Winston, however, did little mathematics during the ten years of her marriage. She taught a little in the summers at the University of Kansas, attended one meeting of the Chicago Section, and published her translation of David Hilbert's famed address before the Paris Congress. Her daughter wrote of the distractions in May Winston's life during this time: "In those days there were many family obligations and my mother unexpectedly had her ailing mother-in-law and one or another of *her* daughters in the house all the time. They were not at all sympathetic to May's taking study time to herself every day, and the atmosphere made the whole thing difficult. My father, I know, felt badly about this." Beshers felt this strain contributed to Newson's sudden death in 1910, at the age of 49, from a rheumatic heart. Caroline N. Beshers to Della Dumbaugh Fenster, 17 February 1990.

[65] Archibald, ed., 1:108. E. H. Moore was rated first among mathematicians in James McKeen Cattell's 1903 poll of American Scientists. Cattell published these results in the fifth edition of *American Men of Science*.

[66] Archibald, ed., 1:146. Despite Moore's progressive views, he was still keenly aware of gender differences in academia. In a letter to President Harper just before the University of Chicago opened, Moore wrote, "Don't you agree that we ought to have no woman instructor, at least no one without a doctor's degree?" This indicated a certain ambivalence toward women in mathematics and reflects the higher standard to which women were generally held at this time. Nevertheless, Moore entertained the idea. See E. H. Moore to W. R. Harper, May 10, 1892, University of Chicago Archives, William Rainey Harper Papers, Box 14, Folder 15. (We thank the University of Chicago for permission to quote from its archives.) The research of Green and LaDuke showed that Chicago went on to award more than twice as many doctorates to women in mathematics prior to 1940 as any other institution. They attributed this achievement to "both the high overall mathematics productivity of Chicago

and the absence of institutional and personal biases against women." See Green and LaDuke, Pre-1940 Ph.D.'s, 18–19.

[67] Rossiter, 88.

[68] Pengra became Mrs. A. R. Crathorne. Based on the faculty at Wisconsin and on her thesis topic, On functions connected with special Riemann surfaces, in particular those for which $p = 3, 4, 5$, it seems likely that either Ernest Brown Skinner or Linnaeus Wayland Dowling served as Pengra's advisor.

[69] See *Bull. NYMS* **3** (1894): 148; and *Bull. AMS* **6** (1900): 466. For more on the life of Ruth Gentry, see Betsey S. Whitman, Women in the American Mathematical Society before 1900: Part Two, *Association for Women in Mathematics Newsletter* **13** (1983): 7–9, esp. 7.

[70] See *Bull. AMS* **6** (1900): 267; *Bull. AMS* **7** (1901): 289; and *Bull. AMS* **11** (1905): 449. Anne Bosworth later became Mrs. Theodore Moses Focke.

[71] For evidence of the participation of Moore and Osgood in these affairs, see *Bull. AMS* **11** (1905): 218; *Bull. AMS* **9** (1903): 508; and *Bull. AMS* **11** (1904): 166.

[72] For Syke's participation, see *Bull. AMS* **11** (1905): 218; and *Bull. AMS* **12** (1906): 267. For Smith's, see *Bull. AMS* **11** (1904): 166.

[73] See *Bull. AMS* **11** (1904): 111; and *Bull. AMS* **12** (1906): 224, respectively.

[74] See *Bull. AMS* **6** (1900): 170. For an interesting account of Brooks' short career, see Rossiter, 15–16. Brooks' career began brilliantly with two substantial papers on radioactivity with Ernest Rutherford at McGill University, a period of study at Cavendish Laboratory in England, and an instructorship in physics at Barnard College. This all came to an abrupt end when Brooks became engaged in 1906 and attempted to bypass the societal rule of, as Rossiter put it, "resignation upon marriage." Barnard held fast to its views against the combination of home duties and college responsibilities. Ultimately, a still single Brooks resigned. Her resignation was not by force but, rather, as a result of the uncomfortableness of a broken engagement to a Columbia graduate student. She never did physics again.

[75] Bessie Growe Morrison, Removal of Any Two Terms From a Binary Quantic by Linear Transformations, *American Journal of Mathematics* **23** (1901): 287–296.

[76] Relative to the present study, this latter group included G. C. Alden, A. F. Becker, H. D. Buckingham, Mary Emma Byrd, Susan Jane Cunningham, M. E. Decherd, M. F. Gould, Fanny Hitchcock, Anna Johnson [later Pell and later Wheeler], Myrtle Knepper, R. A. Pesta, S. F. Richardson, Adelaide Smith, May E. G. Waddell, and M. S. Walker. (On Anna Johnson, though, see [45] above.) For more information on Anna Johnson, see Louise S. Grinstein and Paul J. Campbell, Anna Johnson Pell Wheeler (1883–1966), 241–246, in *Women of Mathematics*. For details on Mary Emma Byrd and Susan Jane

Cunningham, see Whitman, Women in the AMS before 1900: Part One, 11–12.

[77] For the definitions of the various regions see Table 4 of our companion paper in this volume. Although the numbers here are admittedly small, the percentage reflecting the birth places of the women match up remarkably closely to the comparable data for the full "active" sample space, namely, Northeast 42.8%, South 5.6%, Midwest 30.0%, West 1.9%, Canada 5.9%, and foreign 13.8%.

[78] Compare the corresponding statistics for the broader community in Table 11 of ibid.

[79] Clearly, many women also pursued careers outside of academics. For more information on the career options available to professional women, see Rossiter, Chapter 3, 51–72; Welch, 4; and Jane M. Bancroft, Occupations and Professions for College-Bred Women, *Education* **5** (1885): 486–95. In 1884, Bancroft received her doctorate in history from Syracuse. Martha Foote [Crowe] followed suit a year later with a doctorate in English. These two women were among those who actively sought co-education at leading universities and appropriate employment for educated women. Bancroft's article addresses the fundamental question of what a nineteenth-century woman would do with a college degree.

[80] For more details on Charlotte Cynthia Barnum, see Whitman, Women in the AMS before 1900: Part Two, 7.

[81] Isabel Maddison, Charlotte Angas Scott; An Appreciation, *Bryn Mawr Alumni Bulletin* **12** (1932): 9–12, quoted in Kenschaft, Charlotte Angas Scott, 103. Naturally, not all students held Scott in the highest regard. Kenschaft cites one student's doodle as evidence to the contrary: S is for Scott/Superior Scott/She is kind in the main/If you have any brain/But if you have not/Superior Scott! [p. 103].

[82] Mabel Young, Marion Stark, and Helen Russell, Helen A. Merrill, '86, *Wellesley Magazine* (July 1949): 353–354, quoted in Claudia Henrion, Helen Abbott Merrill (1864–1949), 147–151, p. 149–150, in *Women of Mathematics*.

[83] Isabel McLaughlin Stephens, We Shall Never Find Your Equal, *Smith Alumnae Quarterly* (Aug. 1935).

[84] Whitman, Women in the AMS before 1900: Part Two, 8–9.

[85] Whitman, Mary Francis Winston Newson, 577.

[86] Smith College Archives. We thank Maida Goodwin for her help in locating information on Ruth Goulding Wood. Ms. Goodwin shared a special interest in this project since her grandmother was Ruth Wood's first cousin.

[87] Isabel Maddison, *Bryn Mawr Alumnae Bulletin* (1951): **14**. Fanny Hitchcock, a member of the mathematical community over the period from 1891 to 1906, spent her career at the University of Pennsylvania. Upon her death in 1936, she designated that part of her estate go to the University's Trustees to

"provide for the professional, technical, and vocational training of women, either in the regular departments of the University now open to women, or in a separate department as the Trustees may deem best, ... provided that such departments shall, in all cases, be open to men on the same terms as to women." Whitman, Women in the AMS before 1900: Part Two, 8. Note the careful wording of Hitchcock's gift. See Rossiter, 47, for an account of a misuse of funds intended for the advancement of women's education at the University of Pennsylvania in 1887.

[88] This supports all but one of Rossiter's conclusions with regard to the progress of women in science at this time. Contrary to her overall findings, the women in our study were restricted neither in membership nor activity within the AMS. See Rossiter, 313–314.

[89] Rossiter, xvi.

ELEMENTS

OF

ANALYTICAL GEOMETRY

AND OF THE

DIFFERENTIAL AND INTEGRAL

CALCULUS.

BY ELIAS LOOMIS, A.M.,

PROFESSOR OF MATHEMATICS AND NATURAL PHILOSOPHY IN THE UNIVERSITY OF
THE CITY OF NEW YORK, AUTHOR OF "A TREATISE ON ALGEBRA," "ELEMENTS
OF GEOMETRY AND CONIC SECTIONS," "ELEMENTS OF PLANE AND SPHER-
ICAL TRIGONOMETRY, WITH THEIR APPLICATIONS TO MENSURATION,
SURVEYING, AND NAVIGATION," "RECENT PROGRESS OF
ASTRONOMY," ETC., ETC.

NEW YORK:
HARPER & BROTHERS, PUBLISHERS,
82 CLIFF STREET.
1851.

代微積拾級卷六

米利堅羅密士譔　英國　偉烈亞力　口譯

海寗　李善蘭　筆述

代數幾何六

論撱圜

撱圜亦圓錐曲線之一平面曲線也周之各點距二

定點之和恒等二定點乃曲線之二心

如圖吧呃爲二定點設吧點繞呃點任至何處距

呃吧二點之和恒等卽成撱圜以吧呃爲二心吧

代數賣合彼　卷六　一

Title page of Alexander Loomis's *Calculus*
1851 American edition (left) and translated into Chinese
by Li Shan-lan and Alexander Wylie in 1859 (right)

Mathematical Exchanges Between the United States and China A Concise Overview (1850–1950)

Dianzhou Zhang*

*Department of Mathematics, East-China Normal University,
Shanghai, People's Republic of China*

and

Joseph W. Dauben

*Herbert H. Lehman College, CUNY, and
Ph.D. Program in History, The Graduate Center,
City University of New York*

A major influence on the development of Chinese mathematics during the period 1850–1950 came from the United States. The first Chinese translation of a calculus textbook, the first Chinese Ph.D. in mathematics, the first director of the first Chinese Mathematical Institute, and the first graduate program in mathematics in China—were all associated in one way or another with American mathematicians or involved funds established for Chinese foreign-exchange students to go to the United States. These funds, largely in the form of Boxer Indemnity Scholarships, enabled young Chinese to study in American colleges and universities. On the other hand, Chinese mathematicians began to make original contributions of their own to mathematics in the United States, especially after World War II.

INTRODUCTION

Ancient Chinese mathematics, with its own long and impressive history, entered a period of stagnation and decline after the 14th century A.D. Not long thereafter, however, Western mathematics reached China when Matteo Ricci (1552–1610) and Xu Guang-qi (1562–1633) translated the first

*Supported by a grant from the Wang Foundation, Hong Kong.

six Books of Euclid's *Elements* into Chinese in 1607 [1]. At the beginning of the Qing dynasty, the Emperor Kong-Xi (1654–1722) learned a considerable amount of Occidental mathematics from European missionaries, among them Jean François (1654–1707), Joachim Bouvet (1656–1730), and Ferdinand Verbiest (1623–1688). Unfortunately, this interest in Western Mathematics was only an isolated example. From the middle of the 18th century to the Opium War (1840), the Chinese intelligentsia basically subscribed to a doctrine of "back to the ancients." Some Chinese mathematicians in their textual research argued that the algebra Kong-Xi had learned from Western sources was equivalent to the Chinese method of "Tian Yuan Shu," which provided a means of solving equations that dates back to the 14th century. In any case, from about 1750 on, Chinese interest in Western mathematics was virtually nonexistent [Li Di 1984, 254, 277, and 300].

Slightly more than a century later, the English Missionary Alexander Wylie (1815–1887) cooperated with the self-taught Chinese mathematician Li Shan-lan in translating the rest of Euclid's *Elements* into Chinese in 1858 [2]. The following year they also translated an American textbook on calculus into Chinese, and this was the beginning not only of renewed interest in Western mathematics, but of direct U.S. influence upon Chinese mathematics itself.

As is well known, British forces opened the door of the Qing Government to Western commerce and trade as a direct result of the Opium War (1840). Consequently, European influences in China were earlier and more extensive in the 19th century than those exerted by Americans. It is all the more surprising, therefore, that ultimately, the major influence on modern Chinese mathematics should have come from the United States through an unusual combination of events that went far beyond the well-known efforts of Christian missionaries.

I. ELIAS LOOMIS (1811–1889)

The first American to have any considerable effect upon mathematical developments in China (and East Asia generally) was Professor Elias Loomis, although he never visited China and seems to have had few if any contacts with mathematicians there. From 1844 to 1860, Loomis was professor of mathematics and natural philosophy at the City College of New York. Later he accepted a position at Yale University.

Loomis made his most important scientific contributions in the field of meteorology. In 1846 he published the first "synoptic" weather map using a new method of representing meteorological data. In the years to follow, his methods exerted a profound influence on the formulation of theories to predict storms. Most of his scientific achievements were of a practical, rather than a theoretical, nature. In his own time, however, Loomis was better known for the publication of a large number of textbooks on mathematics, astronomy, and meteorology, rather than for his own original scientific investigations [Kutzbach 1973, 487].

One of the most influential of these textbooks appeared in 1851: *Elements of Analytical Geometry and of the Differential and Integral Calculus* [3]. Only eight years later, the Chinese translation of this book was made. It is this Chinese translation that is of special interest for the subsequent history of modern mathematics in China.

Alexander Wylie and Li Shan-lan

Alexander Wylie (1815–1887) arrived in Shanghai from England in 1847, and shortly thereafter began to cooperate with the Chinese mathematician Li Shan-lan in translating a number of mathematical works from English to Chinese in the 1850s. Before sailing from England Wylie had already begun to study Chinese, and in the 30 years he lived in China, he not only became fluent in the language, but also learned something of Manchu and Mongolian. Wylie at the same time acquired an extensive knowledge of Chinese literature, and was the author of many books, pamphlets, and articles both in English and Chinese. At first, he was salaried by the British and Foreign Bible Society, and was in charge of the press of the London Missionary Society. About 1863 or 1864 the Society appointed him as its agent in China. Thereafter, in addition to his work in the central office, he traveled widely. Under his direction, several foreign and a number of Chinese colporteurs were employed, making it possible to cover a large part of the Empire in carrying out the Society's mission to broadcast the Scriptures. In 1877, because of failing eyesight, Wylie retired [La fourette 1929, 265, 437].

Meanwhile, Li Shan-lan (1811–1882) was to become the most important Chinese mathematician of the 19th century, having studied Chinese mathematics by himself. He independently proposed the "Jian Zhui Shu"—a calculating trick for integrating polynomials as well as $1/x$. In addition, Li developed a combinatory theory along the lines of earlier Chinese mathematics and wrote a famous book, ⟨ *Do Ji Bi Lei* ⟩ (The Study in Piles), in which some combinatory identities appear that are still of interest today.

In 1852, in cooperation with Alexander Wylie, Li Shan-lan began to study Western mathematics systematically. Meanwhile, according to Joseph Edkins [4], during the period 1847–1852, "Wylie studied the Sung dynasty arithmetic with the help of Li Shan-lan, a native mathematical author of high attainments" [Edkins quoted in Wylie 1897, 5–6]. When the Mathematics Board of the Tong Wen Guan (the Foreign Language Institute established in Beijing by the Qing government in 1868, and later merged into Metropolitan University) appointed its first professor of mathematics, it chose Li Shan-lan. Equally impressive, Li was the only Chinese professor in the faculty of science at the time [Li Yan and Du Shiran 1987, 256–259; Horng 1991, 97–104].

Li Shan-lan and Alexander Wylie first cooperated on completing the work Matteo Ricci and Xu Guang-qi had begun by translating books VII–XIII plus the two spurious books XIV and XV of Euclid's *Elements* [Heath 1956, 51]. After finishing another joint translation—this time of De Morgan's *Algebra*—they went on to translate the textbook Elias Loomis had written on calculus into Chinese: 〈 *Dai Wei Ji Shi Ji* 〉 (Elements of Analytical Geometry and of the Differential and Integral Calculus). This was an especially significant undertaking because it was the first book on calculus to appear in China, published in 1859 by Muo Hai Publishing House. This firm, founded by Dr. Walter Henry Medhurst (1796–1857), seems to have employed Alexander Wylie to serve as editor and some sort of overall manager.

An important feature of the Li–Wylie translation was a list of 330 English–Chinese mathematical terms they included as an appendix. Needless to say, defining technical terms for a first translation into a totally unrelated foreign language system is an extremely difficult task, more so than translating from Latin to English or even French to Russian, for example. In the case of the list of Chinese–English technical terms drawn up by Li and Wylie, not only was this the first such list of terms for modern mathematics in Chinese, but most of these Chinese terms subsequently became standard in Chinese mathematics and are still in use today [Li Di 1984, 354].

The influence of Li and Wylie's translation of Loomis' *Calculus* was soon felt well beyond China. Before the middle of the last century Japanese scholars still learned most of their mathematics from China. According to Yoshio Mikami, an early authority on the history of mathematics in Japan, "The first material introduction of European mathematics into Japan was certainly Alexander Wylie's translation of Loomis' *Calculus* which was brought to Japan soon after its publication" [Mikami 1913, 178].

Many Japanese mathematicians, including the famous Hagiwara Teisuke [5], had only the Li–Wylie translation of Loomis as a guide to learning calculus. "The Chinese translation of Loomis was perhaps the best source of information he could avail himself of" [Mikami 1913, 256]. Barely 13 years later, the Japanese mathematician Fukuda Riken wrote a book ⟨*Dae Wei Ji Shi Ji Yi Gie*⟩ (Commentary on the Li–Wylie translation of Loomis' Calculus), in which he explained in detail for Japanese readers how to understand the Chinese translation. This is why Japanese mathematicians today use most of the same Chinese characters for the calculus as do the Chinese [Li Di 1984, 384]. It is no exaggeration, therefore, to say that Loomis has exerted an extraordinary influence throughout East Asia on the development of modern mathematics, wherever it is practiced, thanks to the Li–Wylie translation of his calculus.

In addition, several other mathematical textbooks by Loomis were also translated into Chinese with the help of two American missionaries, Calvin Wilson Mateer (1836–1908) and A. P. Parker (1850–1924). These were ⟨Xing Xue Bei Zhi⟩ (Elements of Geometry), translated by Zou Li-wen and C. W. Mateer in 1895; ⟨Ba Xian Bei Zhi⟩ (Trigonometry), translated by Xie Hong-lai and A. P. Parker in 1893; and ⟨Dai Xing He Can⟩ (Elements of Analytical Geometry), translated by Xie Hong-lai and A. P. Parker in 1893 [6].

As for the translation of Loomis' *Calculus*, why should an Englishman, Alexander Wylie, have chosen the work of an American author, Elias Loomis, for translation? This choice is especially surprising considering the state of mathematics in general in the United States at the time. Until the end of the 19th century, there were no outstanding American mathematicians of European calibre. By mid-century, even the best colleges in the United States still did not provide a scientific education comparable to the best available in Europe [7]. Why, therefore, should Loomis' textbooks have been taken as models in China? Why not pick a popular text by a European author, among whom fundamental works by prominent mathematicians were readily available? The answer, presumably, must lie with Wylie himself. It may be that he did not read—or have access to—English or European texts. More likely, the fact that for the 1840s Loomis was writing at an introductory level (rather than at the research level) made his text an appropriate choice for Wylie and Li. Loomis could be recommended as an excellent educator [8], and, pedagogically, his textbook of 1851 on calculus (the one translated into Chinese) proved to be very successful, selling 25,000 copies by 1874 [Loomis 1874, Preface].

As Loomis explained in the first edition (1851) of his book:

It was written, not for mathematicians, nor for those who have a peculiar talent or fondness for the mathematics, but rather for the mass of college students of average abilities, ... although I do not claim for it any originality, it appears to me that I have here developed it in a more elementary manner than I have before seen it presented, except in a small volume by the late Professor Ritchie, of London University. ... I have accordingly given special attention to the development of the fundamental principle of the Differential Calculus, and shall feel a proportionate disappointment if my labors shall be pronounced abortive [Loomis 1851, Preface].

Li Shan-lan in his Chinese preface to the translated edition, also explained the reason why he and Wylie had chosen the textbook by Loomis for translation:

Loomis, a famous American mathematician, combined algebra, differential and integral calculus into a whole book, its title reflecting all of these, and set up an orderly system. This book is of benefit to the students. Mr. Wylie knew and appreciated this very much. He made great efforts to obtain this book, and asked me to translate it into Chinese with him [Li, Shan-lan 1859, Preface].

Loomis' textbooks were certainly very popular in the United States. They were elementary, practical and easily understood by beginners. These were undoubtedly overriding considerations when Alexander Wylie came to choose a textbook for translation into Chinese.

Unfortunately, few historians of mathematics have paid much attention to Loomis and his works, and most histories of mathematics have entirely overlooked the fact that Loomis was especially influential in East Asia during the last century. For example, the otherwise informative biography of Loomis written by Gisela Kutzbach for the *Dictionary of Scientific Biography* never mentions the Li−Wylie translation [Kutzbach 1973, 487]. It may well be that Loomis himself did not know that the 1851 edition of his calculus book had been translated into Chinese. He did not, at least, mention the Chinese translation in his new preface to the subsequently revised edition of the book in 1874, well after the Chinese translation had appeared.

Other standard works on the history of American mathematics also fail to mention Loomis (and consequently overlook as well his influence upon China and Japan). For example, D. E. Smith and J. Ginsburg in their *A History of Mathematics in America before 1900* [Smith and Ginsburg 1934] do not mention E. Loomis at all, nor does J. V. Grabiner in her article, Mathematics in America: the First Hundred Years [Grabiner 1977, 9−23].

A more recent study by George Rosenstein, specifically devoted to the subject of textbooks: The Best Method. American Calculus Textbooks of the Nineteenth Century, also fails to consider the merits of Loomis' work we have referred to [Rosenstein 1989]. It may well be that Loomis was not so important for the history of mathematics in America, but in the history of mathematical development internationally, especially in Asia, Loomis was a figure of considerable influence.

II.   THREE UNIVERSITIES: CORNELL, HARVARD, AND CHICAGO

While Loomis was an indispensable factor in introducing modern Western mathematics to students in China, America also proved significant for institutional reasons as well. Above all, after 1900 the United States represented the major foreign influence upon the development of 20th-century Chinese mathematics. Consider, for example, the following list of China's important "firsts": (1) The first Chinese professors of mathematics who studied in the West, Qin Fen (Harvard, 1909) and Zheng Tung-sun (Cornell, 1910); (2) The first Chinese Ph.D. in Mathematics, Hu Min-fu (Harvard, 1917); (3) The first director of China's National Mathematical Institute of China, Chiang Li-fu (Harvard Ph.D., 1918); (4) The first mentor who supervised the first graduate student (S. S. Chern) to earn a Masters degree in Mathematics in China, Sun Tan (Chicago Ph.D., 1928); and (5) The first Chinese expert on number theory, Yang Ke-chuan (Chicago, 1928), and the first topologist, Kiang Tsai-Han (Harvard, 1930).

ZHENG TUNG-SUN (Tsen Tze-fan, 1887–1963), was a pioneer of mathematical education in Chinese colleges and universities. He studied at Cornell University as an undergraduate student during 1907–1910 (at the end of the Qing Dynasty, just before the founding of the Republic of China in 1911). At the time, very few Chinese chose to study mathematics seriously, but among those who did, Zheng was the most proficient. After one year of graduate work at Yale University, he returned to China, and taught mathematics at several colleges. In particular, Zheng became the first Chinese professor of mathematics at the famous Qing Hua College in 1920 (before Zheng, all professors of mathematics at Qing Hua were from the United States). When Qing Hua College was renamed Qing Hua University in 1926, Zheng was the founder of the Department of Mathematics and worked there until his retirement in 1952. This was a crucial period in the history of science education in China, and thus Zheng exerted considerable early influence upon modern Chinese mathematics [Ke Jiu 1989, 42]. Zheng had many excellent students; among them were Hwa Loo-Keng and Chern Shing-Shen (later, Chern married Shi-Ning Zheng, the daughter of Zheng Tung-Sun).

QIN FEN (1887–1971) graduated from Harvard University in 1909 (where he studied astronomy and mathematics). Qin also returned to China where he was professor of mathematics at Jiang Nan College (1910–1912), Jiao Tung University (1912–1915), and Beijing University (1915–1919). After 1919, Qin gave up teaching to become an officer at the National Education Board and National Economic Council. As a political activist, he was unable to contribute as much time to Chinese mathematics as did Zheng [Perleberg 1954, see entry for Chin Fen].

HU MIN-FU (1891–1927) arrived in the United States in 1910, received his B.A. degree from Cornell University in 1914, and his Ph.D. degree from Harvard University in 1917. Hu was the first Chinese to receive a Ph.D. in mathematics. His doctoral thesis, Linear Integro-Differential Equations with a Boundary Condition, was published the following year in the *Transactions of the American Mathematical Society* [Hu 1918, 363–402]. This was also the first Chinese research paper on modern mathematics to be published in an academic journal.

Hu was an excellent student at Cornell, and was elected to membership in two prestigious honor societies, Phi Beta Kappa and Sigma Xi [9]. In 1914, a group of Chinese students studying at Cornell founded the first Chinese scientific society—Chinese Sciences Agency—and established the first Chinese scientific journal, *Ke Xue* (*Science*). Hu was one of the leading members of this group. After returning to China in 1918, he was instrumental in creating the famous Da Tong university in Shanghai, and continued to contribute regularly to the Chinese Sciences Agency and *Ke Xue* [Hu 1989, 225].

Because Zheng Tung-sun and Hu Min-fu both graduated from Cornell, and due to the subsequent success of *Ke Xue* in the early part of this century in China, Cornell University is especially well known today through the Chinese scientific community.

CHIANG LI-FU (1890–1978) went to the University of California at Berkeley in 1911, and later earned his Ph.D. from Harvard University in 1918. His dissertation, The Geometry of a Non-Euclidean Line-Sphere

Transformation, was written under the supervision of J. L. Coolidge. Nearly 30 years later, when the National Mathematical Institute of China was founded in 1946, Chiang was elected its first director.

Chiang constituted a "one-man" department of mathematics at Nan Kai University, in Tianjin, beginning in 1919. Not surprisingly, he had to teach all of the mathematics courses by himself. Later, many students who studied with Chiang achieved considerable success. Some became prominent leaders in Chinese mathematics (for example, S. S. Chern and T. H. Kiang). To this day, the Nan Kai Mathematical Institute commands a strong reputation in China for mathematics.

Qing Hua University

In 1930, the first graduate school was established in Beijing at Qing Hua University (earlier spelling: Tsing Hwa). Of the four professors who comprised the Department of Mathematics, two had their doctorates from the University of Chicago, and one from Cornell University, while the fourth had studied in France at the University of Paris [Chern 1989, 31]. They were (in addition to Zheng Tung-sun already mentioned above):

SUN TAN (SUN GUANG-YUAN, 1900–1984, Ph.D. Chicago, 1928). Sun's dissertation (Projective Differential Geometry of Quadruples of Surfaces with Points in Correspondence) was directed by E. P. Lane. China's first graduate student, S. S. Chern, earned his Masters degree under the supervision of Professor Sun in 1935 (Chern was enrolled as a graduate student at Qing Hua in 1930 having received his B.A. degree from Nan Kai University). Chern describes his own case as follows: "I went to Qing Hua University in 1930 to study with D. Sun, because Sun was one of the few Chinese mathematicians who continued to do research after his dissertation" [Chern 1989, 31].

YANG KE-CHUAN (also known as YANG WU-ZHI, Ph.D. 1928). Yang was born in 1896, and arrived in the United States in 1923. After studying at Stanford University for three quarters, he became a graduate student at the University of Chicago. Yang received his Ph.D. in 1928 with a dissertation, Various Generalizations of Waring's Problem, which was directed by L. E. Dickson. His was also the first Chinese Ph.D. in modern algebra and number theory. For a long period in the 1930s and 1940s, Yang served as Chairman of the Department of Mathematics at Qing Hua. Meanwhile, he

guided many gifted students in studying number theory, including Hwa
Loo-Keng (see below), who is the best-known of these.

HSIONG CHING-LAI (1893–1969) was founder and Chairman of the
Department of Mathematics at Qing Hua University from 1926 to 1930.
Hsiong received his Masters Degree in France in 1921 and, somewhat
later, his National Science Doctorate from the University of Paris in 1933
[Mo You 1987, 41–42]. His dissertation, a contribution to the theory of
functions of a complex variable, Sur les fonctions entières et les fonctions
meromorphes d'ordre infini, was published in the *Journal de Mathématiques pures et appliquées* [Hsiong 1935, 233–308].

Several other notable mathematicians also returned to China after
studying abroad. Among the 14 Chinese Ph.D.s in mathematics awarded
before 1930, nine were from the United States, two from France, two from
Germany and one from Japan [Yuan 1963]:

Hu Min-fu (1917, Harvard)	Chiang Li-fu (1918, Harvard)
Wong Bing-chin (1922, Berkeley)	Yu Ta-wei (1922, Harvard)
Wei Si-luan (1925, Göttingen)	Chu Kun-ching (1927, Göttingen)
Sun Tan (1928, Chicago)	Yang Ke-chuan (1928, Chicago)
Tchao Tsin-yi (1928, Lyon)	Chen Kien-kwong (1929, Tohoku)
Fan Wei-kwok (1929, Lyon)	Kiang Tsai-han (1930, Harvard)
Liu Chin-nien (1930, Harvard)	Liu Shu-ting (1930, Michigan)

As these scholars returned from Europe and the United States (especially from Harvard), a new indigenous generation of mathematicians was
about to take root in China.

American Influence on Chinese Mathematics

Schematically, the chain of influence representing the connection between
American mathematicians and Chinese students and scholars might roughly
be diagrammed as follows:

(1) In Geometry:

J. L. COOLIDGE → CHIANG LI-FU → S. S. CHERN

E. P. LANE → SUN TAN → S. S. CHERN

S. S. Chern was taught by Chiang Li-fu at Nan Kai University (1926–1930)
and guided by Sun Tan at the Graduate School of Qing Hua University

(1931–1934). Chern was first invited by Oswald Veblen and Herman Weyl to spend several years at the Institute for Advanced Study, Princeton, from 1943 to 1945. While there, his work served to inaugurate the field of global differential geometry. In 1949 Chern went on to accept a position at the University of Chicago, where his supervisor, Sun Tan, had received his Ph.D. Subsequently, Chern has earned a strong reputation for his work in geometry. Moreover, he has been instrumental in training many Chinese students who have gone on to establish themselves as well-known mathematicians of the first rank, including S. T. Yau (a Fields Medal winner), Wu Wen-jun (Topologist, Chairman of the Chinese Mathematical Association, 1983–1988), and Liao Shan-Tao (a member of the Third World Academy of Sciences), among many others. Today, many Chinese mathematicians have important research positions, specializing in the area of differential geometry pioneered by Chern.

(2) In Algebra and Number Theory:

L. E. DICKSON → YANG KE CHUAN → HWA LOO-KENG (HUA LUO-GENG)

Dr. Yang Ke-chuan played an important role in promoting algebra and number theory in China. After receiving his Ph.D. in mathematics at the University of Chicago, he became a Professor of Mathematics at the Graduate School of Qing Hua University in 1929 until 1949. A year later, a paper by Hwa Loo-Keng came to Yang's attention, and made a strong impression. In 1930 Hwa (who was only a shopboy in his home town at the time), had just published an article in *Ke Xue* on the unsolvability of the fifth degree equation by means of radicals. Yang immediately recognized Hwa's mathematical gifts and recommended him to Professor Hsiong Chin-lai, Chairman of Mathematics at Qing Hua. Because Hwa had not been educated at any college or university, he was appointed as a secretary assistant in the Department, and consequently Yang came to exert an early influence on Hwa. Hwa's first important work, "Waring's Problem," was clearly influenced indirectly by L. E. Dickson, with whom Yang had studied at Chicago [Hwa 1934, 361]. Subsequently, Hwa not only came to be known as a specialist of considerable international reputation in algebra and number theory, but he headed as well the first notable Chinese school on number theory.

(3) In Analysis:

M. BÔCHER → HU MIN-FU

H. M. MORSE → KIANG TSAI-HAN

Professor Maxime Bôcher supervised Hu Min-fu's dissertation for the Ph.D. at Harvard in 1917. Hu was working on linear integral and differential equations with boundary conditions, which should have been an auspicious beginning for modern Chinese mathematics in analysis. Unfortunately, Hu died at the age of 36 (drowned in the summer of 1927), too soon to have had any real effect upon analysis in China.

Kiang Tsai-han (also known as Jiang Ze-han, 1902–), was a student of Chiang Li-fu. Kiang received his Ph.D. at Harvard in 1930, where Professor Marston Morse supervised Kiang's dissertation. Kiang went on to become the first topologist in China. His dissertation, On an aspect of Marston Morse's critical point theory, was published in the *American Journal of Mathematics* [Kiang 1932, 92–109 and 657–666].

Upon returning to China, Kiang was made Professor at Beijing University, where he served as Chairman of the Department of Mathematics for twenty years (1932–1952). Thanks to his teaching and research, the topological tradition of American mathematics was transmitted in a direct way to China. Kiang supervised a number of students, and had an especially favorable influence on the future of topology in China. He also prompted other Chinese analysists to study Marston Morse's critical point theory. Kiang and Jiang Bo-ju, one of Kiang's students, made notable contributions to the theory of "fixed point class" in the 1960s. In 1985, Jiang Bo-ju was elected to the Third World Academy of Science for his work on topology (especially his research on the "Nelson Number" of a mapping).

III. A KEY FACTOR: BOXER INDEMNITY SCHOLARSHIPS

At the beginning of this century, the Yi He Tuan Uprising, directed largely against foreigners and Chinese Christians, was put down in August of 1900 when the combined armies of eight countries (Russia, Britain, Germany, France, the United States, Italy, Japan, and Austria) occupied Beijing. Better known as the Boxer Rebellion (in Chinese, "Yi He Tuan" means an organization consisting of amateurs in "righteous harmony of fists"—hence the term "boxer"), the foreign troops were sent to liberate the besieged foreign legations in Beijing. This, in turn, eventually forced the Qing government to sign the Boxer Protocol, which called for a Boxer Indemnity demanding that China pay considerable sums, a total of 450,000,000 taels —about $332,000,000)—to each of the foreign governments in 39 years [Adams 1940, 84; Twitchett and Fairbank 1980, 125–127]. It was "an exorbitant sum that crippled the Chinese government and from any point of view can be stigmatized as a high point of imperialistic exploitation," [Hunt 1972, quoted in Twitchett and Fairbank 1986, p. 383; Dai 1963].

Of the $24,440,000 to be paid to the United States, the U.S. Congress in 1908 returned that part of the indemnity in excess of individual claims against actual property losses during the uprising. This amounted to nearly $12,000,000, slightly less than half of the total indemnity. The remainder was to be used for educating Chinese students in the United States, and this proved to be a potent mechanism for the support and encouragement of Chinese higher education [Twitchett and Fairbank 1986, 383].

The first group of 47 Boxer indemnity students was sent by the Chinese government to the United States in 1909. A year later, the number jumped to 70, but slipped slightly to 63 in 1911. By 1929 the total number of Chinese students who had studied in the United States, thanks to Boxer Indemnity Scholarships, was 1286 [Twitchett and Fairbank 1986, 383–384].

Only one of the first group of 47, however, was chosen to study mathematics. This was Wang Ren-fu, who received his B.A. from Harvard in 1913 [Chen Xue-xun 1986, 732]. He then returned to China where he became a professor of mathematics at Beijing University, as well as a member of the first Council of the Chinese Mathematical Society when it was founded in 1935 [Li Di 1984, 401].

The Second Wave

In 1910, the second group of Boxer Indemnity Scholars arrived in the United States. This group had to pass a nationwide examination in China before receiving Indemnity Scholarships. Among this group were many gifted students, including Hu Shih, who originally studied agriculture but later became interested in the work of John Dewey and became a student of philosophy. Hu was a writer, poet, educator, and also worked at the Chinese Embassy in Washington, D.C. (1940). Later he served as President of the Academia Sinica (in Taiwan).

In this second group, several chose to study mathematics. Chao Yuan-ren and Hu Min-fu both went to Cornell, where they received their B.A. degrees in 1914. Later Chao left mathematics and went on to become a founder of modern Chinese linguistics. Hu Min-fu, as already mentioned, was the first Chinese Ph.D. in mathematics. Among students in the third group of indemnity scholars, Dr. Chiang Li-fu studied at the University of California at Berkeley, arriving there in 1911.

Most Indemnity Scholars from these three groups were interested primarily in engineering, a few in science, and a very few in mathematics.

According to data from the first two groups, from a total of 117 scholars, 76 studied engineering, 15 were in humanities and management, 12 in agriculture and medicine, 12 in science (four in mathematics), with two in other fields [Chen Xue-xun 1986, 732–735].

As for other countries, after its revolution in 1917, the Soviet Union renounced its Boxer Indemnity. Great Britain also established Boxer Indemnity Scholarships, but only much later than the United States. Not until February of 1923 did the Chinese government send a representative to London to discuss the possibility of remitting Boxer Indemnity Funds to China for educational purposes. No agreement was reached, however, until 1930, whereupon the first group of British Boxer Indemnity Scholars from China went to England in 1933 [10]. Various lesser amounts returned from British, French and Italian Boxer indemnity funds were also used in part for educational purposes [Twitchett and Fairbank 1986, 387].

Qing Hua University and the Boxer Indemnity

An important and direct result of the availability of American Boxer Indemnity Funds was the establishment of Qing Hua College and (later) Qing Hua University in Beijing. In fact, Qing Hua has been called the "Indemnity College" or "Indemnity Monument" [Chen 1986, 282]. At first, the Chinese government used Indemnity Funds to finance Qing Hua College, which it had intended primarily for training students who would then be sent to the United States as undergraduates. Later, in order to send scholars directly to American graduate schools, the college was renamed Qing Hua University in 1926, but its financing still came from the American Boxer Indemnity. That same year, the Chinese government renewed the nationwide Qing Hua open examinations for Boxer Indemnity Scholarships which continued until 1949.

Thanks to the financial support made possible through the Boxer Indemnity funds, many aspiring Chinese students were thereby able to study in the United States (and some in Europe) after studying at Qing Hua. Among these were Kiang Tsai-han (1926), S. S. Chern (1935) and the later Nobel prize winner in physics, C. N. Yang (1946). These Chinese graduate students were dedicated not only to the promotion of science in general both in China and the United States, but to the development of mathematics as well [Zhou 1989, 21].

Moreover, because the funds remitted by the United States to China for these educational purposes were a dependable source of income, Qing Hua soon became a university of superior quality in the 1930s. The graduate program in mathematics, for example, was established in 1930. By

World War II, in fact, this graduate school had become the center for teaching and research in mathematics for all of China.

Beijing University and the Boxer Indemnity

Beijing University (Beijing Daxue, known popularly as Bei Da, according to its abbreviation in Chinese), founded in 1912, was the first and most reputable modern national university in China. In its early years it was actually influenced more by Japan than by the United States. From 1901 to 1909, about 34,280 Chinese students studied in Japan [Twichett and Fairbank 1980, 351]. Among those was Feng Zu-xuen who earned his B.A. degree at Kyoto Empire University and headed the group of mathematics at Beijing University in 1912. Unfortunately, due to initial lack of support from the Chinese government, financial backing for Bei Da was inadequate and the academic level slowly decreased (especially apparent in the Faculty of Science). In the 1920s, payments were only made occasionally to the University, and professors received only a part of their salaries. Virtually no significant research in the sciences was done at Bei Da in this period.

After 1930, however, the United States began to exert a much greater influence on the development of Bei Da. This was due primarily to the efforts of Hu Shih, Dean of the Faculty of Liberal Arts at Bei Da in 1931, who had received his Ph.D. from Columbia University in 1917. In 1931, Hu was one of the members of the Management Committee of the China Foundation for the Promotion of Education and Culture, founded in 1924, which took charge of the remainder of the United States Boxer Indemnity. This Management Committee agreed to provide $200,000 to Bei Da every year from 1931 to 1935. At the same time, Kiang Tsai-han, who had received a Boxer Indemnity Scholarship in 1926 and received his Ph.D. at Harvard in 1930, became one of the two professors of mathematics at Bei Da (the other was Feng Zu-xuen). Subsequently, he headed the Department of Mathematics for nearly 20 years. As a result of this generous financial support from the China Foundation for the Promotion of Education and Culture, along with the concerted efforts of the faculty members, Bei Da regained its former reputation in China, as did the Department of Mathematics [Kiang 1990, 74].

When the People's Republic of China was created in 1949, the Boxer Indemnity Scholarships came to an end. Thus ended for a time an extraordinary period of Chinese−American interaction which proved especially beneficial in the development of an indigenous Chinese higher mathematics.

IV. D. E. SMITH AND LI YAN

Prior to World War II, the American educator David Eugene Smith (1860–1943) was easily the most prolific—and the most prominent—historian of mathematics in the United States. Smith was a professor of mathematics at Columbia University from 1901 until his retirement in 1926. He published about 130 books (many of these were elementary textbooks), 100 reviews, and 155 essays and monographs. In the early 1910s, when Smith called attention to the work of the Indian mathematician Mahaviracharya, he discovered that China and other Oriental countries deserved a place in the history of mathematics that previously had not been sufficiently recognized [11]. Smith, in collaboration with Yoshio Mikami, published the *History of Japanese Mathematics* in 1914.

Somewhat earlier, in 1905, Smith had already shown an interest in Chinese mathematics, and asked a missionary A. P. Parker (then in China), to search for older Chinese books on mathematics, to be sent to Smith in New York City. As an American living in Shanghai, Parker was also among those who helped to translate several of the mathematical textbooks by Loomis into Chinese (as already mentioned). In a letter to Smith of May 2, 1905, Parker explained as follows:

> When the Jesuits came they published from time to time a good many mathematical works based on the knowledge of the subject then prevalent in Europe. These works have done very much to stir up the Chinese to the study of the subject. But these books are now being set aside and are being substituted by modern works, translated by missionaries, which are being used in all the schools of China that have been established to teach Western education. [All letters cited here are from the D. E. Smith collection, Professional Correspondence, University Archives, Rare Book and Manuscript Library, Butler Library, Columbia University, New York City, U.S.A.].

Smith replied to this letter on June 5, from which it is clear that he was only interested in early mathematical works printed in China: "If, however, you do not succeed in finding the older books, I do not wish to trouble you to send the modern ones."

However, the catalogue of ancient Chinese mathematical treatises acquired by Smith over the years (compiled by and available for consultation in the archives of Columbia University, New York City) includes 32 books covering many of the key ancient Chinese mathematical works printed in China before 1900 [Appendix III].

Apparently, Smith intended to make a substantial commitment to studying the history of ancient Chinese mathematics. He sent a letter (dated January 15, 1913) to Feng Foo, secretary of the Commercial Publishing House at Shanghai, saying that:

> You mentioned the fact that you are in a position to get hold of old books and documents on Chinese mathematics. If you can get hold of any manuscripts written more than one hundred years ago or any books that were written before 1800, I should be glad to have an option on them. I am anxious to secure a first-class library of early Chinese mathematics, because I expect to take up in a larger way the history of Chinese mathematics in the next few years.

At about this same time, Smith also began to correspond with Li Yan (1892–1963). Li was one of two prominent pioneers in Chinese history of mathematics (the other was Qian Bao-zong (1892–1974)). Although Li Yan studied at the Tang Shan Railway Engineering College, he did not graduate. Instead, he became a clerk at the Long Hai Railway Office in 1913. Thus he was posted to a remote station where his free time was spent studying and writing about ancient Chinese mathematics. At the time, no one else in China was doing research on this subject from a modern point of view. Li was the only one.

On January 23, 1915, when Li was 22 years old, he wrote a bold letter to D. E. Smith: "I venture to take the liberty of writing you to ask your help in the completion of my 'History of Chinese mathematics'." He went on in his letter to explain that he hoped to provide a "comprehensive view of our Chinese mathematics," from its earliest beginnings down to the present time. Li hoped that Smith would work with him to improve the book he intended to publish. Consequently, he sent an outline of his manuscript to Smith for comments and improvements. Smith replied warmly to Li on March 2, 1915:

> As perhaps you know, I got out a History of Japanese Mathematics with Mr. Mikami of Tokyo about a year ago. ... I have been hoping to get out a history of Chinese mathematics along the same lines, and last Spring I communicated with the learned Jesuit Father, Père Vanhee of Liège, who has written so extensively upon the subject. ... It occurs to me, however, that the three of us might possibly collaborate. ... If, however, you care to get your material into English and send it to me with this arrangement in view, I shall be glad to look it over and take it up with Father Vanhee as soon as I am able to communicate with him. ... You might consider this matter, and if you decide to make the translation and send it to me, it will have my best attention.

The next letter (written by Li to Smith on May 31, 1915, from Foochow) expressed Li's gratitude, admitting that "to collaborate on the work, of course, is the best plan." The only trouble Li foresaw was the possibility that perhaps he was not up to the demands this collaboration would entail.

Li's manuscript on the history of mathematics in China was in three volumes! The first was devoted to Ancient Mathematics, the second to Medieval and the third to Modern Mathematics. Li, of course, was well aware that "It is indeed the hardest task to translate into English the specific terms that have been extensively used in Ancient China without any loss of the true meaning." Along with this letter, Li enclosed an outline of his *History of Ancient Chinese Mathematics* [cf. Appendix I] for Smith's comments.

Smith received Li's letter within a month, and replied on July 7, explaining to Li that when it came to publication, three volumes would be too much. Smith also told Li that he had written to a colleague at the Vatican in Rome about Vanhee. Smith was concerned because he had received no word from Vanhee for some time. With no news of his whereabouts, Smith suggested (in a letter dated November 4) that Vanhee might have been killed during World War I. Actually, Smith was mistaken about this. Four years later, he finally received a letter from Vanhee, who wrote on February 24, 1919 to explain:

> I was taken a prisoner and put in close confinement during 9 months without any visits, any books, without necessary food and clothing and finally condemned to four years hard work!!! ... I never have thought the Germans as a nation were so cruel, so barbarian, and so false, you will never realize what we have had to suffer since five years.

Several months later, Smith decided to write a letter to the "New York Times" (April 28) enclosing a copy of Vanhee's letter. Smith explained that "Father Vanhee is a Jesuit scholar of wide reputation and is, perhaps, the best European authority on the history of Chinese mathematics." two days later, *The New York Times* published the two letters [Smith 1919]. Indeed, Vanhee continued to produce papers on the history of Chinese mathematics throughout the 1920s, including two published in the *American Mathematical Monthly* [Vanhee 1926, 492–495 and 502–504].

Li Yan's History of Chinese Mathematics

In response to Li Yan's interest in collaborating with Smith to produce an historical survey of Chinese mathematics (cf. Appendix I), Smith drew up his own outline of the work he envisioned, and mailed it to Li on November 4, 1915:

> History of Chinese Mathematics
> by David Eugene Smith and Y. Li
> Introduction (Chapters 1–4)
> Part One: Ancient Mathematics (3000–200 B.C.)
>   Chapters 1–12.
> Part Two: Medieval Mathematics (200 B.C.–600 A.D.)
>   Chapters 1–6.
> Part Three: Modern Mathematics (600 A.D.–1900)
>   Chapters 1–5.

(A more detailed version of Smith's outline is given in Appendix II.)

Several years passed before Li was ready with a Chinese version of the manuscript, which he then sent to Mao Yi-sheng for translation into English (Li must have realized that his own command of English was not up to the demands of doing this himself).

Mao Yi-sheng (Mao Thomson, 1896–1989) was to become a famous engineer in China, specializing in bridge construction. He was also a leader among those who supported the cause of modern science and technology in China. Mao had been a classmate of Li at the Tang Shan Railway Engineering College, and later received his Masters degree in Engineering from Cornell University in 1917. Although he found it very difficult to translate Li's manuscript, he finally sent an English version to Smith on June 25, 1917.

Smith eagerly examined Mao's translation of Li's work, but was disappointed with what he found. This prompted a letter to Li on July 25, 1917, the last letter from Smith of which there is any record:

> It seems to me that what we need chiefly is the translations of problems and solutions from the older works, so that we shall thus get the reader in touch with the original material. ... Now, in your manuscript there are almost no translations from any of the ancient works mentioned. There is nothing from the Yih-King [Y Jing or I-ching], although the text relating to the ho-tu [He Tu] and the lo-shu [Luo Shu] would probably be very interesting.
>
> Western readers want the translation of certain typical parts, not merely a statement of the contents of the books.
>
> ...Unless you feel like sending a large amount of exact translation from the chief mathematical works of China, and more exact information, it would not be worth while for us to proceed further [12].

In retrospect, it is a great pity that the collaboration between Smith and Li was not more successful. Nevertheless, Smith remained interested in the history of Chinese mathematics. Li Yan may also have benefitted from his correspondence with Smith, since he eventually published the first important Chinese treatise on the history of Chinese mathematics (in Chinese) in 1919 [Li Di 1984, 410].

V. CHINESE MATHEMATICIANS DEDICATED TO AMERICAN MATHEMATICS

Modern Chinese mathematics quickly began to mature in the 1930s. Despite the hiatus in most mathematical activity during the disruptive years of World War II, as soon as it was over, a large number of mathematicians left China for the United States, where they pursued research in a wide variety of areas. After 1949, in response to the newly proclaimed People's Republic of China, some of these scholars returned and subsequently became leaders of contemporary Chinese mathematics in the PRC. Those who stayed in the United States became Chinese-Americans, and many of these have made substantial contributions of their own to American mathematics. Recently, a large number of graduate students in mathematics departments across the United States have come from mainland China, Taiwan, or Hong Kong. Of the 933 Ph.D.s in mathematics awarded in 1989–1990, 227 were given to graduate students from China or to students of Chinese descent [13]. What follows are brief descriptions of several of the most prominent Chinese-American mathematician who have achieved international reputations in this century.

Chern Shiing-Shen

Among Chinese-American mathematicians, the most prominent is undoubtedly S. S. Chern. Chern was born on October 26, 1911, in Jiaxing, China. After finishing his undergraduate studies at Nan Kai University in Tianjin, he received his Masters degree from Qing Hua University in 1934.
  Chern then went to Germany where, two years later, he received his Ph.D. from Hamburg University in 1936. While in Europe he took the opportunity to visit Paris, where he met Elie Cartan. Every two weeks Chern was invited to Cartan's home to discuss their mutual interests in mathematics. During the Second world War, and especially during the difficult period of the war with Japan, Chern was professor of mathematics at South-West Associate University (During World War II, three universities—Qing Hua, Bei Da, and Nan Kai—merged into a single institution in

Kunming, China). As a result of correspondence with Oswald Veblen in the 1930s, both Herman Weyl and Oswald Veblen were familiar with Chern's works. Despite the war, Chern was invited to visit the Institute for Advanced Study at Princeton during 1943–1945, where he did basic work on modern differential geometry. In 1946, he was made Acting Director of the National Institute of Mathematics in Shanghai.

Chern returned again to the United states after the war, in 1949, when Marshall Stone asked Robert Hutchins, President of the University of Chicago, to offer Chern a professorship. Stone also added a most interesting threat that proved effective:

> If the appointment were not made, I would not be a candidate for reappointment as chairman when my three years term expired. ... Happily, the protest was successful [Stone 1976, 185].

And as Robert Osserman has said:

> For a single most decisive factor contributing to the rebirth of geometry in America, I would propose the move of Chern to the United States from China in the late forties [Osserman 1989, 517].

S. S. Chern has been the leading figure in global differential geometry. His ground-breaking discovery of characteristic classes (now known as Chern classes) was the turning point that set global differential geometry on a course of tumultuous development. The field has blossomed under his leadership, and his results, together with those of his numerous students, have influenced the development of topology, algebraic geometry, complex manifolds, and most recently, gauge theories in mathematical physics [*Notices of the American Mathematical Society* 1984, 7].

Chern was awarded the Chauvenet Prize in 1970, the U.S. National Medal of Science in 1975, and the Steele Prize in 1983 [14]. The importance of Chern's research was again recognized when he received the Wolf Prize in 1983–1984 for his contribution to global differential geometry [15]. He was elected a Member of the U.S. National Academy of Science in 1961, and served as the Vice President of the American Mathematical Society during 1963–1964. In addition to serving as the first Director of the Mathematical Sciences Research Institute in Berkeley, California (1981–1984), he has also been Director of the Nan Kai Mathematical Institute at his old university (Nan Kai University) at Tianjin, since 1985. A bibliography of the publications of S. S. Chern and a list of his Ph.D. students were published in *Shiing-Shen Chern. Selected Papers*, New York: Springer-Verlag, 1978 and [Chern 1989]; see also [Donald 1985, 33–40].

Yau Shing-tung

S. T. Yau was born in Guangdong, China, on April 4, 1949. After graduating from the Zhong Wen Da Xue (Chinese University) in Hong Kong, he enrolled as a graduate student at the University of California, Berkeley. Under the supervision of S. S. Chern, Yau received his Ph.D. in 1971. Subsequently, he solved a series of important problems in differential geometry, including Calabi's Conjecture, as well as the Positive Mass Conjecture.

In 1981, Yau was awarded the Osward Veblen Prize in geometry from the American Mathematical Society and the John J. Carty Medal from the National Academy of Sciences [16]. The following year he won a Fields Medal, perhaps the highest honor that can be given to a mathematician, for his work on differential geometry [*Notices of the American Mathematical Society* 1981, 162; 1983, 571]. He taught at Stanford University and the University of California, San Diego, and is currently Professor of Mathematics at Harvard University (since 1987).

Lin Chia-chiao

C. C. Lin, born in Fujian on July 7, 1916, received his B.A. degree from Qing Hua University in 1937 and his Masters degree from the University of Toronto in 1941. In 1944 he earned his Ph.D. at the California Institute of Technology. His research on turbulence theory and astronomy has been especially successful. As a well-known applied mathematician, Lin was elected to membership in the U.S. National Academy of Sciences in 1962. Today he is Professor of Applied Mathematics at the Massachusetts Institute of Technology [17].

Chow Wei-Liang

W. L. Chow, born October 1, 1911, in Shanghai, was an undergraduate at the University of Chicago in 1931. He went on to study in Germany, where he earned his Ph.D. at the University of Leipzig in 1936. In 1950 he was appointed Professor of Mathematics at the John Hopkins University (U.S.A.), where he established himself as a famous algebraic geometer. Owing to Chow's contributions to algebra and algebraic geometry, Peter Lax included Chow (along with Chern) in his list of illustrious immigrant-mathematicians to the United States [Lax 1977, 132], [17].

Fan Ky

Fan was born September 19, 1914 in Hangzhou. He was educated at Beijing University, and then went to France where he received his Ph.D. from the University of Paris in 1941. Later, coming to the United States in 1945, he became an assistant to John Von Neumann, and worked at the Institute for Advanced Study in Princeton from 1945 to 1947. Today he is a distinguished analyst at the University of California, Santa Barbara. Fan Ky was also Director of the Institute of Mathematics in Taiwan, from 1978 to 1984. When he retired in 1985, the Ky Fan Assistant Professorship was established at Santa Barbara the following year.

Other Chinese-American mathematicians have also made a notable contribution to American mathematics:

Wang Hsien-chung (1918–1978), born in Beijing, earned his M.A. from the South-West Associate University in 1944. He then went to England, where he was awarded a Ph.D. at Manchester University in 1948. After 1950, he taught at a number of universities in the United States. In 1958, he was appointed Professor of Mathematics at Northwestern University, and from 1966, he has taught at Cornell University. Wang is best-known for his work on the theory of Lie Groups, and was invited to give a plenary lecture entitled "Some geometrical aspects of coset spaces of Lie groups" at the International Congress of Mathematicians at Edinburgh in 1958.

Chung Kai-lai was born in Shanghai, 1917. After graduating from South-West Associate University in Kunming, he arrived in the United States in 1945, earning his Ph.D. at Princeton in 1950. Chung went on to establish himself as a prominent American expert on probability theory in the postwar period. Today he is Professor of Mathematics at Stanford University.

Wang Hao, born in Jinan, Shandong, in 1921, was educated at South-West Associate University. He earned his M.A. at Qing Hua University in 1945. In the same year, Wang went to Harvard University where he received his Ph.D. in 1948. Wang is especially well-known for his many works on mathematical logic, and in 1983 he was awarded the "Milestone Award" [18]. Today Wang is Professor of Mathematics at Rockefeller University in New York City.

Lam Tsit-yuan was born on February 6, 1942, in Hong Kong. He received a B.A. from the University of Hong Kong in 1963 and a Ph.D. from Columbia University in 1967. In 1968 he joined the faculty of the University of California, Berkeley, and was promoted to the rank of Professor in 1976. In 1982, Lam was awarded the Steele Prize of the

American Mathematical Society for his book, *Algebraic Theory of Quadratic Forms* (1973) and four of his papers (*K* and *K*—an introduction to *K*-Theory, Ten lectures on quadratic forms over fields, Serre's conjecture, and The theory of ordered fields) [*Notices of the American Mathematical Society* 1982, 505].

Siu Yum-Tun was born in Guangdong in 1943. After graduating from Hong Kong University in 1963, he came to the United States where he received his Ph.D. from Princeton University in 1966. Y. T. Siu has a well-known reputation for his works on the theory of functions of several complex variables. In 1983 he was an invited plenary speaker at the International Congress of Mathematicians held in Warsaw, Poland. Today Siu is Professor of Mathematics at Harvard University.

VI.   MATHEMATICIANS IN CHINA

Among mathematicians who studied in the United States and then re-turned to China, these too have made important contributions in a variety of ways to mathematics in China, as well as internationally.

Hwa Loo-keng

Hwa Loo-keng (Hua Luo-geng 1910–1985) was the most prominent math-ematician of the new generation in China. Indeed, he was a legendary figure. As a village shopboy, he had no degree, but in 1931 he was made a faculty member   at Qing Hua University. In 1936–1937 Hwa went to Cambridge, England, were he studied analytical number theory. Hwa's contributions to Waring's Problem, additive number theory, group theory, and the theory of complex functions of several variables are among his most original. After World War II, Hwa was invited to the Institute for Advanced Study at Princeton for two years (from 1946–1948), and thereafter he was made full professor at the University of Illinois (1948–1950). In 1951 he returned to China, where he became Director of the Institute of Mathematics in Beijing. From 1950 to 1983 he also served as Chairman of the Chinese Mathematical Society. Hwa was elected a foreign member to the U.S. National Academy of Sciences in 1982, and to membership in the Academy of Third World Countries in 1983.

Hsu Pao-lu

Hsu Pao-lu (1910–1970) was a founder of modern mathematical statistics. Hsu received his B.A. from Qing Hua University in 1935, and then went to England to study mathematical statistics with R. A. Fisher. His contribution to multivariate analysis was both fundamental and elegant. He stayed in the U.S. during 1946–1947, and published a series of articles on statistics in various American journals. Hsu played an important role in the formation of multivariate statistical analysis at the time. Chung Kai-Lei, a student of Hsu, later played an important role in the area of probability and statistics in the United States [Jiang Ze-han 1979, 12]. In 1979, the *Annals of Statistics* published a set of papers in memory of Hsu.

CONCLUSION

Chinese mathematicians have followed a long and difficult path on their way to achieving international recognition for their many contributions to contemporary mathematics, both at home and abroad. Here discussion has been limited primarily to the importance and significance of Chinese-American contacts, but this should not be taken to minimize the value of links between Europe and China as well. Nevertheless, for reasons that are at once historical, economic and political, it seems likely that Chinese-American relations will continue to dominate the future of mathematics in China, and mutually enrich scientific exchanges between both countries well into the next century.

In fact, Cheng Min-de, Professor of Mathematics at Beijing University and head of the Mathematical Board of the Chinese Science Foundation, said at the opening ceremony of a recent conference on "Prospects for Chinese mathematics in the 21st Century":

> The updated situation of mathematics can be described as follows: the United States and Soviet Union hold a leading position, Western Europe follows closely, Japan is trying hard to catch up, and China is an unknown. If its potential is realised, Chinese mathematics is destined to have a bright future. Mathematics may be the first scientific area in China to come up to the advanced world standard [Cheng 1988].

In just this spirit, S. S. Chern, on the occasion of the 30th anniversary of the founding of *Teaching Mathematics*, a popular Chinese mathematical journal, offered the following prophetic words of congratulation which he inscribed in a presentation copy of the journal (in Chinese), as follows:

"A great nation of mathematics in the 21st century!"

Professor Chern also pointed out that his hopes for a "Great Nation" looked ahead to a time when there would be an indigenous Chinese mathematical school, promoting mathematical exchanges with other countries on an equal and independent footing [Chern 1989, 312].

For China, this is no idle expectation. But to reach this goal, increased international cooperation is necessary, including the continuation of exchanges between China and the United States. In the 19th century, the roots of such exchanges survived despite the Yi He Tuan uprising against foreign influences in China. It may be ironic in retrospect, but nevertheless, it was the positive use to which the reparations demanded by the Boxer indemnity were put, that subsequently made it possible for so many of China's youngest and brightest minds to come to the United States for their first advanced training in Western mathematics.

NOTES

[1] The standard, pioneering work on the history of Chinese mathematics is volume III of Joseph Needham's *Science and Civilization in China* [Needham 1959], although Yoshio Mikami's earlier book, *The Development of Mathematics in China and Japan* should also be mentioned [Mikami 1913]. More recent work, including surveys of Chinese mathematics, are [Li Di 1984], [Li Yan and Du Shiran 1987], and [Martzloff 1988].

[2] This translation included as well Books XIV and XV, which were not written by Euclid [see Heath 1956, Vol. I, 5].

[3] Loomis [1851]. Later Loomis divided the original book into two parts: both *Elements of Analytical Geometry* and *Elements of the Differential and Integral Calculus* appeared in 1874.

[4] Joseph Edkins (1823–1905), the English missionary, in conjunction with Li Shanlan, translated the *Theory of Conic Section* and *An Elementary Treatise on Mechanics* (by William Whewell). The later, published in 1859, is the earliest Chinese translation dealing with Newton's mechanics.

[5] Hagiwara Teisuke (1828–1909) was the last great mathematician of the old Japanese school. For biographical details about Hagiwara, see [Mikami 1913, 252].

[6] All of these were translated from volumes in *Loomis' Series of Text Books*, which included 14 in all: *Elementary Arithmetic, A Treatise on Arithmetic, Elements of Algebra, A Treatise on Algebra, Elements of Geometry, Trigonometry and Tables, Elements of Analytical Geometry, Elements of Differential and*

Integral Calculus, Elements of Nature Philosophy, Elements of Astronomy, Practical Astronomy, Recent Progress of Astronomy, A Treatise on Astonomy, and *A Treatise on Meteorology.* The entire series was published by Harper and Brothers, New York, in 1874.

[7] On the subject of mathematics in America in the 19th century, its content and quality, see the studies by [Smith 1934] and [Gabiner 1977].

[8] A biography of Loomis published in 1892 mentions that "The Rev. J. McClintock commended his textbook in mathematics as a model of neatness, precision and practical adaptation to the wants of the student" [*The National Cyclopaedia of American Biography*, VII (1938), 233.]

[9] According to the records for Hu Min-fu, Department of Manuscripts and University Archives, Cornell University, Ithaca.

[10] Howard L. Boorman, ed. *Biographical Dictionary of the Republic of China*, New York and London: Columbia University Press, vol. 2 (1968), 14 (see also Chern 1989, 34).

[11] A biography of D. E. Smith appears in *The National Cyclopaedia of American Biography*, James T. White and Company, vol. E (1937–1938), 138.

[12] This is the last letter in the Smith–Li folder of correspondence in the D. E. Smith Collection, University Archives, Butler Library, Columbia University, New York City. If Smith ever received a reply from Li Yan, it has not been preserved.

[13] *American Doctoral Dissertations (1989–1990),* Ann Arbor, Michigan: University Microfilms International, Bell and Howell Information Company. See also *World Journal* (in Chinese), New York, July 27, 1990.

[14] The Chauvenet Prize is awarded for a noteworthy paper of an expository or survey nature published in English, and of general interest to members of the Mathematical Association of America. Chern received the Chauvenet Prize for his paper, Curves and Surfaces in Euclidean Space, published in the MAA series *Studies in Mathematics*, **4** (1967), 16–56; see also *Notices of the American Mathematical Society* (1970), 117.

[15] The Steele prizes were established in 1970 in honor of George D. Birkhoff, William F. Osgood, and William C. Graustein by P. Steele. The American Mathematical Society awarded one of the 1983 Steele Prizes to S. S. Chern for the cumulative influence of his total mathematical work, but especially for his significance in the development of the field of differential geometry, and the lasting influence he has had upon mathematics through his Ph.D. students. See *Notices of the American Mathematical Society* (1987), 1166.

[16] The Wolf Prize is one of the greatest honors that a mathematician can receive. Wolf Prize winners are selected on the basis of lifelong achievement in mathematics. Since 1978, two mathematicians have been awarded these prizes each year by the Wolf Foundation of Herzlia Bet (Israel). To date: I. M. Gel'fand and C. L. Siegan (1978); J. Leray and A. Weil (1979); H. Cartan and A. N. Kolmogorov (1980); L. Ahlfors and O. Zariski (1981); M. G. Krein and

H. Whitney (1982); S. S. Chern and P. Erdős (1983); K. Kodaira and H. Lewy (1984); S. Eilenberg and A. Selberg (1985–1986); K. Ito and P. Lax (1987); F. Hirzebruch and L. Hörmander (1988); A. Calderon and J. Milnor (1989); E. De George and I. Piatetski-Shapiro (1990) (collected from "News and Announcements," *Notices of American Mathematical Society*).

The Veblen prize was established in 1961 in memory of Professor Oswald Veblen. It is given only once every five years for exceptional research in geometry or topology that has appeared during the preceding five years. The American Mathematical Society awarded the tenth Veblen prize in 1981 to Shing-Tung Yau for his work on nonlinear partial differential equations, for his contributions to the topology of differential manifolds, and for his work on the complex Monge–Ampère equations on compact complex manifolds (see *Notices of American Mathematical Society* (1987), 1163).

The Carty medal is awarded for noteworthy and distinguished accomplishment in any field of science coming within the scope of the Charter of the Academy. Established in 1930, the prize includes a bronze medal and an honorarium, awarded every three years [Siegman 1985, 374].

[17] The biographical information provided here for Lin Chia-chiao, Chow Wei-liang, and Fan Ky is drawn from a variety of sources. These include *Chinese Encyclopaedia* (volume on mathematics) Beijing: The Encyclopaedic Press, 1989; *American Men and Women of Science*, New York and London: R. R. Bowker, 1982; Chern 1989; Li Di 1984; and Mo You 1986.

[18] The Milestone Award is for foundational work in automatic theorem proving. The principal endowment for this prize was made by the Fredkin Foundation and the funds are administered by Carnegie Mellon University. For details see *Notices of the American Mathematical Society*, **30** (1991) 425.

APPENDIX I

Li Yan to D. E. Smith, May 31, 1915.
D. E. Smith Professional correspondence, Archives of Butler Library, Columbia University, New York, NY.
(What follows is a verbatim transcript; no changes have been made to correct or improve the English of this Document)

LI YAN'S HISTORY OF CHINESE MATHEMATICS

VOLUME II ANCIENT MATHEMATICS

Part I. Introductory.
  Chapter I. Origin of Chinese Civilization.

II. Chronological view of Chinese Mathematics.

III. Mathematics in Yih-King (Book on Change)

1. Forward about Yih-King. 2. Relation existed between mathematics and Yih-King. 3. Loh-Shu and Magic Square.

4. Relation Between Mathematics and Loh-Shu.

VI. Relation between mathematics and astronomy as confined to the science of almanack.

V. Relation between mathematics and Chinese philosophy

Part II. General View of Ancient Mathematics

Chapter I. Introductory.

II. One Number.

III. On the Method of Counting.

IV. On Arithmetics.

1. Concept of number. 2. Four fundamental procedures of arithmetics. 3. On fraction. 4. Details on fractions.

5. On decimals. 6. On domination of numbers. 7. On proportion. 8. On evolution and involution.

V. Methods in calculation as employed by ancient mathematicians.

1. By fingure. 2. By slips of wood or cardboard.

3. By Abacus.

VI. On Equations.

VII. On Geometry. 1. Plane. 2. Solid.

VIII. Concept of Modern Analysis in Ancient China.

Part III. Biography of Ancinent Mathematicians.

Chapter I. Lih-Sheu. Biography and Works.

II. Chou-Kon and Shan-Koh.

1. Biography and works. 2. On Chou Peh Soan Chin.

3. On Kieu Chan Soan Shue.

III. Sun-Sze. 1. Biography and works.

2. On Sun Sze Soan Chin.

IV. Conclusion.

Part IV. Civilization of Ancient World.

Chapter I. Brief Note. II. Babylonia. III. Egypt. IV. Greece and Rome. V. India and Arabis.

Part V. Concluded Words About Ancient Chinese Mathematics

Chapter I. On the Method of Measurement in Ancient China.

1. Introductory. 2. Measure (length and area).

3. Measure (volume). 4. Weight.

II. Specific terms explained.

III. Cause of regression of mathematics in Chou and Chin dynasties.

APPENDIX II

Smith's revision of the outline presented in appendix I, as drafted by Li Yan.
D. E. Smith, Professional correspondence, Archives of Butler Library, Columbia
   University, New York, NY.
Dated November 4, 1915:

History of Chinese Mathematics
By David Eugene Smith and Y. Li

INTRODUCTION
I. Origin and Early Development of Chinese Civilization.
   1. Various traditions.
   2. Scientific data.
   3. Probable truth of the matter.
II. Chronology of Chinese mathematics.
   1. Extreme claims of antiquity.
   2. Scientific data as to antiquity.
   3. Probable truth as to antiquity of Chinese mathematics.
   4. Chronological summary, arranged both by dynasties and by dates according to
      the Christian era.
   5. Basis of the three-fold division into ancient mathematics (c. 3000–200 B.C.),
      Medieval (200 B.C.–600 A.D.) and Modern (600–1900 A.D.)
III. Mathematics in 1-Ching.
IV. Relation of ancient mathematics to philosophy and astronomy. Schlegel's
   theory of the signs of the Zodiac.

Part I Ancient arithmetic.
I.   The origin of number. System of counting, of writing numbers.
II.   Ancient arithmetic.
      Two fold division, corresponding to the Greek logistic
      (computation) and arithmetic (theory of numbers). Examples of
      theory of numbers.
III. Method of computation.
      Possible antiquity of the bamboo rods.
      Possible use of knotted cords, like the quipu of Peruvians.
      Uncertainly as to ancient method.
IV.   Biographical sketches of ancient mathematicians.
V.   The ancient mathematical classics. A scientific study of
      their dates. Their contents.
VI.   Early Chinese algebra. The nature, symbols, equations. Its
      antiquity.
VII. Early Chinese geometry. Relation to mensuration. The
      Pythagorean theorem. Comparison with early Egyptian
      geometry.

VIII. Ancient ideas of higher mathematics.

IX. Ancient metrology.

X. The origin and early history of magic square.

XI. Probable effects of intercourse with other countries.

XII. Conclusion as to the ancient civilization and ancient mathematics.

Part II Medieval Mathematics
(200 B.C.–600 A.D.)

I. 200 B.C.–200 A.D. Biographies. Discoveries. Treatises.

II. 200 A.D.–400 A.D. Biographies. Discoveries. Treatises.

III. 400 A.D.–600 A.D. Biographies. Discoveries. Treatises. The development of the theory of the circle.

IV. Reform in metrology.

V. Probable effect of intercourse with other countries.

VI. Conclusion as to the medieval civilization and medieval mathematics.

Part III Modern Mathematics
(600 A.D.–1900)

I. 600–1000 A.D. Biographies. Discoveries. Treatises. Growth of algebra. Mechanical computation. On Ta Yen Ch'in I.

II. 1000–1400 A.D. Biographies. Discoveries. Treatises. Golden age of algebra. Possible effect upon Persian and Italian algebra. On T'ien Yuan and Szu Yuan. On the rise of trigonometry.

III. From 1400–1900 A.D. Biography. Discoveries. Treatises. The information of Occidental mathematics. Influence of Matthew Ricci. Reaction against European mathematics. Magic square and recreations. Typical problems.

IV. Different schools in modern mathematics.

V. Conclusion.

APPENDIX III

Miwa Kai. 1989. *D. E. Smith Collection of Works in Chinese on Mathematics and other Subjects*, Catalogue of Butler Library, Columbia University, New York, NY.

(We list here only authors, titles and press dates of mathematical books which appeared before 1900).

1. *Giu zhang suan ging* (Nine Chapters on the Mathematical Art), 1890.

2. *Giu zhang suan shu xi cao tu shuo* (Nine Chapters on the Mathematical Art, with Commentaries and Diagrams), 1820.

3. *Jian yi suan fa guei chu cuo yao* (Summary of the Simple Abacus Calculation for Division), 1889.

4. *Qin ding si ku quan shu jian ming mu lu* (Summary of the Index of the Complete Library), 1782.

5. *Qin ding si ku quan shu zong mu* (Index of the Complete Library), 1894.

6. *Suan jing shi shu* (Ten Mathematical Manuals), 1890.

7. *Yu zhi shu li ging yun* (Imperial Collected Basic Principles of Mathematics), 1882.

8. *Yu zhi li xiang kao cheng* (Imperial Compendium of Calendrical Science and Astronomy), 1723.

9. *Zhou bei suan jing* (Arithmetical Classic of the Gnomon and the Circular Path of Heaven) Wu Ying Dian Ju Zhen Ben, 1311.

10. *Zhou bei suan jing* (Arithmetical Classic of the Gnomon and the Circular Path of Heaven), Proofread by Bao Shan, 1600(?).

11. Dai Zhen. *Gou gu ge yuan ji* (On the Right-Angled Triangles and Cutting Circle), 1744.

12. *Ibid*. 1890.

13. Euclid. *Gi he yuan ben* (Elements), translated by Matteo Ricci and Xu Guang-qi (Books I–VI), Alexander Wylie and Li Shan-lan (books VII–XV), 1865.

14. Fang Zhong-tong. *Shu du yan* (Treatise in Mathematics), 1890.

15. Mei Ding-giu. *Li suan quan shu* (Complete Book of Calendrical Science), 1723.

16. Mei Jue-cheng. *Mei shi cong shu qi yao* (Collected Works of Mei Family), 1761.

17. Lao Nai-xuan. *Chou suan qian shi* (An Introduction to Rod Computation), 1897.

18. ——. *Chou suan fen fa qian shi* (An Introduction to Rod Computation in special methods), 1898.

19. ——. *Chou suan mung ke* (An Elementary Introduction to Rod Computation), 1898.

20. ——. *Gu chou suan kao shi* (Commentary on Ancient Rod Computation), 1886.

21. Lao Nai-xuan *et al*.. *Do ji chou fa* (Sums of Piles Calculated by Rods), 1900.

22. ——. *Yan Yuan xiao cao* (A Sketch of Computation), 1898.

23. Li Rui. *Li shi suan xue yi shu* (Collected Mathematical Works of Li Rui), 1823.

24. Tan Zai-zi. *Li suan tian ji quan shu* (Complete Books of Calendrical Computation and Astronomy), 1752.

25. John Wallace. *Dai shu xue* (Algebra), translated by John Fryer and Hua Heng-fang, 1897.

26. Xu Chao-jun. *Gao hou meng qiu* (Introduction to Astronomy), 1815.

27. Xu Gui-lin. *Suan Yong* (A View of Computation), 1830.

28. Ruan Yuan. *Chou ren zhuan* (The Compilation of the Biographies of Mathematicians and Astromers), 1896.

29. Zhang Zuo-nan. *Cui wei shan fang shu xue* (Mathematical Works at Cui Wei House), 1821.

30. Zhen Luan. *Wu jing suan shu* (Five Books on the Mathematical Art), 1890.

31. Zhu Shi-jie. *Si yuan yu jian xi cao* (Detailed Analysis of the Precious Mirror of the Four Elements), 1836.

32. ——. *Suan xue qi meng* (The Elementary Introduction to Mathematics), 1871.

REFERENCES

Adams, James Truslow, ed. 1940. *Dictionary of American History*, vol. 3, New York: Charles Scribners's Sons.

Anderson, T. W. *et al.* 1979, Pao Lu Hsu (1910–1970), *The Annals of Statistics*, **7**, No. 3, 467–490.

Chen, Xue–xun. 1986. *Gin Dai Zhong Guo Jiao Yu Shi Liao* (*Historical Materials of Modern Chinese Education*). Beijing: People Education Press (in Chinese).

Cheng, M. D. 1988. Zai 21 Shi Ji Shu Xue Zhan Wong Hui Yi Shang De Bao Gao, *Ke Xue* (*Science*), **40**, No. 4, 1988 (in Chinese).

Chern, S. S. 1989. *Chen Sheng Shen Wen Xuan* (*Selected Writings*), Beijing: Academy Press (in Chinese).

Dai, Xuan-chi. 1964. *Yi He Tuan Yan Giu* (Studies on Yi He Tuan). Taipei, Wen Hai Press (in Chinese).

Donald, J. and Alexanderson, G. L., eds. 1985. Chern Shih-Shing, *Mathematical People*, Birkhäuser, 33–40.

Duren, P., eds. *A Century of Mathematics in America*. Providence, R.I.: American Mathematical Society, 1989.

Grabiner, Judith V. 1977. Mathematics in America: The first Hundred Years, in Tarwater, 9–23.

Greene, Roger S. 1923. "Education in China and the Boxer Indemnities," *Chinese Social and Political Science Review*, **7**(4), 199–207.

Heath, Thomas L. 1956. *The Thirteen Books of Euclid's Elements*, 3 vols., New York, Dover (reprint).

Horng, Wann-Sheng. 1991. Li Shan-lan: The Impact of Western Mathematics in China during the Late 19th Century, Dissertation: City University of New York.

Hsiong, King-lai. 1935. Sur les fonctions entières et les fonctions méromorphes d'ordre infini. *Journal de Mathématiques pures et appliquées*, **14**, 233–308.

Hu Min-fu. 1918. Linear Integro-Differential Equations with a Boundary Condition. *Transactions of the American Mathematical Society*, **19**, 363–402.

Hu Yan-geng. 1989. Hu Min-fu—The pioneer of *Science*, *Ke Xue* (*Science*), **41**, No. 3, 225.

Hunt, Michael H. 1972. The American Remission of the Boxer Indemnity: a Reappraisal. *Journal of Asian Studies*, **31** (3), 539–559.

Hwa Loo-Keng. 1935. On the Representation of integers by the Sums of Seven Cubic Functions. *The Tôhoku Mathematics Journal*, **41**, 361–366.

Kiang Tsai-han. 1932. On the Critical Points of Non-Degenerate Newtonian Potentials. *American Journal of Mathematics*, **54**, 92–109; On the Existence of Critical Points of Green's Function for Three Dimensional Regions, *American Journal of Mathematics*, **54**, 657–666.

——. 1990. Huei Yi Hu Shi De Gi Gian Shi (In Memory of Dr. Hu Shi), *World Journal* (*New York Edition*), (December 20), 74.

Kutzbach, Gisela. 1973. Loomis, Elias. *Dictionary of Scientific Biography.* Scribner's: New York, vol. 8, 487.

La fourette, Kenneth S. 1929. *A History of Christian Missions in China.* New York: The Macmillan Company.

Lax, Peter. 1977. The Bomb, Sputnik, Computers and European Mathematicians, in Tarwater 1977, pp. 129–135.

Li Di. 1984. *Zhong Guo Shu Xue Shi Jian Bian (A Concise History of Chinese Mathematics),* Shenyang: Liao Ning People's Press (in Chinese).

Li Shan-lan. 1859. Preface to *Dai Wei Ji Shi Ji* (in the Library of East-China Normal University, Shanghai).

Li Yan and Du Shiran. 1987. *Chinese Mathematics, a Concise History.* Oxford: Clarendon Press.

Loomis, Elias. 1851. *Elements of Analytical Geometry and of the Differential and Integral Calculus.* New York: Harper and Brothers Publishers.

——. 1874. *Elements of Differential and Integral Calculus,* New York: Harper and Brothers Publishers.

Martzloff, Jean-Claude. 1988. *Histoire des Mathématiques Chinoises.* Paris: Masson.

Mikami, Yoshio. 1913. *The Development of Mathematics in China and Japan.* New York: G. E. Stechert Co., rep. New York: Chelsea Publishing Company, 1961, 176–178, 301.

Mo You and Xu Shen. 1986. *Xian Dai Zhong Guo Shu Xue Shi Hua. A Concise History of Modern Chinese Mathematics.* Nanning: Guang Xi People Press (in Chinese).

Needham, Sir Joseph. 1959. *Science and Civilization in China.* III, Part I, Cambridge: Cambridge University Press.

Osserman, Robert. 1989. The Geometry Renaissance in America: 1938–1988, in Duren *et al.,* Part II, vol. 2, 513–526.

Perleberg, Max. 1954. *Who's Who in Modern China.* Hong Kong: Ye Olde Printerie.

Rosenstein, George. 1989. The best Method. American Calculus Textbooks of the Nineteenth Century, in Duren *et al.,* vol. 3, 77–110.

Siegman, Gita. 1985. *Awards, Honors and Prizes,* Detroit: Gale Research Company, vol. I, 374.

Smith, David Eugene and Jekuthied Ginsberg. 1934. *A History of Mathematics in America before 1900,* Chicago: The Mathematical Association of America.

——. 1919. A Belgian's Testimony on the Bad Treatment of Professors By the Germans. A letter to the Editor of the New York Times. *The New York Times,* April 30, 1919, 10.

Stone, Marshall H. 1976. Reminiscences of Mathematics at Chicago, in Duren *et al.,* vol. II, 183–190.

Tarwater, D., ed. 1977. *The Bicentennial Tribute to American Mathematics: 1776–1976.* Washington D.C.: The Mathematical Association of America.

Twitchett, Denis and John K. Fairbank. 1980. *The Cambridge History of China*, **11** (Late Ch'ing, 1800–911). Cambridge: Cambridge University Press.

——. 1986. *The Cambridge History of China*, **13** (Republican China 1912–1949). Cambridge: Cambridge University Press.

Vanhee, Louis and S. J. Brussels. 1926. The Great Treasure House of Chinese and European mathematics (translated and shortened by D. E. Smith), *American Mathematical Monthly*, **33**, 502–504.

——. 1926. Napier's Rods in China. *American Mathematical Monthly*, **32**, 494–497.

Wylie, Alexander. 1897. Jottings on the science of Chinese Arithmetic, in A. Wylie, *Chinese Researches*, 159–194.

Yang, Tsui-Hua. 1991. *Patronage of Sciences: The China Foundation for the Promotion of Education and Culture*, Taiwan, R.O.C.: Academia Sinica (in Chinese).

Yuan, Tung-li. 1963. *Bibliography of Chinese Mathematics, 1918–1960*, Washington D.C.: Tung-li Yuan.

Zhou, Pei-yuan. 1989. Reminiscences of teacher Tong-Sun. *Memorial Booklet for Zheng Tong-Sun*, Nanjing: Jiang Su Education Press (in Chinese).

Notes on the Contributors

Joseph W. Dauben is Professor of History and the History of Science at Herbert H. Lehman College of the City University of New York. He also serves as a member of the graduate faculty of the Ph.D. Program in History at the Graduate Center, CUNY, where he directs the Specialization in History of Science program. From 1976–1986 he served as editor of *Historia Mathematica*, an international journal for the history of mathematics published by Academic Press, and from 1986–1993 was chairman of the International Commission on History of Mathematics. His first book, *Georg Cantor, His Mathematics and Philosophy of the Infinite*, was published by Harvard University Press in 1979; it has recently been reprinted in paperback by Princeton University Press (1990). A Chinese edition translated by Zheng Yu-Xin (Nanjing University) appeared in 1989.

Professor Dauben first went to China as a visiting scholar jointly sponsored by the U.S. National Academy of Sciences and the Academia Sinica in Beijing for four months in 1989. During the fall semester of 1991 he was a member of the Institute of History, National Tsing-Hua University, Taiwan. Currently he is working on an English translation of selections from the *Ten Classics of Ancient Mathematics*, a project sponsored by the U.S. National Endowment for the Humanities, in collaboration with colleagues in Beijing, Taipei, and Singapore.

Della Dumbaugh Fenster is completing her graduate work in the history of mathematics at the University of Virginia. Her dissertation is entitled "Leonard Eugene Dickson and His Work in the Theory of Algebras." Her more general interests lie in the emerging American mathematical community and the role of women in this development, in the early history of algebra in the United States, and in the role of mentoring in mathematics.

Ivor Grattan-Guinness has both a doctorate (Ph.D.) and higher doctorate (D.Sc.) in the history of science from London University. He is Reader in Mathematics at Middlesex University, England. He was editor of the history of science journal *Annals of Science* from 1974 to 1981, and from 1979 to 1992 he was the founder-editor of the journal *History and Philosphy of Logic*. From 1986 to 1988, he was the president of the British

Society for the History of Mathematics. His latest book is *Convolutions in French Mathematics, 1800–1840*, 3 volumes (1990, Birkhäuser Basel and Berlin). He is the editor of the *Companion Encyclopaedia of the History and Philosophy of the Mathematical Sciences* (Routledge, London, 1993). He is presently writing a general history of mathematics for the Fontana History of Science series, and is also working on the history of mathematical logic and the foundations of mathematics from 1870 to 1930.

Jesper Lützen teaches the history of mathematics and exact sciences at the Mathematics Institute of Copenhagen University. He is a member of the editorial boards of *Historia Mathematica* and *Archive for History of Exact Sciences*. His research interests center on nineteenth-century analysis and mathematical physics. He is the author of *The Prehistory of the Theory of Distributions* (New York: Springer, 1982) and *Joseph Liouville 1809–1882. Master of Pure and Applied Mathematics* (New York: Springer, 1990).

Karen Hunger Parshall is Associate Professor of Mathematics and History at the University of Virginia. Her book with David E. Rowe, *The Emergence of the American Mathematical Research Community 1876–1900*, will appear in 1994, and she is currently finishing a critical edition of selected correspondence of the nineteenth century British mathematician, James Joseph Sylvester, and his circle, to be published by B. G. Teubner. She has just stepped down as the book review editor of *Historia Mathematica* to assume the positioon of the journal's managing editor.

Volker Peckhaus took his doctorate in 1990 from the University of Erlangen-Nürnberg, where he is presently Wissenschaftlicher Assistent in the Institut für Philosophie. He is a member of the editorial board of *Historia Mathematica* and author of *Hilbertprogramm und Kritische Philosophie. Das Göttinger Modell interdisziplinarer Zusammenarbelt zwischen Mathematik und Philosophie* (Göttingen: Vandenhoeck & Ruprecht, 1990). His research interests are concerned with the history of formal logic and foundations as well as methodological problems in the history of science.

Walter Purkert studied mathematics and the history of science at Leipzig University. His research has touched on many aspects of the history of mathematics during the nineteenth and twentieth centuries. He is the author, with H. J. Ilgauds, of *Georg Cantor 1845–1918*, Vita Mathematica, vol. 1 (Basel: Birkhäuser, 1987). He has served as an associate editor of *Science Networks* and as a member of the editorial board of *Historia Mathematica* and the Executive Committee of the International Commission on the History of Mathematics. He is currently engaged in archival research documenting the life and work of the Bonn mathematician Felix Hausdorff.

Rossana Tazzioli studied mathematics at the University of Genoa and took her doctorate in mathematics at the University of Bologna. Her main research interests are the history of mathematics during the nineteenth century, particularly differential geometry, mathematical physics, and related fields. She is the author of "Ether and Theory of Elasticity in Beltrami's Work," *Archive for History of Exact Sciences* 46(1) (1993):1–37.

Peter Ullrich is Hochschuldozent in the Fachbereich Mathematik of the Westfälische Wilhelms-Universität in Münster. Within pure mathematics, his research interests lie in rigid analytic geometry. His interests in the history of matheematics have focused on complex analysis in the nineteenth and twentieth centuries. He is the editor of Karl Weierstrass's lectures entitled *Einleitung in die Theorie der analytischen Funktionen* (Dokumente zur Geschichte der Mathematik, vol. 4, Wiesbaden, Vieweg, 1988) based on the notes taken by Adolf Hurwitz.

Zhang Dian-Zhou is Professor of Mathematics at East China Normal University in Shanghai. He is a specialist in functional analysis, and is particularly interested in using the experience of the history of contemporary mathematics as it may pertain to the future development of teaching and research in China. Professor Zhang is the author of numerous works on the history of Chinese and Western mathematics, and is an authority on the history of modern mathematics in particular. His most recent book, *A Concise History of Chinese Mathematics in the 20th Century*, the first of its kind, will be published at the end of the year.

Professor Zhang spent two years in the United States in 1989–1991, supported by a fellowship from the K. C. Wang Foundation of Hong Kong. As a Visiting Professor in the Ph.D. Program in History, The Graduate Center, CUNY, he spent much of this time working with Professor Joseph Dauben on the paper that is printed in this volume of *History of Modern Mathematics*. He also spent part of 1992 as a Research Professor at the Mathematical Sciences Research Institute in Berkeley, California.

Professor Zhang is currently organizing the next ICMI–China Regional Conference on Mathematical Education, which will meet in Shanghai August 16–20, 1994. He and Professor Dauben are also working on a study of international cooperation in research related to the history of ancient Chinese mathematics in the early twentieth century.

ISBN 0-12-599663-2

90040

9 780125 996631